D1117287

The Desert Grassland

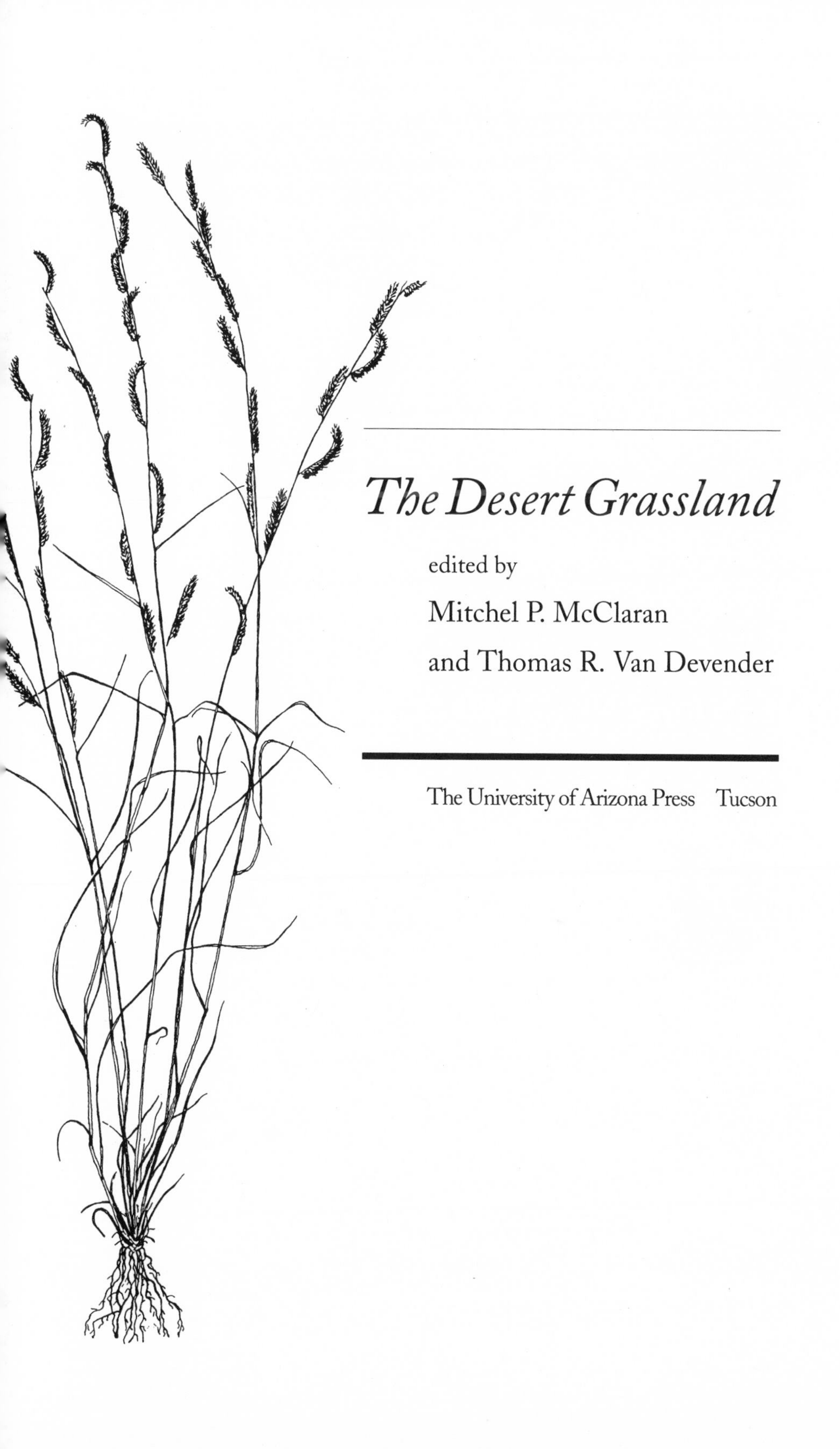

The Desert Grassland

edited by

Mitchel P. McClaran

and Thomas R. Van Devender

The University of Arizona Press Tucson

This book was published in cooperation with the Arizona-Sonora Desert Museum, a private, nonprofit museum that interprets the natural history of the Sonoran Desert region primarily through living, outdoor exhibits.

The University of Arizona Press

♾This book is printed on acid-free, archival-quality paper.
Manufactured in the United States of America

99 98 97 96 95 5 4 3 2 1

Library of Congress Cataloging-in-Publication Data
The desert grassland/edited by Mitchel P. McClaran and Thomas R. VanDevender.
p. cm.
Includes bibliographical references and index.
ISBN 0-8165-1580-8 (cloth)
1. Deserts—Southwestern States. 2. Deserts—Mexico 3. Grasslands—Southwestern States. 4. Grasslands—Mexico. 5. Desert ecology—Southwestern States 6. Desert ecology—Mexico. 7. Grassland ecology—Southwestern States. 8. Grassland ecology—Mexico. I. McClaran, Mitchel P., 1956- . II Van Devender, Thomas R.
QH104.5.S6D47 1995
574.5′2652′0979—dc20 59-327
CIP

British Cataloguing-in-Publication Data
A catalogue record for this book is available from the British Library.

Contents

Preface vii

1 Desert Grasslands and Grasses 1
Mitchel P. McClaran

2 Desert Grassland, Mixed Shrub Savanna, Shrub Steppe, or Semidesert Scrub? The Dilemma of Coexisting Growth Forms 31
Tony L. Burgess

3 Desert Grassland History: Changing Climates, Evolution, Biogeography, and Community Dynamics 68
Thomas R. Van Devender

4 Landscape Evolution, Soil Formation, and Arizona's Desert Grasslands 100
Joseph R. McAuliffe

5 The Role of Fire in the Desert Grasslands 130
Guy R. McPherson

6 Diversity, Spatial Variability, and Functional Roles of Invertebrates in Desert Grassland Ecosystems 152
Walter G. Whitford, Gregory S. Forbes, and Graham I. Kerley

7 Diversity, Spatial Variability, and Functional Roles of Vertebrates in the Desert Grassland 196
Robert R. Parmenter and Thomas R. Van Devender

8 Human Impacts on the Grasslands of Southeastern Arizona 230
Conrad J. Bahre

9 Revegetation in the Desert Grassland 265
Bruce A. Roundy and Sharon H. Biedenbender

Appendix 1: Common and Scientific Plant Names 305

Appendix 2: Common and Scientific Invertebrate Names 313

Appendix 3: Common and Scientific Vertebrate Names 317

List of Contributors 323

Index 325

Preface

With this book, the desert grassland will become more accessible and therefore more personal. The chapters in this volume will pique and enrich the layperson's curiosity, supplement the college student's lectures and laboratories, provide an accessible reference for scholars, and assist land managers responsible for desert grasslands. The writing style is targeted at a sophisticated lay audience with some background in natural history, anthropology, biology, ecology, and geology. The college student and scholar will appreciate the contemporary interpretations of older works, and the land manager will gain a quick and provocative insight into the mechanics, diversity, and human use of desert grasslands.

This book tells the story of a diverse and rich biotic community that defies conventional definitions of grassland. It is a story of a place where soil, landforms, animals, and plants are closely tied but also surprisingly independent. A story that informs us that many of the grass taxa present today were also present in the Ice Age forests that existed there before the climate dried and the trees vanished. A story that describes how the intensity and frequency of fire can influence the flora and fauna present. A story that describes how humans—from Amerindians to contemporary ranchers, public land managers, and real estate developers—have changed the relative abundance of woody and herbaceous species and introduced new plants and domesticated animals. It is a story that concludes with a review of the attempts, sometimes failed and sometimes successful, to reestablish plants in the desert grassland where overgrazing, drought, farm abandonment, and increased noxious plant density have occurred. Finally, it is a comprehensive and contemporary story of the desert grassland, and the authors provide fresh and invigorating voices for their interpretations.

This book is the product of an idea that was hatched during the 1991 Arizona–Sonora Desert Museum Board of Trustees retreat at Westward Look Resort in Tucson, Arizona. As a new and enthusiastic trustee, I suggested that the museum could foster greater learning and appreciation for the significance of its new habitat exhibits if the opening of each new exhibit was accompanied by a symposium and book that described the state of knowledge about each habitat. Fortunately, my wild idea fell on the attentive ears, dedicated souls, and creative minds of David Hancocks and Carol Cochran, the museum's executive director and director of education, respectively. It is because of their interest and fortitude that the symposium was convened in September 1992 in Tucson during the opening month of the museum's new Desert Grassland Habitat exhibit.

Tom Van Devender became involved in the planning of the symposium in the winter of 1991–92, and his broad knowledge of scholars working in the desert grassland and his wealth of personal knowledge helped bring together the group that presented their work at the symposium. When Tom, Carol, and I began editing the manuscripts submitted by those speakers, we concluded that a contribution covering the role of fire and an introductory chapter were sorely needed. I garnered the help of Guy McPherson to write the chapter on fire, and I wrote the introductory chapter. Except for those two chapters, however, all the contributions reflect the verbal presentations at the symposium.

Each chapter was reviewed by scholars familiar with the grassland so that we could present a book of great integrity. The following reviewers provided excellent suggestions that improved the book immensely: Tony L. Burgess, Reginald Chapman, Scott A. Elias, Robert P. Gibbens, Carlton H. Herbel, Charles Hutchinson, Hiram Johnson, Howard E. Lawler, S. Clark Martin, Paul S. Martin, Guy R. McPherson, Phil R. Ogden, Carl A. Olson, Phil A. Pearthree, Yar Petryszyn, Gerald G. Raun, Richard Reeves, Wade Sherbrooke, E. Lamar Smith, Henry A. Wright, and Richard Young.

Tom Van Devender compiled the appendixes of common and scientific names used in the book. John R. Reeder and Charlotte Reeder assisted him with grass taxonomy, and Carl A. Olson helped with the invertebrate taxonomy.

Monte Bingham, the master draftsman for the book, provided all the figure drawings unless otherwise noted. Sandi Lehman assisted with word processing, and Nancy Lyon standardized the table formats throughout.

Editing this book has been at various times a struggle, a joy, and a source of pride, but it has always been a source of inspiration because it has enabled me to learn more about the desert grassland in the last eighteen months than I could have otherwise in five years.

Mitchel P. McClaran

The Desert Grassland

I

Desert Grasslands and Grasses

Mitchel P. McClaran

The term *desert grassland* challenges conventional perceptions of grasslands because the terms *desert* and *grassland* evoke very different impressions. The dominant plants on desert grassland are not always grasses; they can be shrubs or small trees as well (see Burgess, this volume). The climate is only slightly cooler and wetter in desert grassland than in more typical deserts, and the difference is largely imperceptible in June, when temperatures exceed 40°C and the last measurable rainfall occurred in March. The desert grassland achieves the lowest total biomass production of all North American grassland types (Sims and Singh 1978b). The soils are very low in organic matter and often contain high amounts of calcium carbonate (Schmutz et al. 1991).

The mixed grass-shrub vegetation commonly called desert grassland grows in the basins and valleys that skirt the hills and mountain ranges of southwestern North America (Brown 1982a,b; Burgess, this volume; Dick-Peddie 1993; Schmutz et al. 1991). The North American desert grassland reaches its northern limit in Arizona, New Mexico, and Texas, but the majority of its 500,000 km² (Lauenroth 1979) extends 1,500 km to the south through 13 Mexican states from Sonora to Puebla (fig. 1.1; Schmutz et al. 1991). The desert grassland occurs at elevations between 1,100 and 1,800 m

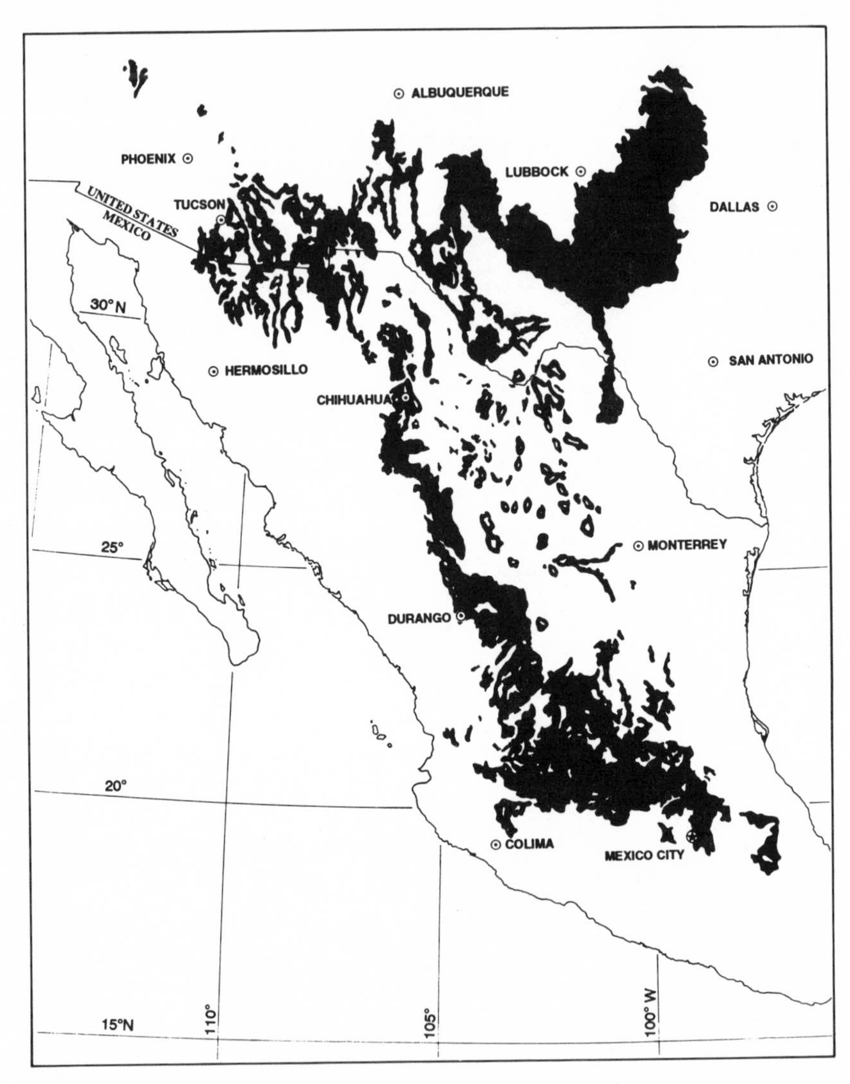

Figure 1.1. Geographic distribution of desert grasslands in southern and southwestern North America. Provided by F. W. Reichenbacher, Southwestern Field Biologists, Tucson, Arizona, December 1993.

in the United States, and as high as 2,500 m in Mexico (Brown 1982a,b; Schmutz et al. 1991). There is remarkable climatic, geologic, edaphic, and biotic diversity across this vast area and elevational range, but there is a consistent theme of dry, hot, sunny climates with short, unimposing vegetation consisting of grasses, forbs, shrubs, small trees, and succulents.

Grasses are a significant food source for wild and domestic animals, and they anchor and stabilize the scant and precious desert grassland soil. The ability of grasses to withstand frequent defoliation and maintain vigor is the result of several morphological and physiological traits unique to the grass family. Unfortunately, the appreciation of this remarkable versatility is often stifled by the intricacies of grass morphology and the resulting difficulties of species identification.

In this introductory chapter I discuss the general climate and vegetation characteristic of the desert grassland with the intent of describing what this vegetation looks like, when it grows, and how the growth pattern is linked to the temperature and rainfall regimes. Brief descriptions of four desert grassland areas illustrate the spatial variation and the relationship between climate and vegetation. In the latter part of the chapter I discuss grass morphology, taxonomy, and growth to lay the foundation for a discussion of grass regrowth following defoliation. An understanding of just what a desert grassland is and the significance of grass's ability to tolerate defoliation should be sufficient preparation for a rewarding exposure to the remainder of this book.

Desert Grassland

Climate

Of all the North American grasslands, the desert grassland has the driest and the least cloudy and frost-prone climate (Schmutz et al. 1991; Sims and Coupland 1979; Sims et al. 1978). The average annual precipitation on desert grasslands in the United States ranges from 230 to 460 mm, while desert grasslands in some parts of Mexico receive up to 600 mm (Brown 1982a,b; Schmutz et al. 1991). The portion of the total annual precipitation that falls between May and October increases from less than 50 percent in the northwest near Prescott, Arizona, to 75 percent in Texas and eastern New Mexico, and 90 percent in central Mexico. In the United States, about 50 percent of the desert grassland area receives less than 250 mm annually (Martin 1975), and the year-to-year variation in precipitation (as measured by the coefficient of variation) is greater there than in all other areas except deserts (Hidy and Klieforth 1990).

The seasonal rainfall pattern is strongly influenced by the migration of the Bermuda high-pressure cell and the subtropical Pacific jetstream (Neilson 1986, 1987). From April through August, the Bermuda cell moves across

Texas, New Mexico, and into eastern Arizona, and the subtropical jetstream migrates north. The Bermuda cell brings moisture from the Gulf of Mexico that mixes with moisture from the Pacific Ocean and Gulf of California being transported by the subtropical jetstream. Because rainfall is greatest at the western edge of the Bermuda cell, warm-season rainfall commences in June–July in the eastern portion of the grassland and is greatest in the west from late July to mid-August. Some late-summer moisture falls in the east as the Bermuda cell retreats in September. August rainfall is most reliable in southeastern Arizona, June–July rainfall is more reliable in Texas and New Mexico, and winter rainfall is most common in the far northwestern grasslands (Martin 1975; Neilson 1986, 1987).

The average annual temperature is 13–16°C, and the desert grassland commonly experiences 75 freezing nights each winter and 20 summer days hotter than 40°C (Brown 1982a,b; Schmutz et al. 1991; Sims et al. 1978). The low frequency of frost, high solar radiation, and warm daytime temperatures in the winter extend the potential growing season (as measured by a 15-day consecutive daily mean greater than 4.4°C) to more than 300 days (Sims et al. 1978).

The combination of low rainfall, warm temperatures, and high solar radiation also creates a significant moisture deficit: the amount of annual potential evapotranspiration (combination of transpiration from plants and evaporation from the soil) is regularly two to three times greater than the rainfall (Sims et al. 1978). This moisture deficit does more to restrict the actual growing season than temperature does. The water deficit also results in infrequent soil leaching and the buildup of calcium carbonate, which can create a dense layer of caliche (McAuliffe, this volume; Schmutz et al. 1991).

Vegetation

Vegetation in the desert grassland is difficult to characterize because so many species grow in the vast space covered by this grassland. This variety is magnified by the discontinuous distribution of desert grassland (fig. 1.1; Burgess, this volume; McAuliffe, this volume) and the resulting mosaic of grassland with desertscrub at lower elevations and with evergreen-oak woodland and chaparral at higher elevations. This intermingling and natural fragmentation of vegetation results in the presence of taxa that also occur in desertscrub vegetation at lower, warmer, drier elevations as well as taxa from the evergreen-oak woodland and chaparral at higher, colder, wetter elevations.

This natural fragmentation of desert grassland must be considered when evaluating the effects of fragmentation caused by human agriculture, housing, and road development (Bahre, this volume). Increased fragmentation can lead to a depauperate fauna and flora, both through isolation and reduced carrying capacity (Saunders et al. 1991) and also as the result of changes in surface water drainage and fire patterns (McPherson, this volume). Because the desert grassland is naturally fragmented, however, especially compared with grassland in the Great Plains, creative research designs and judicious analysis will be required to distinguish human-caused impacts on biodiversity from the influence of natural fragmentation.

I briefly treat this spatial and biotic variety by describing four desert grassland areas, three in Arizona and one in New Mexico, that exemplify the vegetational variety of the grassland. All four areas are open to the public and have been the sites of significant research efforts.

Jornada Experimental Range. The vegetation on the 78,000-ha Jornada Experimental Range is a classic example of desert grassland. On the sandy and gravelly upland soils, black grama (*Bouteloua eriopoda*), mesa dropseed (*Sporobolus flexuosus*), and bush muhly (*Muhlenbergia porteri*) grasses grow with woody plants such as honey mesquite (*Prosopis glandulosa*), creosotebush (*Larrea divaricata*), snakeweed (*Gutierrezia sarothrae*), and soaptree yucca (*Yucca elata*; fig. 1.2). Tobosa (*Hilaria mutica*), burrograss (*Scleropogon brevifolius*), and alkali sacaton (*Sporobolus airoides*) grasses grow along with tarbush (*Flourensia cernua*) and creosotebush on the clay bottomland soils (fig. 1.2). Recent vegetation classifications call this vegetation desert grassland (Dick-Peddie 1993), semidesert grassland (Brown 1982b), and Chihuahuan desert grassland (Schmutz et al. 1991).

The average annual rainfall there is 230 mm, and about 55 percent of it falls between July and September; the average temperature is about 15°C (Sims et al. 1978). The effective growing season, the time when soil moisture and temperatures are favorable, is between July and September. Most of the area lies between 1,200 and 1,400 m elevation.

The Jornada is 35 km north of Las Cruces, New Mexico. It was established in 1912 and initially administered by the United States Department of Agriculture's Bureau of Plant Industry to improve livestock grazing management. Currently, the Department of Agriculture's Agricultural Research Service office in Las Cruces administers the Jornada. Current research efforts continue to include livestock grazing management, but the scope has broadened to include more basic ecological studies as well.

Figure 1.2. Examples of vegetation on the Jornada Experimental Range, southern New Mexico, September 1993. A. The tall succulent soaptree yucca (*Yucca elata*), black grama (*Bouteloua eriopoda*), and the short shrub snakeweed (*Gutierrezia sarothrae*) dominate this area of sandy-gravelly soil on a gentle slope.

Its lengthy tenure as a research facility has fostered the study of long-term vegetation changes on the Jornada. Descriptions, culled from early land surveys, of historic increases in honey mesquite, creosotebush, and tarbush, and decreases in grass dominance (Buffington and Herbel 1965) have been supplemented with long-term experiments and measurements of plant populations (Gibbens and Beck 1988; Gibbens et al. 1993; Hennessey et al. 1983). Three major findings emerge from these studies: (1) significant increases in shrub cover since 1858 and the establishment of extensive mesquite dunelands; (2) dramatic swings in grass populations in response to droughts in the 1930s and 1950s, when tobosa and burrograss populations recovered from droughts but black grama did not; and (3) the negative influence of small animals on the establishment of grasses and shrubs.

From 1970 to 1972, the Jornada was the desert grassland site for the International Biological Programme (IBP). The IBP described seasonal changes in biomass, energy dynamics, and the physical environment in set-

B. On a clay-soil bottomland site, the dominant plants are the evergreen shrubs creosotebush (*Larrea divaricata*) and tarbush (*Flourensia cernua*), and the sod-forming grasses burrograss (*Scleropogon brevifolius*) and tobosa (*Hilaria mutica*).

tings throughout the world to explain the factors controlling biological productivity (Sims et al. 1978). The results of this research can be used to compare the characteristics of the desert grassland with other grasslands and biotic communities.

Many of the initial long-term studies have been maintained, and in addition the Jornada became a Long Term Ecological Research (LTER) site, sponsored by the National Science Foundation, in 1980 (Franklin et al. 1991). LTER research is designed to facilitate the long-lasting efforts needed to interpret the mechanisms responsible for vegetation dynamics (McClaran et al. 1995).

Santa Rita Experimental Range. The 20,000-ha Santa Rita Experimental Range, 40 km south of Tucson, Arizona, is near the western edge of the desert grassland. The elevation, which varies from 900 to 1,300 m, strongly influences the amount of annual rainfall (250-500 mm/yr); and all elevations receive 50–60 percent of their annual total between July and September (Anable 1989). The mean annual temperature is about 16°C at 1,300 m, and is higher at lower elevations (Martin 1966). Most of the Santa Rita Range is lower than the Jornada, and the resulting warmer temperatures support more species with Sonoran Desert affinities such as blue and foothill paloverdes (*Cercidium floridum* and *C. microphyllum*) and saguaro (*Carnegiea gigantea*). The vegetation is a mixture of grasses, forbs, shrubs, small trees, and cacti (fig. 1.3). The dominant grasses include Arizona cottontop (*Digitaria californica*), Lehmann lovegrass (*Eragrostis lehmanniana*), slender grama (*Bouteloua repens*), and Rothrock grama (*B. rothrockii*). Dominant shrubs and small trees include burroweed (*Isocoma tenuisecta*), snakeweed, and velvet mesquite (*Prosopis velutina*). Cane cholla (*Opuntia spinosior*), chainfruit cholla (*O. fulgida*), and variable prickly pear (*O. phaeacantha*) are among the dominant cacti. The mixture of vegetation on the Santa Rita has recently been referred to as semidesert grassland (Brown 1982b), Sonoran Desert grassland at the lower elevations (Schmutz et al. 1991), and high desert bunchgrass at the higher elevations (Schmutz et al. 1991).

The Santa Rita Experimental Range was established in 1903 as the first range experiment station in the world. It was administered by the United States Department of Agriculture's Forest Service until 1989, when title was transferred to the state of Arizona and administration was assumed by the University of Arizona College of Agriculture (Anable 1989; Martin 1966). Early research efforts there were aimed at stabilizing the livestock industry through improved grazing management, revegetation, and brush control, but the scope has broadened to include more basic ecological studies as well.

Photographs taken on the Santa Rita Range in the late nineteenth and early twentieth centuries are the foundation for a series of subsequent, or repeat, photographs of the same area. These repeat photographs illustrate the extent and magnitude of vegetation changes (Bahre 1991, this volume; Hastings and Turner 1965; Martin and Reynolds 1973; McAuliffe, this volume). In combination with long-term livestock grazing exclosures and a network of rain gauges, these photographs show that velvet mesquite has increased on both grazed and ungrazed areas, that burroweed and cacti populations express dramatic fluctuations that are not completely related to grazing or weather patterns, and that areas with a shallow clay or argillic soil horizon have not experienced similar mesquite, cactus, or burroweed dynamics.

Some of the earliest and most thorough grazing system research in the desert grassland was conducted on the Santa Rita Experimental Range, culminating in the development of the Santa Rita grazing system (Martin

Figure 1.3. Among the dominant plants on the Santa Rita Experimental Range at 1,000 m elevation (September 1993) are the short shrub burroweed (*Isocoma tenuisecta*; foreground) and short trees (background) including blue paloverde (*Cercidium floridum*) and velvet mesquite (*Prosopis velutina*). The dominant cacti include cane cholla (*Opuntia spinosior*), chainfruit cholla (*O. fulgida*), and variable prickly pear (*O. phaeacantha*). The dominant perennial grasses include Arizona cottontop (*Digitaria californica*) with the long white inflorescence, bush muhly (*Muhlenbergia porteri*), and pappusgrass (*Pappophorum vaginatum*). Scattered plants of the summer annual grass six-weeks needle grama (*Bouteloua aristidoides*) are also visible.

and Severson 1988), which uses three pastures to provide an ungrazed or rest period of 12 consecutive months in two out of three years. Ongoing grazing system research is evaluating the application of short, intense grazing periods followed by long periods of rest.

A series of elegant experiments and field studies determined the impact of velvet mesquite on soil nutrients and grass growth on the Santa Rita Range (Tiedemann and Klemmedson 1973a,b; Tiedemann et al. 1971). The higher concentration of several nutrients, including nitrogen, under the tree canopy has been described as an island of fertility (Garcia-Moya and McKell 1970), and some perennial grasses produce more biomass under trees or in shade when they are planted in soil taken from beneath a velvet mesquite canopy.

The recent spread and increasing dominance of Lehmann lovegrass is the latest installment in the chronicle of historic vegetation change on the Santa Rita Range. Introduced in the 1930s from southern Africa as a potential source of forage and erosion control, it was seeded on only 200 ha before seeding was stopped in the 1970s (Anable et al. 1992). The lovegrass now occurs on about 90 percent of the Santa Rita Range (Anable et al. 1992) and is the most common perennial grass on about two-thirds of that area. Repeat photography, livestock exclosures, and long-term vegetation composition plots have facilitated studies of its spread. Lovegrass does not require grazing in order to spread and increase, nor does it increase with increasing grazing pressure, although native grass density does decrease with greater grazing pressure (McClaran and Anable 1992). Unexpectedly, the density of two native grasses, Arizona cottontop and Rothrock grama, was unrelated to the density of lovegrass or the length of time that lovegrass had been growing in a given area (Van Deren 1993).

Appleton-Whittell Research Ranch Sanctuary. The 3,300-ha Appleton-Whittell Research Ranch Sanctuary lies 100 km southeast of Tucson near the town of Elgin, Arizona. At nearly 1,500 m elevation, the mean annual temperature is approximately 12°C. The mean annual precipitation is about 400 mm, with about 60 percent occurring as rain between July and September. Frost and snow are more common here than on the other areas discussed because the Research Ranch is higher. The cooler temperatures and greater rainfall preclude the presence of plants with strong desert affinities, while species with oak woodland affinities are common. Vegetation on the rolling hills includes blue grama (*Bouteloua gracilis*), hairy grama (*B. hirsuta*), plains lovegrass (*Eragrostis intermedia*), scattered Emory oak (*Quercus emoryi*), Arizona white oak (*Q. arizonica*), and sotol (*Dasylirion wheeleri*). In the seasonally flooded bottomland, big sacaton (*Sporobolus wrightii*) dominates (fig. 1.4). Plains grassland (Brown 1982a) and high desert sod-grass (Schmutz et al. 1991) are other names applied to this type of desert grassland.

The Research Ranch is a combination of private and public land. The public land includes state land administered by the Arizona State Land Office, and the federal land is administered by the United States Department of Agriculture Forest Service. Until 1969, these lands were grazed by livestock through a permit arrangement with the public land institutions

Figure 1.4. Looking from the gravelly ridges across the clay soil bottomland on the Appleton-Whittell Research Ranch Sanctuary (September 1993). The dominant plants on the ridges are short perennial grasses, including blue grama (*Bouteloua gracilis*), hairy grama (*B. hirsuta*), and plains lovegrass (*Eragrostis intermedia*). The short succulents in the foreground are beargrass (*Nolina microcarpa*, right) and sotol (*Dasylirion wheeleri*, left). The tree is Emory oak (*Quercus emoryi*). Big sacaton (*Sporobolus wrightii*) is the dominant plant on the bottomlands.

(Bahre 1977). In 1969, the Appleton family terminated its livestock operation and established a livestock-free sanctuary that included both the privately and publicly owned parts of the Research Ranch. The National Audubon Society acquired the property in 1980 and continues to manage it as a natural history research facility and nature reserve.

Studies of bird predation on grasshoppers (Bock et al. 1992a), interspecific bird competition (Bock et al. 1992b), and factors influencing plant species abundance in grassland marshes, or ciénegas (Fernald 1987), have been conducted there. As an ungrazed area surrounded by ongoing livestock grazing, the Research Ranch offers the opportunity to study the changes that follow livestock removal and serves as a benchmark for comparison with nearby grazed areas (Bock and Bock 1993; Brady et al. 1989).

Empire-Ciénega Resource Conservation Area. The rolling hills and seasonally flooded clay bottomlands, perennial creek, and ciénega of the Empire-Ciénega Resource Conservation Area are situated between the Santa Rita and Whetstone Mountains approximately 75 km southeast of Tucson, and just north of the town of Sonoita at 1,300–1,400 m elevation. Because the Empire-Ciénega Conservation Area is only slightly lower, warmer, and drier than the nearby Appleton-Whittell Research Ranch Sanctuary, the vegetation composition of the grasslands in the two areas is quite similar. The most conspicuous difference is a greater abundance of velvet mesquite on the Empire-Ciénega area (fig. 1.5). In addition, cottonwood (*Populus fremonti*) and velvet ash (*Fraxinus pennsylvanica* var. *velutina*) form intermittent gallery forests along Ciénega Creek.

Privately owned cattle ranches occupied the area from the late 1800s until they were purchased in 1960 for subdivision and home development, a proposition that never materialized. In the mid-1970s the area was purchased for mineral and water rights, and in 1988 the land was acquired by the federal government. It is now administered by the United States Department of the Interior's Bureau of Land Management (BLM) office in Tucson. The area is managed to provide recreational opportunities, wildlife habitat, historical preservation and interpretation of the ranch facilities, and continued livestock grazing.

Compared with the other grassland sites, very little research has been conducted here; however, the greater accessibility provided under BLM administration presents convenient opportunities for the interested visitor to experience the desert grassland. In addition to the natural and cultural features of the Empire-Ciénega area, visitors witness an award-winning livestock grazing operation. The rancher who has leased the grazing rights since the mid-1980s was named Range Manager of the Year in 1989 by the Arizona Section of the Society for Range Management in recognition of his progressive practices of frequent relocation of livestock to provide rest periods for grazed plants and his conscientious erosion control efforts on the dirt road system.

Vegetation Growth and Climate

The amount and timing of vegetation growth are mainly controlled by rainfall, plant physiology, and soil characteristics (Brown 1982a,b; Burgess, this volume; Martin 1975; McAuliffe, this volume; Schmutz et al. 1991; Sims and Coupland 1979). Total (above- and belowground) plant production av-

Figure 1.5. On the rolling uplands on the Empire-Ciénega Resource Conservation Area (September 1993), scattered velvet mesquite (*Prosopis velutina*) interrupts the grass-dominated aspect. The dominant grasses are blue grama (*Bouteloua gracilis*), hairy grama (*B. hirsuta*), sideoats grama (*B. curtipendula*), sprucetop grama (*B. chondrosioides*), and plains lovegrass (*Eragrostis intermedia*). The tall forb at the right front is camphorweed (*Heterotheca subaxillaris*). Light-colored cudweed (*Gnaphalium* sp.) dominates the middle of the photograph.

erages about 250–350 g/m^2 in the desert grassland (Martin 1975; Schmutz et al. 1991; Sims and Singh 1978b). Only about 0.16 percent of the usable solar radiation is converted to plant products (Sims and Singh 1978b). These production and solar conversion values are the lowest of all North American grasslands (Sims and Singh 1978b), and they are largely the result of low rainfall, high evapotranspiration, and shallow soils.

Variation in the timing and amount of rainfall is particularly important in favoring species with different physiological characteristics. For example, the deeper root systems of woody species give them a prolonged growing season relative to grasses and other herbs (Burgess, this volume). Although most shrubs and succulents grow in the spring, evergreen species such as burroweed, creosotebush, and tarbush can grow whenever temperature and soil moisture conditions are suitable. Deciduous species such as honey and velvet mesquite and catclaw (*Acacia greggii*), however, do not grow when their leaves are absent in the winter and early spring (Schmutz et al. 1991).

There is a strong relationship between the summer rainfall pattern and grass growth: typically, more than 90 percent of grass growth occurs in July–September (Martin 1975; Sims and Singh 1978a). The period of fastest growth, between late July and August, is later than that of other North American grasslands (Sims and Singh 1978a).

An example from the Jornada Experimental Range illustrates the insignificance of winter moisture for total plant production. Rainfall was 183 and 324 mm, and production was 379 and 282 g/m^2 in 1971 and 1972, respectively (Sims and Coupland 1979). Thus, more total rainfall did not result in greater plant production. In 1972, much of the additional rain fell in the winter months when most grasses are largely inactive (Sims et al. 1978).

A few grasses do grow in the winter and early spring months. Squirreltail (*Elymus elymoides*) and New Mexico feathergrass (*Stipa neomexicana*) (Gurevitch 1986) are the most conspicuous spring-growing grasses, and they are especially abundant when winter moisture is above average. In addition, Arizona cottontop, bush muhly, Lehmann and plains lovegrass, sideoats grama (*Bouteloua curtipendula*), and the threeawns (*Aristida* spp.) often have green leaves in winter and produce some growth in spring (Martin 1975; Schmutz et al. 1991).

C_4 and C_3 Photosynthetic Pathways

The relatively reliable summer precipitation and warm temperatures in the desert grassland result in the dominance of grasses with the C_4 photosynthetic pathway. C_4 plants have a greater photosynthetic capacity at higher temperatures and use water more efficiently than plants that use the more typical C_3 photosynthetic pathway. The proportion of C_4 grass species is higher in the desert grassland than in all other North American grasslands (French 1979); more than 95 percent of all grass production in the desert grassland is from species with the C_4 pathway (Sims et al. 1978).

The C_4 and C_3 photosynthetic pathways differ in their mode of acquiring CO_2 from the atmosphere: C_4 plants have a preliminary stage that uses a four-carbon acid (hence C_4) to enrich CO_2 concentrations to nearly three times that occurring in the cells of C_3 plants (Long and Hutchin 1991).

All plants eventually use a three-carbon acid (hence C_3) in the energy-generating Calvin-Benson cycle, but C_4 plants have an enriched intracellular CO_2 concentration during the Calvin-Benson cycle (Hattersly and Watson 1992).

C_4 plants have a greater photosynthetic capacity at higher temperatures because the greater CO_2 concentration permits the stomata (leaf pores) to remain closed longer, thus reducing transpiration water loss. In addition, C_4 plants are more efficient at higher temperatures because the increased CO_2 concentration reduces the impact of photorespiration. During photorespiration, O_2 attaches to the primary photosynthesis enzyme, ribulose biphosphate carboxylase (RubisCO). In some respects, O_2 can be viewed as competing with CO_2 for space on the RubisCO enzyme (Long and Hutchin 1991). The competitive advantage of O_2 increases as leaf temperatures increase, but the higher CO_2 concentration in C_4 plants reduces that advantage (Long and Hutchin 1991).

The C_4 photosynthetic pathway was discovered in a grass species (Kortschack et al. 1965), and about half of all grass species worldwide use that pathway (Hattersly 1992). But C_4 photosynthesis is not the exclusive pathway in any grass subfamily, and one species even has both C_3 and C_4 individuals (Hattersly and Watson 1992). This variety in the presence/absence of the C_4 pathway suggests that it has evolved at least four times in the family and that there have been several reversions from C_4 back to C_3 photosynthesis (Hattersly and Watson 1992).

Grasses

The grass family, Poaceae, with some 10,000 species and 650–900 genera (Smith 1993), ranks behind only the sunflower (Asteraceae), legume (Fabaceae), and orchid (Orchidaceae) families (Watson 1990) in ubiquity. Members of the family are found in nearly every type of environment from tropical to Arctic and alpine regions, and from wet marsh and meadow habitats to dry deserts. The significance of their profusion may be exceeded by the economic importance of grasses, which are used for cultivated crops, natural forage for grazing animals, and building materials (Smith 1993).

Grass Morphology

Among the distinctive morphological characteristics of the grasses are the inflorescence of spikelets containing one to many florets (flowerlike units);

achenelike seeds often referred to as a caryopsis or grain; usually hollow, round stems with swollen nodes; and alternate, ranked leaves (Smith 1993).

A grass plant can be described as a group of tillers (shoots) that originate from the base of the plant (root crown). Each tiller contains several phytomers (leaf and stem units) that are stacked on each other (Briske 1991; Etter 1951). Each phytomer consists of a node, internode, leaf blade and sheath, collar, ligule, and axillary bud (fig. 1.6). The node is the spot on the leaf base where the leaf sheath arises and surrounds, partially or completely, the lower part of the internode. The leaf blade originates at the top of the sheath and projects away from the internode. The leaf sheath and blade are joined at the collar, and the ligule occurs on the inside of the collar. The axillary bud is a secondary meristem (point of cell division, differentiation, and growth) that can produce new tillers, or rhizomes or stolons that can form new plants. Rhizomes (underground tillers) and stolons (aboveground tillers) originate only on the axillary bud in the lowest phytomer of a tiller. Not all species produce rhizomes or stolons. The new plants that emerge from the rhizome or stolon are produced asexually and are characteristic of sod-forming grasses such as the stolon-bearing black grama (fig. 1.7) and rhizome-bearing tobosa.

When the internodes and leaves are fully developed and the plants are tall, the series of phytomers is commonly called a culm or stem (e.g., Gould 1951).

The reproductive structures of grasses are markedly different from those of most flowering plants. One to several florets are contained within a spikelet, and spikelets are arranged in an inflorescence at the top of a tiller (fig. 1.7). The floret contains the sexual organs—the female ovary and stigmas (typically two) and the male stamens (typically three)—and three vegetative characters: the lodicules (typically two), palea, and lemma. By analogy, the lodicules are similar to the perianth (petals and sepals) of more typical flowers, the palea is similar to a prophyll (the first leaf on an axillary branch), and the lemma is viewed as a modified leaf (Clayton 1990). Functionally, the lodicules swell and separate the lemma and palea, enabling the exertion and exposure of the stigma and stamens for fertilization. Wind is almost exclusively the mechanism of pollen transport, but some tropical grass species are visited by insects (Soderstrom and Calderon 1971). Lodicule swelling lasts for only 10–60 minutes before the lemma and palea again cover and protect the ovary, but the stigma and/or stamens remain exposed (Clayton 1990). The stigma and stamens are not always exerted simultaneously.

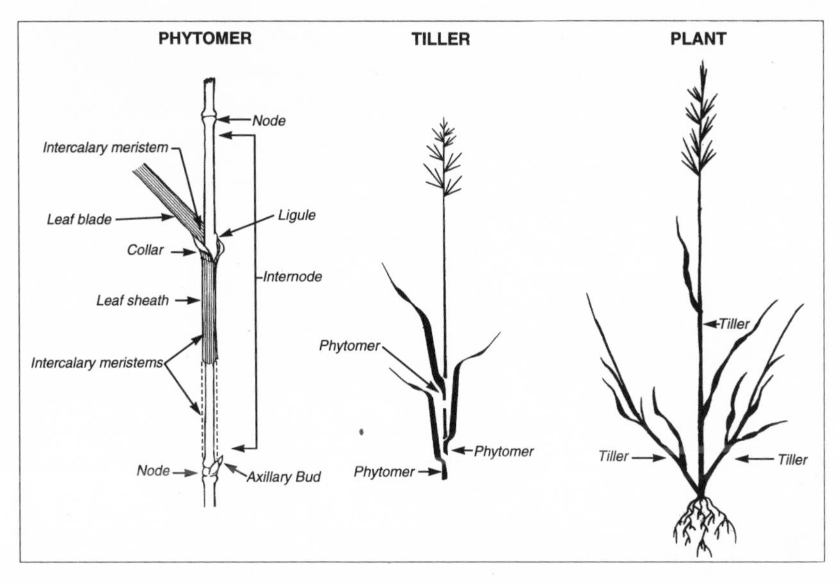

Figure 1.6. Grass plants are assembledges of tillers, and tillers are assembledges of phytomers. Each phytomer consists of a node, internode, leaf blade and sheath, collar, ligule, axillary bud, and three intercalary meristems. Adapted from Briske 1991; and Etter 1951.

Spikelets contain at least one floret, a rachilla, and typically two glumes (fig. 1.7). The glumes are empty scales at the base of the spikelet, and the florets are arranged on the branchlike rachilla above the glumes. The description and classification of grass inflorescences focus on the arrangement of spikelets along the branchlike rachis (Clayton 1990). Inflorescences in the grass family vary from simple panicles, racemes, and spikes to complicated combinations of spikelike racemes (Allred 1982).

The presence of unisexual florets or spikelets is a very simple and common variation of this floral arrangement. For example, sterile florets are found in gramas (*Bouteloua* spp.; fig. 1.7), and sterile spikelets are present in bluestems and beardgrasses (*Andropogon* spp., *Bothriochloa* spp.).

The diaspore, the dispersal unit for the grass seed, can vary from a simple grain as in big sacaton, to the floret as in threeawns, the spikelet as in gramas, or the entire inflorescence as in squirreltail.

Allred and Columbus (1988) developed a floral formula for grasses as an aid to species identification. They emphasized the number of glumes, florets, and reduced (unisexual or sterile) florets; the site of disarticulation (where the seed or floret separates from the inflorescence); and the inflor-

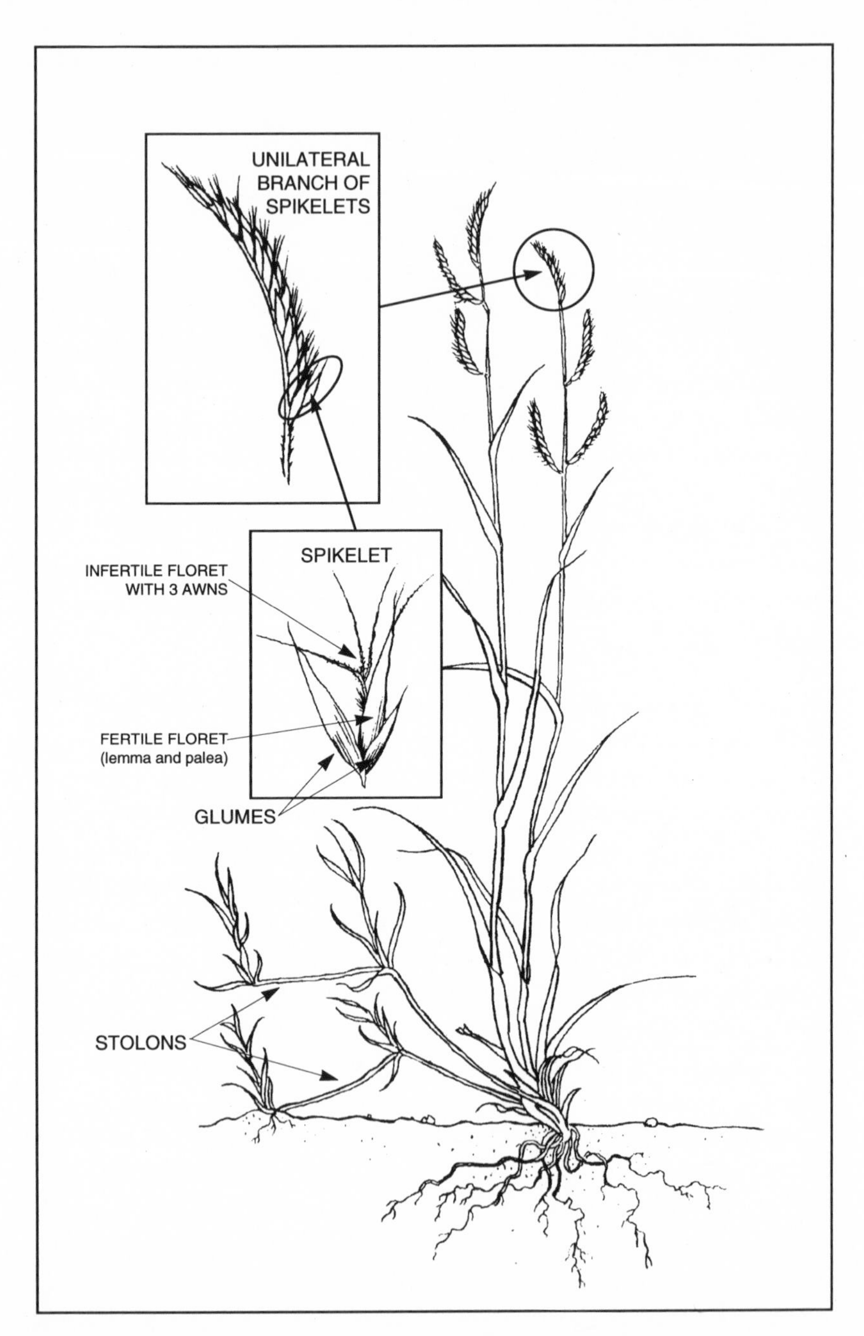

Figure 1.7. Black grama (*Bouteloua eriopoda*) reproduces asexually with stolons (extravaginal tillers) and sexually with florets. The florets (one fertile and one infertile) are arranged in spikelets, with 12–20 spikelets on a unilateral branch and unilateral branches of spikelets arranged in an inflorescence.

escence type. The formula includes an account of the number of nerves and awns on the glumes and floret (usually the lemma). Their formula approach is extremely helpful because these traits form the basis of most species identification guides.

As a final note on sexual reproduction in grasses, it is interesting to examine the energetic benefits of the grass flower's reduction to inconspicuous florets and its reliance on wind pollination. In their summary of grass physiological studies, Redman and Reekie (1982) showed that up to 60 percent of the energy devoted to seed production originates from photosynthesis within the inflorescence and the flag leaf (the first leaf below the inflorescence). In contrast, the conspicuous flowers with colorful petals and fragrant aromas of other plant families have very little, if any, photosynthetic capacity. In essence, the energetics of grass seed production benefits from the lack of investment in showy petals and attractive aromas as well as sustained productive capacity from photosynthesis within the inflorescence.

Grass Taxonomy and Identification

Taxonomic treatments of the grass family include subdivisions at the subfamily and tribal levels because of the large number and variety of species. Although floret and spikelet characteristics weigh heavily in taxonomic treatments at the genus and species level, biochemical, cellular, and embryonic characteristics are often included in the schemes that differentiate subfamilies and tribes. In general, the level of agreement among taxonomists is greatest at the species and genus levels, and lowest at the subfamily and tribal levels (Renvoize and Clayton 1992; Watson and Dallwitz 1992). To the ecologist, natural historian, or land manager, the different systems for classifying tribes and subfamilies are largely irrelevant because each approach uses nearly identical criteria for distinguishing genera and species (Watson 1990).

Several published sources identify the grass species and genera found in the desert grassland. Some sources emphasize floral characteristics (Allred 1993; Beetle and Johnson-Gordon 1991; Gould 1951, 1975; Hitchcock and Chase 1950), and some use only vegetative characteristics to determine species identity (Barnard and Potter 1984; Copple and Pase 1978). Identification using vegetative characteristics emphasizes ligule, leaf, phytomer, and tiller morphology. The vegetative identification guides can be very useful because floral parts are absent during most of the year.

Grass Growth and Regrowth Following Defoliation

Understanding the pattern of grass growth provides a basis for comprehending how members of this family can tolerate defoliation (removal of biomass by grazing animals, fire, or clipping). All plant growth occurs at meristems, areas of cell division, morphological differentiation (genesis of different plant structures such as leaves, internodes, and inflorescences), and cell growth. Phytomer differentiation into node, leaf blade and sheath, internode, and axillary bud occurs in the apical meristem that is located at the base of the tiller (figs. 1.6 and 1.8A; Briske 1991). There are three intercalary meristems—areas of cell division and growth but not differentiation—and they are located at the base of the leaf sheath, leaf blade, and internode (fig. 1.6; Dahl and Hyder 1977). Intercalary meristem activity ceases in the leaf blade when the ligule is formed, and it ceases in the leaf sheath when the ligule is exposed. There is no regular pattern for cessation in the internode intercalary meristem (Briske 1991).

Typically, many phytomers are produced before the internodes begin to elongate, which keeps the apical meristem near the base of the tiller (fig. 1.8B). At this stage, only leaf blades and sheaths are extended above the tiller base. Elongation of the internodes increases the height of the tiller and elevates the apical meristem above the tiller base (fig. 1.8C). Tiller extension by internode elongation is usually initiated by differentiation of the inflorescence in the apical meristem; however, tall asexual tillers can also be produced. All phytomer differentiation ceases in the apical meristem after inflorescence differentiation, and therefore all subsequent vegetative growth within the tiller occurs from the growth of existing cells (Briske 1991). Increased day length is known to initiate inflorescence differentiation (Briske 1991).

New tillers originate from axillary buds located above the node at the base of the phytomer (fig. 1.6). New tillers may emerge from the leaf sheath by appearing near the top of the sheath (intravaginal tillers) or through the sheath base (extravaginal tillers). Intravaginal tillers typically appear from the phytomer nearest the ground surface (the oldest phytomer) and give the plant a very bunchlike, or caespitose, appearance. Tillers originate from axillary buds on elevated phytomers in some grasses, however, resulting in a diffuse branching pattern and bushy appearance. Diffuse branching is common in two desert grasses: bush muhly and Arizona cottontop. Burgess (this volume) refers to these as suffrutescent grasses. Rhizomes and stolons (below- and aboveground asexual reproductive tillers;

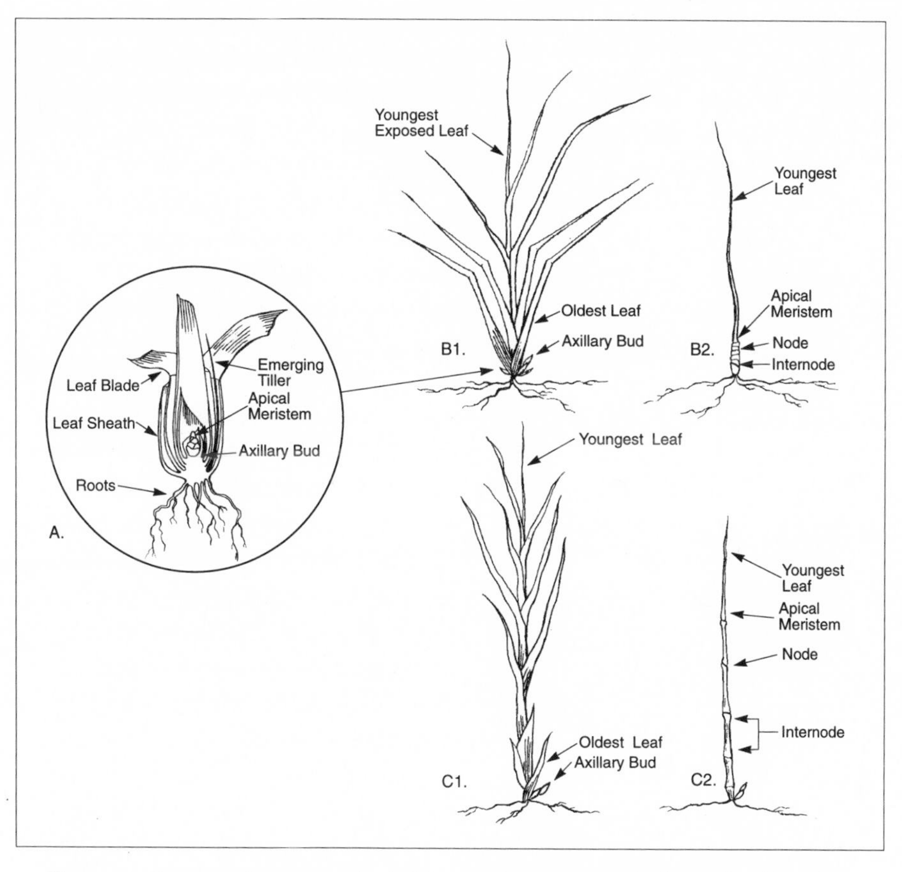

Figure 1.8. Patterns of grass growth. A. The apical meristem at the base of a tiller where the components of phytomers are being differentiated. New tillers (intravaginal) develop from axillary buds. Adapted from Jewiss 1972. B1. Leaf arrangement when the apical meristem is near the base of the tiller. B2. With all but the youngest leaf removed from B1, the short internodes and basal location of the apical meristem are obvious. C1. Leaf arrangement when the apical meristem is elevated near the top of the tiller after the internodes have elongated. C2. With all but the youngest leaf removed from C1, the elongated internodes and elevated position of the apical meristem are obvious. Adapted from Dahl and Hyder 1977.

fig. 1.7) emerge as extravaginal tillers from the base of the lowest phytomer (Briske 1991).

It has been suggested that recruitment of new tillers from axillary buds in the basal phytomer is stimulated by an increase in the red : far-red light wave ratio (Deregibus et al. 1985), changes in growth hormone levels (Harrison and Kaufman 1980), and removal of the apical meristem (Richards et al. 1988). A shift to greater red light exposure can occur if the plant cover shading the base of the tiller is removed. An increase in the number of tillers occurred when red light was experimentally increased with optical fibers at the base of grass plants (Deregibus et al. 1985). Tillering response to apical meristem removal is not consistent among grass species. Richards et al. (1988) compared the tillering response of grasses whose apical meristem was removed with a treatment in which the apical meristem and 85 percent of the foliage were removed. Subsequent tillering was greater in two temperate grasses when they were defoliated than when only the apical meristem was removed. Three tropical grasses (including the desert grassland species tanglehead, *Heteropogon contortus*), however, exhibited a similar tillering response to both treatments. These results suggest that apical dominance is more important in controlling tillering in the tropical species, and absolute defoliation is more important in the temperate species.

Although the number of tillers per plant appears to be unaffected by defoliation, the tillers are recruited more slowly through the growing season in defoliated plants than in undefoliated ones (Butler and Briske 1988; Olsen and Richards 1988).

Regrowth potential following defoliation can occur at any or all of the three types of meristems: apical, intercalary, and axillary. The relative contribution of each to regrowth is a function of the severity of the defoliation and the season in which it takes place (Briske 1991). If the defoliation occurs early in the growing season or early in the development of a tiller before the apical meristem is elevated, then only leaf blades and sheaths are defoliated and the apical meristem remains intact. In this case, additional phytomers continue to emerge from the intact apical meristem. If the intercalary meristems are not removed, the blades, sheaths, and internodes will continue to elongate. If the blade tip is defoliated but the intercalary meristem remains, the blade will continue to grow from that meristem but the leaf tip will retain a cut appearance (Hyder 1974). Regrowth is limited to new tiller development from axillary buds when defoliation occurs late in the season and the elevated apical meristem is removed, when floral differentiation has been initiated in the apical meristem, or when the level of defoliation is so severe that the apical meristem is removed when it is still at the base of the tiller (Briske 1991).

For a long time it was thought that the concentration of carbohydrates in the grass crown (the basal, aboveground portion of the grass) at the time of defoliation controlled regrowth by providing a reserve of energy, but more recent studies have shown that up to 99 percent of the carbohydrates used for regrowth originate in the remaining undefoliated plant tissue. The carbohydrate reserves in the grass crown are equivalent to only three days of photosynthesis (Davidson and Milthrope 1966; Richards and Caldwell 1985). Therefore, the amount of plant tissue remaining after defoliation is a better predictor of regrowth potential than carbohydrate reserves in the crown.

The pattern of carbohydrate allocation to axillary meristems following defoliation combines with meristem removal and amount of leaf area remaining in determining regrowth potential. Richards and Caldwell (1985) compared the regrowth of two grasses with very similar meristem locations and found that the species allocating more carbohydrates to axillary meristems showed greater regrowth, even though the total available carbohydrates were no different.

Desert Grass Demography

New plant recruitment in desert grassland grasses is from seeds, rhizomes, or stolons. The average longevity of plants following seedling emergence or stolon or rhizome establishment is about two years, but the probability of survival increases from less than 30 percent in the first two years to more than 60 percent for the next four years (Canfield 1957; Wright and Van Dyne 1976).

On the Jornada Experimental Range, increased livestock grazing decreased the survival rates of black grama and red threeawn (*Aristida longiseta*) on sandy soils, but mesa dropseed survival increased with greater grazing intensity on loamy soils (Wright and Van Dyne 1976). The depression of initial recruitment, seedling survival, and maximum longevity by livestock grazing were greatest for sideoats grama, tanglehead, and wolftail (*Lycurus setosus*) on the Santa Rita Experimental Range (Canfield 1957).

In the desert grassland, maximum longevity varies from 28 years for black grama to 4 years for Rothrock grama (Canfield 1957; Wright and Van Dyne 1976). Longevity is highly variable among desert grassland locations; the maximum age of black grama was only 14 years on the Santa Rita Experimental Range (Canfield 1957).

Summary

The desert grassland does not conform to parochial interpretations of grasslands. Because it experiences the hottest, driest, and sunniest climate of all North American grasslands, it has one of the lowest levels of primary production and rates of solar energy conversion. Furthermore, the potential abundance of shrubs and small trees is unusual for grasslands. The period of peak growth is later than in other North American grasslands and coincides with the greatest period of rainfall in June–September. As a result of this late-season growth period, the proportion of grasses with the C_4 photosynthesis pathway is greater than that found in other North American grasslands.

The relative proportions of grasses and shrubs in the desert grasslands have changed dramatically during the last century; contemporary changes also include the arrival and spread of introduced species. Evaluating the impact of human-caused fragmentation of species richness (biodiversity) will require extraordinary efforts because the naturally fragmented spatial distribution of the desert grassland is already very significant. These historic and current vegetation changes, and the fragmented spatial distribution, can be observed by visiting the four desert grassland areas described above, all of which permit public access.

The grass family is the fourth largest plant family, but its members are probably more noteworthy for their unique morphology and physiology, which translates into an ability to withstand repeated defoliation. Their three types of meristems are the basis for this regrowth potential. Severity of defoliation and the season in which it occurs are the most important factors determining the expression of this regrowth potential, the former because the remaining plant tissue provides the majority of carbohydrates that fuel regrowth; the latter because late in the growing season the apical meristem is elevated, and if it is defoliated, regrowth is limited to the remaining intercalary or axillary meristems.

Desert grassland grasses are relatively short-lived, and the influence of defoliation on longevity is not uniform among species. Maximum longevity is on the order of one to several decades, but this varies greatly among locations. Finally, all grasses do not respond to weather patterns or defoliation in the same way.

The ecologist, natural historian, or land manager able to identify the different grasses can begin to appreciate and systematically observe the variety of behaviors exhibited by the desert grassland grasses. Grass identification guides generally focus on differences in reproductive morphology, but the guides that focus on vegetative characteristics can be useful during the majority of the year when floral parts are absent.

Acknowledgments

Carol Cochran, G. R. McPherson, and Thomas R. Van Devender provided constructive comments on an earlier draft of this chapter. Katherine Sanchez assisted with acquiring information on the four desert grassland areas. Robert P. Gibbens assisted with photographs at the Jornada Experimental Range. Monte Bingham created the line drawings.

Literature Cited

Allred, K. W. 1982. Describing the grass inflorescence. Journal of Range Management 35:672–675.

———. 1993. A Field Guide to the Grasses of New Mexico. New Mexico State University Agricultural Experiment Station, Las Cruces.

Allred, K. W., and J. T. Columbus. 1988. The grass spikelet formula: an aid to teaching and identification. Journal of Range Management 41:350–351.

Anable, M. E. 1989. Alien plant invasion in relation to site characteristics and disturbance: *Eragrostis Lehmanniana* on the Santa Rita Experimental Range, Arizona, 1937–89. M.S. thesis, University of Arizona, Tucson.

Anable, M. E., M. P. McClaran, and G. B. Ruyle. 1992. Spread of introduced Lehmann lovegrass (*Eragrostis lehmanniana* Nees.) in southern Arizona, USA. Biological Conservation 61:181–188.

Bahre, C. J. 1977. Land-use history of the Research Ranch, Elgin, Arizona. Journal of the Arizona Academy of Science 12(suppl. 2):1–32.

———. 1991. A Legacy of Change: Historic Impact on Vegetation in the Arizona Borderlands. University of Arizona Press, Tucson.

Barnard, C. M., and L. D. Potter. 1984. New Mexico Grasses: A Vegetative Key. University of New Mexico Press, Albuquerque.

Beetle, A. A., and D. Johnson-Gordon. 1991. Gramineas de Sonora. Secretaria de Agricultura y Recursos Hidraúlicos, Comisión Técnico Consultiva de Coeficientes de Agostadero, Hermosillo, Sonora, Mexico.

Bock, J. H., and C. E. Bock. 1993. Cover of perennial grasses in southeastern Arizona in relation to livestock grazing. Conservation Biology 7:371–377.

Bock, C. E., J. H. Bock, and M. C. Grant. 1992a. Effects of bird predation on grasshopper densities in an Arizona grassland. Ecology 73:1706–1717.

Bock, C. E., A. Cruz, M. C. Grant, C. S. Aid, and T. R. Strong. 1992b. Field experimental evidence for diffuse competition among southwestern riparian birds. American Naturalist 140:815–828.

Brady, W. W., M. R. Stromberg, E. F. Aldon, C. D. Bonham, and S. H. Henry. 1989. Response of a semidesert grassland to 16 years of rest from grazing. Journal of Range Management 42:284–288.

Briske, D. D. 1991. Developmental morphology and physiology of grasses. Pp. 85–108 *in* R. K. Heitschmidt and J. W. Stuth (eds.), Grazing management: an ecological perspective. Timber Press, Portland.

Brown, D. E. 1982a. Plains and Great Basin grasslands. *In* D. E. Brown (ed.), Biotic communities of the American Southwest—United States and Mexico. Desert Plants 4:115–121.

———. 1982b. Semidesert grassland. *In* D. E. Brown (ed.), Biotic communities of the American Southwest—United States and Mexico. Desert Plants 4:123–131.

Buffington, L. C., and C. H. Herbel. 1965. Vegetational changes on a semidesert grassland range from 1858 to 1963. Ecological Monographs 35:139–164.

Butler, J. L., and D. D. Briske. 1988. Population structure and tiller demography of the bunchgrass *Schizachyrium scoparium* in response to herbivory. Oikos 51:306–312.

Canfield, R. H. 1957. Reproduction and life span of some perennial grasses of southern Arizona. Journal of Range Management 10:199–203.

Clayton, W. D. 1990. The spikelet. Pp. 32–51 *in* G. P. Chapman (ed.), Reproductive versatility in the grasses. Cambridge University Press, Cambridge, England.

Copple, R. F., and C. P. Pase. 1978. A vegetative key to some of the common Arizona range grasses. U.S. Department of Agriculture Forest Service General Technical Report RM-53.

Dahl, B. E., and D. N. Hyder. 1977. Developmental morphology and management implications. Pp. 258–290 *in* R. E. Sosebee (ed.), Rangeland plant physiology. Society for Range Management, Denver.

Davidson, J. L., and F. L. Milthrope. 1966. Leaf growth in *Dactylis glomerata* following defoliation. Annals of Botany 30:173–184.

Deregibus, V. A., R. A. Sanchez, J. J. Casal, and M. J. Trlica. 1985. Tillering reposes to enrichment of red light beneath the canopy in a humid natural grassland. Journal of Applied Ecology 22:199–206.

Dick-Peddie, W. A. 1993. New Mexico Vegetation: Past, Present, and Future. University of New Mexico Press, Albuquerque.

Etter, A. G. 1951. How Kentucky bluegrass grows. Annals of the Missouri Botanical Garden 38:293–375.

Fernald, A. S. 1987. Plant community ecology of two desert marshes in southeastern Arizona: Babocomari cienega and Canelo Hills cienega. M.S. thesis, University of Colorado, Boulder.

Franklin, J. F., C. S. Bledsoe, and J. T. Callahan. 1991. Contributions of the Long-Term Ecological Research Program. Bioscience 40:509–523.

French, N. R. 1979. Principal subsystem interactions in grasslands. Pp. 173–190 *in* N. R. French (ed.), Perspectives in grassland ecology: results and applications of the US/IBP Grassland Biome Study. Springer-Verlag, New York.

Garcia-Moya, E., and C. M. McKell. 1970. Contribution of shrubs to the nitrogen economy of a desert-wash plant community. Ecology 51:81–88.

Gibbens, R. P., and R. F. Beck. 1988. Changes in basal area and forb densities over a 64-year period on grassland types of the Jornada Experimental Range. Journal of Range Management 41:186–192.

Gibbens, R. P., K. M. Havstad, D. D. Billheimer, and C. H. Herbel. 1993. Creosotebush vegetation after 50 years of lagomorph exclusion. Oecologia 94:210–217.

Gould, F. W. 1951. Grasses of the Southwestern United States. University of Arizona Press, Tucson.

———. 1975. The Grasses of Texas. Texas A&M University Press, College Station.

Gurevitch, J. 1986. Restrictions of a C_3 grass to dry ridges in a semiarid grassland. Canadian Journal of Botany 64:1006–1011.

Harrison, M. A., and P. B. Kaufman. 1980. Hormonal regulation of lateral bud (tiller) release in oats (*Avena sativa* L.). Plant Physiology 66:1123–1127.

Hastings, J. R., and R. M. Turner. 1965. The Changing Mile: An Ecological Study of Vegetation Change with Time in the Lower Mile of an Arid and Semiarid Region. University of Arizona Press, Tucson.

Hattersly, P. W. 1992. C_4 photosynthetic pathway variation in grasses (Poaceae): Its significance for arid and semiarid lands. Pp. 181–212 *in* G. P. Chapman (ed.), Desertified grasslands: their biology and management. Academic Press, New York.

Hattersly, P. W., and L. Watson. 1992. Diversification of photosynthesis. Pp. 38–116 *in* G. P. Chapman (ed.), Grass evolution and domestication. Cambridge University Press, Cambridge, England.

Hennessey, J. T., R. P. Gibbens, J. M. Tromble, and M. Cardenas. 1983. Vegetation changes from 1935 to 1980 in mesquite dunelands and former grasslands of southern New Mexico. Journal of Range Management 36: 370–374.

Hidy, G. M., and H. E. Klieforth. 1990. Atmospheric processes affecting the climate of the Great Basin. Pp. 17–45 *in* C. B. Osmond, L. F. Pitelka, and G. M. Hidy (eds.), Plant biology of the basin and range. Springer-Verlag, New York.

Hitchcock, A. S., and A. Chase. 1950. Manual of the Grasses of the United States. U.S. Department of Agriculture Miscellaneous Publication 200. Reprint. Dover Books, New York.

Hyder, D. N. 1974. Morphogenesis and management of perennial grasses in the U.S. Pp. 89–98 *in* Plant morphogenesis as a basis for scientific management of range resources. U.S. Department of Agriculture Miscellaneous Publication 1271.

Kortschack, H. P., C. E. Hartt, and G. O. Burr. 1965. Carbon dioxide fixation in sugarcane leaves. Plant Physiology 40:209–213.

Lauenroth, W. K. 1979. Grassland primary production: North American grasslands in perspective. Pp. 3–24 *in* N. R. French (ed.), Perspectives in grassland ecology: results and applications of the US/IBP Grassland Biome Study. Springer-Verlag, New York.

Long, S. P., and P. R. Hutchin. 1991. Primary production in grasslands and coniferous forests with climate change: an overview. Ecological Applications 1:139–156.

Martin, S. C. 1966. The Santa Rita Experimental Range: a center for research on improvement and management of semidesert rangelands. U.S. Department of Agriculture Forest Service Research Paper RM-22.

———. 1975. Ecology and management of southwestern semidesert grass-shrub ranges: a status of knowledge. U.S. Department of Agriculture Forest Service Research Paper RM-156.

Martin, S. C., and H. G. Reynolds. 1973. The Santa Rita Experimental Range: your facility for research on semidesert ecosystems. Journal of the Arizona Academy of Science 8:56–67.

Martin, S. C., and K. E. Severson. 1988. Vegetation response to the Santa Rita grazing system. Journal of Range Management 41:291–295.

McClaran, M. P., and M. E. Anable. 1992. Spread of introduced Lehmann lovegrass along a grazing intensity gradient. Journal of Applied Ecology 29:92–98.

McClaran, M. P., P. Ang, A. Capurro, D. Deutchman, D. Shafer, and J. Guarini. 1995. Interpreting explanatory processes for time series patterns: lessons from three times series. Pp. 465–482 *in* T. M. Powell and J. H. Steele (eds.), Ecological time series. Chapman & Hall, New York.

Neilson, R. P. 1986. High-resolution climatic analysis and Southwest biogeography. Science 232:27–34.

———. 1987. Biotic regionalization and climatic controls in western North America. Vegetatio 70:135–147.

Olsen, B. E., and J. H. Richards. 1988. Annual replacement of the tillers of *Agropyron desertorum* following grazing. Oecologia 76:1–6.

Redman, R. E., and E. G. Reekie. 1982. Carbon allocation in grasses. Pp. 195–231 *in* J. R. Estes, R. J. Tyrl, and J. N. Brunken (eds.), Grasses and grasslands: systematics and ecology. University of Oklahoma Press, Norman.

Renvoize, S. A., and W. D. Clayton. 1992. Classification and evolution of the grasses. Pp. 3–37 *in* G. P. Chapman (ed.), Grass evolution and domestication. Cambridge University Press, Cambridge, England.

Richards, J. H., and M. M. Caldwell. 1985. Soluble carbohydrates, concurrent photosynthesis and efficiency in regrowth following defoliation: a field study with *Agropyron* species. Journal of Applied Ecology 22:907–920.

Richards, J. H., R. J. Mueller, and J. J. Mott. 1988. Tillering in tussock grasses in relation to defoliation and apical meristem removal. Annals of Botany 62:173–179.

Saunders, D. A., R. J. Hobbs, and C. R. Margules. 1991. Biological consequences of ecosystem fragmentation: a review. Conservation Biology 5:18–32.

Schmutz, E. M., E. L. Smith, P. R. Ogden, M. L. Cox, J. O. Klemmedson, J. J. Norris, and L. C. Fierro. 1991. Desert grassland. Pp. 337–362 *in* R. T. Coupland (ed.), Natural grasslands: introduction and Western Hemisphere. Ecosystems of the World 8A. Elsevier, Amsterdam.

Sims, P. L., and R. T. Coupland. 1979. Producers. Pp. 49–72 *in* R. T. Coupland (ed.), Grassland ecosystems of the world: an analysis of grasslands and their uses. Cambridge University Press, Cambridge, England.

Sims, P. L., and J. S. Singh. 1978a. The structure and function of the ten western North American grasslands. II. Intra-seasonal dynamics in primary producer compartments. Journal of Ecology 66:547–572.

———. 1978b. The structure and function of the ten western North American grasslands. III. Net primary production, turnover and efficiencies of energy capture and water use. Journal of Ecology 66:573–597.

Sims, P. L., J. S. Singh, and W. K. Lauenroth. 1978. The structure and function of the ten western North American grasslands. I. Abiotic and vegetational characteristics. Journal of Ecology 66:251–285.

Smith, J. P., Jr. 1993. Poaceae [Gramineae]. Pp. 1218–1303 *in* J. C. Hickman (ed.), The Jepson manual: higher plants of California. University of California Press, Berkeley.

Soderstrom, T. R., and C. E. Calderon. 1971. Insect pollination of tropical rainforest grasses. Biotropica 3:1–16.

Tiedemann, A. R., and J. O. Klemmedson. 1973a. Effects of mesquite on physical and chemical properties of the soil. Journal of Range Management 26:27–29.

———. 1973b. Nutrient availability in desert grassland soils under mesquite (*Prosopis juliflora*) trees and adjacent open areas. Soil Society of America Proceedings 37:107–111.

Tiedemann, A. R., J. O. Klemmedson, and P. R. Ogden. 1971. Response of four perennial southwestern grasses to shade. Journal of Range Management 24:442–447.

Van Deren, K. J. 1993. The influence of invasive Lehmann lovegrass on two native grasses in the semi-desert grassland. M.S. thesis, University of Arizona, Tucson.

Watson, L. 1990. The grass family, Poaceae. Pp. 1–31 *in* G. P. Chapman (ed.), Reproductive versatility in the grasses. Cambridge University Press, Cambridge, England.

Watson, L., and M. J. Dallwitz. 1992. The Grass Genera of the World. CAB International, Wallingford, England.

Wright, R. G., and G. M. Van Dyne. 1976. Environmental factors influencing semidesert grassland perennial grass demography. Southwestern Naturalist 21:259–274.

2

Desert Grassland, Mixed Shrub Savanna, Shrub Steppe, or Semidesert Scrub?

The Dilemma of Coexisting Growth Forms

Tony L. Burgess

I have been uncomfortable with the term *desert grassland* since my first exposure to this patchy vegetation of grass, shrubby trees, succulents, rosette plants, and subshrubs—a distinctive species assemblage unlike that of any other North American landscape. Because the main economic utility of this vegetation type is livestock production, we tend to exploit it in the same manner as other grasslands, but an economic focus devalues the presence of the other growth forms in the vegetation called desert grassland. Some desert grassland areas have been fairly stable, but many have experienced shifts in plant dominance to the point that grasses have become scarce and the vegetation has been completely altered.

In this chapter I explore the concepts unifying, and sometimes dividing, the landscapes and communities called desert grassland. I offer explanations of how the different growth forms coexist and how the vegetation can change dramatically. The structure and dynamics of desert grasslands can be understood as direct or indirect consequences of rainfall, soil, fire, and herbivory. After offering a brief review of comparable vegetation on other continents, I discuss how dynamic mixtures of various growth forms have created some confusion in the terms used to describe these North American landscapes. I conclude with a paradigm shift and propose a more fitting name, Apacherian mixed shrub savanna, for these communities.

Distinctive Features of Desert Grasslands

Southwestern desert grasslands typically feature growth forms related to both desertscrub and grassland. Many sites have seen a recent shift from grass dominance to shrub and subshrub dominance. The species present on a given desert grassland site may be fairly constant over time, but the relative abundances of those species can change substantially, propelling the vegetation into a different structure (Westoby et al. 1989). As a consequence, the vegetation at any particular time is likely to be highly dependent on its recent history (Laycock 1991). The following traits are characteristic of desert grasslands:

1. Grasses with suffrutescent, annual, and short-lived herbaceous perennial growth forms are common. *Suffrutescent* describes bushy grasses with perennial stems that may live for several years. A suffrutescent grass (e.g., bush muhly, *Muhlenbergia porteri*) can produce new growth from buds located well above its base, in contrast with a typical grass, which grows from basal buds. Most dominant grasses of desert grasslands, such as blue grama (*Bouteloua gracilis*), are ceaspitose bunchgrasses; but some common species, including curly mesquite grass (*Hilaria belangeri*), are sod forming. Annual and short-lived perennial grasses are often transiently dominant after wet periods.
2. Subshrubs coexist with grasses, and their relative abundances often shift in response to changes in seasonal rainfall, grazing, and the soil surface. Subshrubs (e.g., snakeweed, *Gutierrezia sarothrae*) are woody or partly woody plants usually less than 50 cm tall. They are also called half-shrubs, or chamaephytes in the Raunkiaer plant life form system (Mueller-Dombois and Ellenberg 1974).
3. Succulents that store water (e.g., prickly pear, *Opuntia* sp.) and rosette plants such as sotol (*Dasylirion* sp.) may be conspicuous components of the vegetation. Frequent fires and stable soil moisture regimes tend to exclude these growth forms.
4. Desert grasslands are often distributed in a rather patchy manner over the landscape, unlike the more extensive grasslands on the plains and prairies to the east. The complex landforms and geological mosaics typical of the southwestern Basin and Range Province foster a variety of con-

tacts between woodland, chaparral, grassland, and desertscrub vegetation (D. E. Brown and Lowe 1980; McAuliffe, this volume).
5. Fires are less common in desert grasslands than in grasslands, which have more available moisture and thus greater biomass production (McPherson, this volume).
6. Most desert grasslands receive both winter and summer rainfall. Warm-season rainfall usually exceeds winter rainfall in the southern and eastern areas, but there is an almost equal likelihood of summer and winter rainfall in northern and western Arizona (McClaran, this volume). Desert grasslands usually experience drought in spring and early summer, followed by a midsummer or early autumn peak in rainfall. The southern areas, such as Durango, Mexico, have shorter spring droughts, and summer rains usually start in May (Gentry 1957). In contrast, the plains and Great Basin grasslands to the east and north typically experience spring and autumn rainfall peaks (Neilson 1987).
7. The same climate may support woodland, grassland, savanna, or scrub, depending on the site's soil and topographic position. Different histories of disturbance by fire and livestock add to the complexity of the vegetation mosaic.

Water-Use Strategies

Primary production in desert grassland ecosystems is more constrained by drought than by any other factor. Release from drought comes in distinct pulses of rainfall delivered in varying rhythms and spatial patterns (Noy-Meir 1973). Storm runoff is unevenly distributed over the landscape, and the different soils have different regimes of moisture storage and evaporation. This variable availability of soil moisture is reflected in the variety of coexisting growth forms. Each form represents a different mode of water exploitation, a specialization for using some part of the available soil moisture. Some weather sequences and sites favor one strategy, and some favor another. This gives a rich texture to the landscape but makes it difficult to classify and categorize the complex vegetation patterns.

To interpret differences in water-use strategies it is necessary to understand the dynamics of soil moisture. Most often the moisture is delivered in light showers that do not infiltrate deeply into the soil (Sala and Lauenroth 1982). High daytime temperatures in summer cause rapid evaporation near the soil surface. Shallow soil horizons are wet frequently by light rains, but during the warm season the soil dries very quickly. The result is a highly variable, rapidly changing soil moisture regime near the surface. The deep horizons are seldom wet, but once saturated they lose moisture slowly. Deep soaking by occasional tropical storms in autumn can influence growth for months. Winter moisture delivered gradually at cool temperatures is more likely to penetrate into lower soil horizons than summer rain dropped by brief thunderstorms. In desert grasslands, the deeper parts of the soil profile usually have less available water, but the moisture is more stable than at shallower depths. The exceptions are sites where deep moisture is derived from subterranean aquifers rather than from direct infiltration. Plants with access to stable aquifers are not limited by the factors that shape upland communities, and the resulting phreatophytic riparian vegetation contrasts sharply with surrounding areas.

Plants extract water from the soil by maintaining a diffusion gradient between their internal tissues and the soil (Nobel 1974). Water potential, or pressure, is nearly zero in a saturated soil, by definition. That is, it takes very little pressure to extract water from wet soil. As the soil dries, more pressure is needed to extract water. This condition is expressed in terms of a lower, more negative number. Moist soil might have a water potential of −0.1 megapascals (MPa), whereas soil with a water potential of −2 MPa would be considered dry. If the water potential within the plant is greater (less negative) than the potential in the soil, the roots cannot absorb water. For example, if the internal potential of a barrel cactus (*Ferocactus* sp.) is −0.7 MPa, it cannot take up water from soil with a potential of −1.0 MPa.

A simple classification of water-use strategies can help decipher the vegetative structure of a desert grassland. If one compares root system architecture and the usual variation of a plant's internal water content, sometimes called its characteristic hydrature (Walter and Stadelman 1974), three recognizable types emerge among the plants dominating the southwestern drylands: intensive exploiters, extensive exploiters, and water storers. These divisions are not in strict accord with more traditional growth form schemes (Mueller-Dombois and Ellenberg 1974) because they are based on the combination of a plant's habit with its inferred physiology and life history.

Intensive Exploiters

Intensive exploiters extract a large proportion of their moisture from shallow soil horizons through their dense network of shallow roots (fig. 2.1). An intensive exploiter tracks the soil moisture with its internal water status and appropriates most of the moisture in the shallow soil layers. Intensive

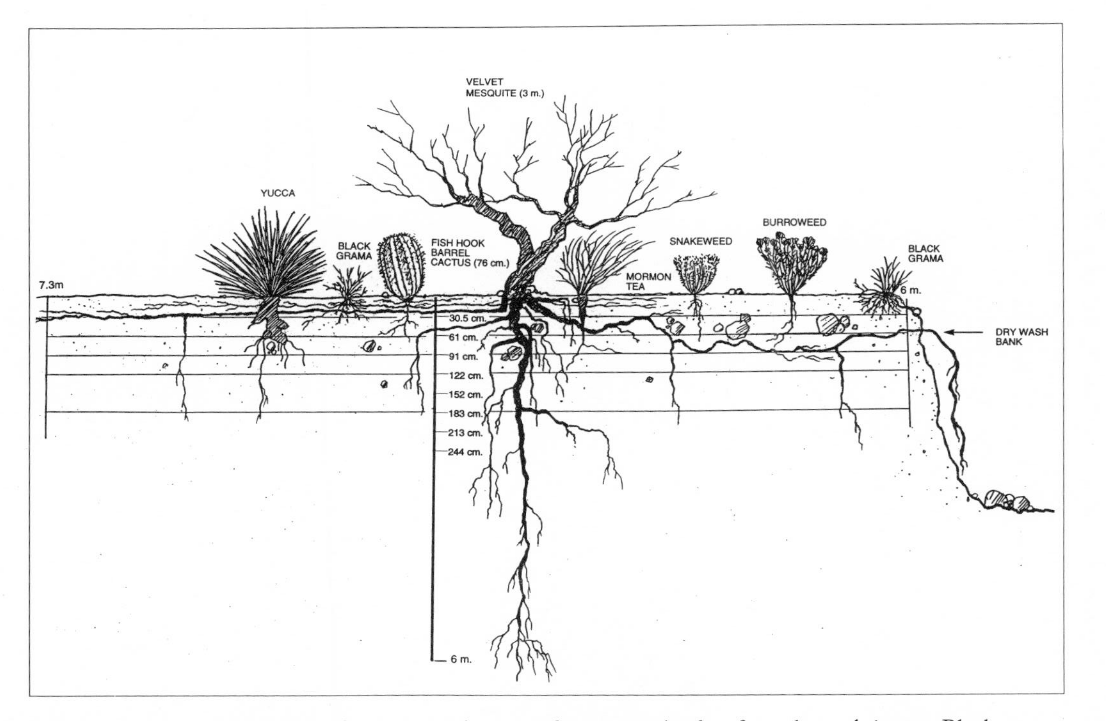

Figure 2.1. Root distribution of common plants in desert grasslands of southern Arizona. Black grama (*Bouteloua eriopoda*) and snakeweed (*Gutierrezia sarothrae*) are intensive exploiters. Velvet mesquite (*Prosopis velutina*) and Mormon tea (*Ephedra* sp.) are extensive exploiters. Burroweed (*Isocoma tenuisecta*) uses an intermediate strategy. Fishhook barrel cactus (*Ferocactus wislizenii*) and yucca (*Yucca* sp.) are water storers. Adapted from Cannon 1911; Cable 1969; and examples from washes in San Xavier District of the Tohono O'odham Nation.

exploiters are very effective competitors for the limited shallow soil moisture. They tend to be very resilient; they recover rapidly from stress or damage and grow rapidly when soil moisture is available. Most intensive exploiters in desert grasslands respond best to summer rainfall, but some species can also use cool-season moisture. Their potential for rapid growth and reproduction allows intensive exploiters to dominate denuded landscapes faster than plants that use other strategies.

Perennial grasses (e.g., tobosa, *Hilaria mutica*) exemplify this strategy. Grasses that dominate semiarid grasslands develop a dense network of roots concentrated in the upper parts of the soil where rainfall penetrates most frequently (Blydenstein 1966; Cable 1969; Sala and Lauenroth 1985). In contrast, the roots of some midwestern tall-grass prairie grasses penetrate several meters into the soil (Weaver 1920, 1954) because water infiltration is deeper on the prairies than in desert grasslands.

Sala et al. (1981) found that the blue grama leaf water potential reflects the water status of the soil occupied by its roots. During the day, leaves transpire while acquiring CO_2 from the atmosphere. They usually lose water faster than the roots can absorb it, and the leaf water potential drops. At night, when transpiration ceases, water absorbed by the roots raises the grass's internal water potential, and by dawn the leaf water potential is almost the same as that of the wettest soil layer occupied by the roots. As the soil dries, the daily maximum water potential of the grass decreases (fig. 2.2). A lower internal water potential gives the leaf less time during the day to open its stomata to take in CO_2 before the internal water level reaches a threshold of dehydration that could damage its tissues. The leaf stomata close as this threshold is approached. If the soil continues to dry out, at some point the grass's active metabolic demands exceed its carbon intake and it must become dormant to survive.

Intensive exploiters respond to brief rainfall events with rapid root growth to absorb soil moisture before it evaporates. Drought-stressed blue grama can rehydrate within one day after receiving as little as 5 mm of rainfall (Sala and Lauenroth 1982). The grass is thus able to briefly resume its acquisition of atmospheric carbon before the soil dries and the plant becomes stressed again (fig. 2.2).

Perennial grasses are typical intensive exploiters, but they are not alone in employing this water-use strategy. Annual grasses and most annual forbs have root architectures and physiologies consistent with the intensive exploiter strategy as well (Cable 1969; Cannon 1911; Harris 1967; Mack and Pyke 1984).

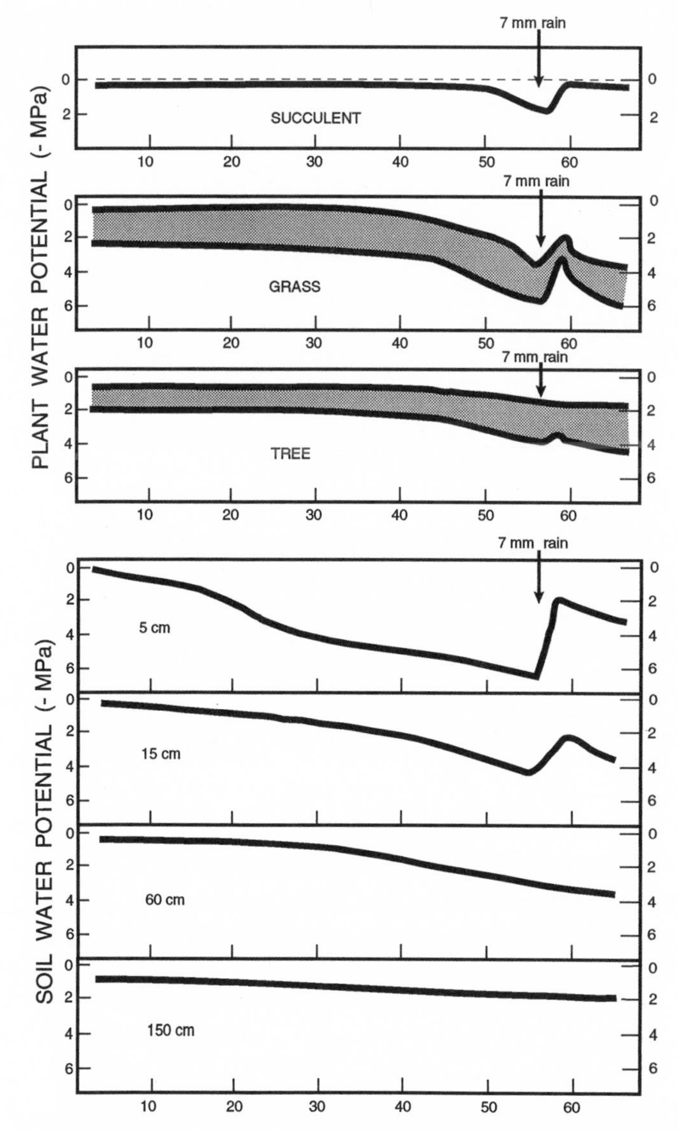

Figure 2.2. Soil and plant water potentials during a drying cycle ended by a light rainfall on day 56. As the soil and plants desiccate, their potentials become more negative. The units of water potential are megapascals (MPa), which indicate the pressure needed to extract water from the soil or plant. The top three diagrams compare internal water potentials among succulent, grass, and tree growth forms. The shaded areas on the grass and tree diagrams represent the daily range of internal water potentials experienced by the plant. The daily range of most succulents is too small to portray. The lower diagrams compare soil moisture conditions at 5, 15, 60, and 150 cm soil depths. Adapted from Cable 1977; Haas and Dodd 1972; Sala and Lauenroth 1982; Sala et al. 1981; and Szarek and Ting 1975.

Many, but not all, subshrubs are or seem to be intensive exploiters. Snakeweed root systems occupy the same soil layers as those of perennial grasses (Weaver 1920) and sometimes appear to compete with the perennial grasses (Campbell and Bomberger 1934; Jameson 1966). Subshrubs differ from most perennial grasses in their pattern of shoot dieback during dormancy. Most grasses die back nearly to the soil surface each fall or winter, and each spring or summer the entire canopy must regrow. Suffrutescent grasses, also known as chamaephytic grasses (Shmida and Burgess 1988), such as black grama (*Bouteloua eriopoda*) and bush muhly are exceptions to this pattern. The shoots of subshrubs and suffrutescent grasses die back only in part, leaving some stems alive to start growth during the next growing season. Subshrubs are the most common perennial growth form in arid ecosystems (Shmida and Burgess 1988), and they often dominate vegetation in semiarid Mediterranean climates (Westman 1981); hence, their presence indicates a strong influence of drought. Snakeweed, burroweed (*Isocoma tenuisecta*), and Wright buckwheat (*Eriogonum wrightii*) are common subshrubs in desert grasslands.

Dwarf shrubs (nanophanerophytes in the Raunkiaer system) differ in having woodier stems that do not die back to the same extent as those of subshrubs. Fairy duster (*Calliandra eriophylla*) and feather dalea (*Dalea formosa*) are dwarf shrubs found in semiarid grasslands. The physiology and root architecture of most subshrubs and dwarf shrubs are poorly understood, and some surprises are certain to be discovered as these species are explored in greater detail.

Perhaps because they produce a leafy canopy more quickly after dormancy ceases, subshrubs and suffrutescent grasses seem to be favored in areas where drought stress is often so severe between rains that intensive exploiters must become fully dormant to survive (Shmida and Burgess 1988). As described above, blue grama can respond quickly to rainfall. After prolonged drought and during winter, however, its stems die back to the crown, and new growth must originate from ground level. If rainfall is scanty, the soil moisture may have evaporated by the time new leaves have developed. A suffrutescent grass such as big galleta (*Hilaria rigida*) can grow new leaves more quickly from stems located well above the soil than it could from basal sprouts (Nobel 1980). Quick regeneration of leaf area may allow subshrubs to use soil moisture that would be lost to herbaceous perennials.

Not all subshrubs are intensive exploiters. Cable (1969) found that burroweed, traditionally thought to compete with perennial grasses, has a taproot that takes up moisture below the soil layers most used by perennial grasses (fig. 2.1). This subshrub depends heavily on winter rainfall that infiltrates more deeply than brief summer showers. Burroweed thus uses soil moisture less available to the shallower roots of intensive exploiter grasses. In this respect it is intermediate between intensive exploiters and extensive exploiters.

Extensive Exploiters

Extensive exploiters have extensive root systems that penetrate a large volume of soil in both shallow and deep layers. Trees and shrubs such as mesquite (*Prosopis* spp.), juniper (*Juniperus* spp.), and tarbush (*Flourensia cernua*) are usually extensive exploiters (fig. 2.1). Parts of the soil that are too deep or have moisture distributed too erratically to support intensive exploiters are sometimes occupied by the roots of extensive exploiters (Walter 1971). Velvet mesquite (*Prosopis velutina*) offers an extreme case; its roots are known to penetrate depths greater than 50 m (Phillips 1963) and to extend 15 m or more beyond the edge of the tree canopy (Cable 1977).

Deep roots use moisture that arrives infrequently, most often during winter, but remains available for several months (fig. 2.2). Extensive root systems contact more stable moisture supplies, like that found beneath buried rocks (Evenari et al. 1971), and their ability to extract water from a large volume of soil compensates for the generally low water potential at deeper levels (Sala et al. 1989). Because their moisture supply is relatively dependable, extensive exploiters often have more stable internal water potentials than intensive exploiters have. They experience drought stress slowly, and they make metabolic adjustments that allow their leaves to continue acquiring CO_2 (Nilson et al. 1983).

Small rainfall events that are critical to the survival of intensive exploiters may be nearly imperceptible to extensive exploiters, producing only minor changes in their internal water status (fig. 2.2). They tend to grow more slowly than intensive exploiters because their woody tissue takes more energy to build. Because their seedlings must rely on shallow soil moisture for their initial growth, extensive exploiters are vulnerable to interference from intensive exploiters, which use the shallow soil moisture and limit the availability of light. Glendening and Paulson (1955) showed that perennial grasses can suppress velvet mesquite seedlings. Seedlings of extensive exploiters rarely become established in the presence of vigorous grasses. To compensate, they tend to be very fecund. A single velvet mesquite can produce over 142,000 seeds in a year (Glendening and Paulson 1955).

Simulation models indicate that woody plants able to persist in association with grasses can show explosive population growth when grasses are temporarily reduced (Walker and Noy-Meir 1982; Walker et al. 1981). Once woody plants become dominant, their long life spans and their ability to extensively use both shallow and deep soil moisture can maintain the landscape in a woodland or scrub state almost indefinitely.

Water Storers

Water storers can store relatively large amounts of water internally that acts as a buffer against the rapid onset of drought stress. Examples of water storers include such stem succulents as prickly pears, fishhook barrel cactus (*Ferocactus wislizenii*), and ocotillo (*Fouquieria splendens*); and rosette or tuft plants such as beargrass (*Nolina* spp.), mescal (*Agave* spp.), sotol, and yucca (*Yucca* spp.). Tuberous-rooted plants such as flameflower (*Talinum aurantiacum*) and purgeroot (*Jatropha macrorhiza*) can also be considered water storers. Plants with this water-use strategy are usually not abundant in grass-dominated vegetation, but they are prominent and diagnostic species in many areas considered to be desert grassland.

Water storers tend to have shallow root systems (fig. 2.1) that can become fully absorptive within hours after a rain (Szarek et al. 1973). The water potential of a water storer is more stable and often substantially higher than that of an intensive exploiter (fig. 2.2); however, once the soil dries to a potential below that of the moist interior of a water storer, only intensive exploiters can continue to absorb soil moisture. Water storers have an advantage over intensive exploiters when soil moisture is transient, as on shallow soils over bedrock and during periods when rainfall is limited to very brief showers.

Many water storers, including prickly pear, cholla, and fleshy-fruited yucca, employ a type of photosynthesis called crassulacean acid metabolism, which allows them to realize a very high water-use efficiency (Kemp and Gardetto 1982; Ludwig et al. 1980; Nobel 1988). They are adversely affected by extreme temperatures and long, hot, unbroken droughts (Burgess and Shmida 1988). Water storers tend to be vulnerable to extreme freezes and are most active during warm weather.

Most water storers grow relatively slowly, probably due to the energy cost of constructing and maintaining nonproductive storage tissue and the loss in light-capturing efficiency associated with maximizing their water-use efficiency. As a result, their seedlings tend to be very vulnerable to competition, fire, and predation. The seedlings often require protection under a nurse plant or rock mulch or a period of exceptionally favorable weather to attain a size and water-storage capacity sufficient for long-term survival. Their population dynamics can be erratic. Tschirley and Wagle (1964) observed and described a 40-year cycle of invasion, increase, and decline of chainfruit cholla (*Opuntia fulgida*) on a grassland site in the Santa Rita Experimental Range.

These three strategies of water use represent different ways of exploiting soil moisture that varies with depth and time. Intensive exploiter populations tend to respond quickly to changes in rainfall regimes. They die during severe droughts but can recolonize rapidly if favorable conditions resume. Extensive exploiters are long-lived unless there is a catastrophic disturbance. Their longevity imposes a slow rate of change on the landscapes they dominate. Water storers also tend to have relatively slow population response times. Vegetation comprising species that have different response times often shows complex and unpredictable dynamic behavior (Allen and Hoekstra 1992).

Factors Influencing Growth Form Coexistence

Climate, soil, fire, and herbivory all influence the coexistence of growth forms in the desert grassland. I introduce each factor separately below and describe possible interactions.

Climate

Climate is the most important factor determining the relative abundance of growth forms. Desert grasslands experience erratic rainfall and variable episodes of drought. Temperatures and rainfall vary considerably over the entire range of the desert grasslands, which is not surprising given their latitudinal and elevational extent over a complex topography. The variable environments occupied by desert grasslands create plant communities that are less stable and support a greater variety of growth forms than most other grasslands.

Desert grasslands receive both winter and summer rainfall. Over the entire region, warm-season rainfall decreases along a gradient from southeast to northwest. Winter storms decrease from north to south (McClaran, this volume; Mitchell 1976; Schmidt 1989). Not only does the proportion of winter to summer rainfall vary regionally, it varies from year to year, although particular seasonal patterns of rainfall may dominate for a decade or more (Webb and Betancourt 1992).

The seasonal distribution of rainfall and the temperature when soil moisture is available can be critical for a plant's success (Stephenson 1990). Under cool conditions soil moisture is available longer, but plant growth is slow. Grasses adapted to respond to warm-season rains are often unable to use winter moisture because they do not grow new leaves during cold weather (Cable 1969). Most of the dominant grasses in desert grasslands employ the C_4 photosynthetic pathway, which functions most efficiently in full sun at high temperatures (Shmida and Burgess 1988; Waller and Lewis 1979). Most subshrubs and cool-season grasses, such as New Mexico feathergrass (*Stipa neomexicana*) and western wheatgrass (*Agropyron smithii*), have the more common C_3 photosynthetic pathway (Gurevitch 1986; Kemp 1983; Kemp and Williams 1980).

Winter rains tend to infiltrate deep into the soil, favoring extensive exploiters, and the cool temperatures favor the growth of subshrubs with C_3 photosynthetic pathways. Neilson (1986) proposed that wet winters produce an abundant crop of annuals that can tie up nutrients and suppress the growth of warm-season grasses during the subsequent summer. A series of years with dry summers and above-average winter rains will produce soil moisture conditions that favor subshrubs and shrubs over perennial warm-season grasses. Conversely, wet summers and falls combined with dry winters and springs can favor warm-season C_4 grasses (Cable 1975).

Between 1915 and 1968 there were only seven years in which study plots on the Jornada del Muerto Range showed black grama seedling establishment. Each of these years had relatively dry winters and springs followed by bursts of late summer or early autumn rain (Neilson 1986). These weather sequences are exceptional during the twentieth century but were fairly common between 1850 and 1900, when the area experienced generally drier winters and wetter summers. To some extent the shift from black grama to mesquite dominance on the Jornada Range can be correlated with changes in the seasonal distribution of rainfall from summer to winter. This would explain why the vegetation shift occurred on sites both grazed and ungrazed by livestock (Hennessy et al. 1983).

Droughts, periods without usable rainfall, have consequences that vary with their intensity. To understand drought effects, it is useful to distinguish between stress and disturbance. Grime (1977) considered stress to be limitation of growth, whereas disturbance is loss of living tissue. During a drought, a grass can avoid excessive desiccation by limiting transpiration, which also curtails CO_2 intake and prevents growth. If the drought continues, dehydration and starvation cause stems and roots to die back, and the effect of the drought changes from stress to disturbance. The location of the threshold between stress and disturbance depends on the physiology and condition of the plant. A drought that would subject a healthy tobosa to severe stress might cause extensive dieback in sand dropseed (*Sporobolus cryptandrus*) (Herbel et al. 1972; Schmutz et al. 1991). Frequent short droughts may allow grasses to prevent establishment of extensive exploiter seedlings, but regular long droughts can eliminate most perennial grasses.

Like drought, temperature extremes can cause stress or disturbance. Severe freezes limit plants more than any other factor except, perhaps, extreme aridity (Woodward 1987). Much of the desert grassland experiences a warm-temperate climate with fairly mild winters punctuated occasionally by severe frosts (D. E. Brown 1982b; McClaran, this volume). Subtropical woody extensive exploiters and succulent water storers are vulnerable to extreme freezes. Unusually low temperatures in 1978, for example, caused extensive dieback of velvet mesquite (Glinski and Brown 1982). There is a profound change in the aspect of desert grasslands in areas where winters are too cold for mesquite and other warm-temperate shrubs (McClaran, this volume).

Soil

Soil conditions convert large-scale patterns of climate into local patterns of soil moisture availability. This small-scale patchiness is exacerbated by the Basin and Range topography, which generates a context for complex, fragmented vegetation. Many landscape patterns and dynamics in desert grasslands can be explained by the effects of soil depth and texture on the availability of water (McAuliffe, this volume; Walter 1971). To demonstrate a few important soil processes, I contrast the behavior of water in three different types of soil materials: sand, clay, and carbonate accumulations.

Sand. The larger pore spaces between sand particles allow water to enter and leave sand more quickly than clay. The capacity of sand to store water is relatively small; thus it saturates rapidly and the rain infiltrates to deep layers, where it evaporates slowly. If one thinks of the rainfall pattern as a signal, then sand transmits that signal into the soil quickly and effectively. In wet climates, sandy soils are considered drier than clay soils because they cannot hold as much water, but the inverse texture effect caused by rapid infiltration allows sand to supply more moisture to plants from a light rainfall than clay does (Noy-Meir 1973).

Clay. Clay soils take longer to absorb water and can hold a larger volume of it than sandy soils. If rainfall occurs as a brief cloudburst, much of the water may run off before it can soak into the tiny pores between clay particles. If the amount of rain is small, even if all of it infiltrates, only the shallowest layers of the soil will be moist because a large volume of water is required to saturate clay. Until the shallow layers are saturated, the water will not move deeper. But once a clay is saturated, it forms a large reservoir of water. In dry climates, clay soil amplifies the dry and wet components of the rainfall signal. Brief downpours scarcely penetrate, whereas saturation from a long, soaking storm can supply water to plants for several months.

Seemingly minor differences in soil texture can have profound effects on the plant community. Knoop and Walker (1985) showed substantial differences in interactions between grass and woody plants on soils with 72 percent versus 87 percent sand in the surface horizon.

Carbonate. Carbonate horizons form where water-soluble minerals, especially calcium carbonate, accumulate in the soil near the limit of moisture infiltration. Over time a carbonate horizon can become an impervious layer of calcrete, or caliche (Gile et al. 1966; Machette 1985). When exposed by erosion, these horizons form calcareous soils similar to those derived from weathering limestone (Gile 1975). Calcrete erodes unevenly to create exposed or slightly buried surfaces that shed water into weathering crevices, resulting in a landscape with patchy water infiltration and retention. Deposits of calcrete or limestone can distort a rainfall signal into very irregular soil moisture patterns. Such conditions promote the coexistence of different growth forms by providing patches suitable only for shallow-rooted intensive exploiters and water storers adjacent to deeper soil pockets where extensive exploiters thrive.

When a soil profile has layers with different textures, infiltrating rainfall becomes segregated into soil moisture held at different water potentials. Sandy surfaces above clay or carbonate subsoils influence interactions between intensive and extensive exploiters in ways that vary with the relative thicknesses of the sand and subsoil horizons. In parts of western South

Africa, grassland forms in areas where heavier soils are covered by a sand veneer. Where the sand layer thins, dwarf shrubs become common (Tinley 1982). Deep sands in Mozambique may support pure grassland or shrub savanna, while nearby sites where calcrete is closer to the surface support a savanna woodland with more woody plant cover (Tinley 1982).

Typical grasslands have a fairly uniform, relatively dense plant cover; there is little bare ground, and the accumulation of plant litter as mulch on the soil surface promotes infiltration and retards evaporation (Beutner and Anderson 1943; Dyksterhuis and Schmutz 1947). When plants are more widely spaced, wind, runoff, and rain splash cause litter to be removed from bare sites and to accumulate only beneath isolated plant canopies. Denuded soil between plants often forms a crust that promotes more runoff, which accelerates the differentiation between bare and vegetated patches (Buffington and Herbel 1965; Gile 1966). Rain tends to flow down the stems of larger trees and shrubs and infiltrate the soil at the stem bases (Pressland 1973). Patch contrast leads to greater accumulation and infiltration of moisture beneath the larger plants, which favors the extensive exploitation strategy of, for example, velvet mesquite (Tiedemann and Klemmedson 1973), which can use scattered pockets of deep soil moisture.

Once an ecosystem crosses the threshold from one state of patchiness to another, positive feedback can propel the landscape in a direction not easily reversed (Walker et al. 1981; Westoby et al. 1989). These thresholds are mostly determined by soil properties. Sand tends to reduce patch contrast, whereas clay and calcrete usually promote it.

Fire

Fire acts as a generalized disturbance event that usually favors intensive exploiters able to rapidly regenerate their canopies. Woody plants, subshrubs, and succulents tend to be reduced by frequent burning (Cable 1967; Heirman and Wright 1973), although some woody and subshrub species are adapted to tolerate recurrent fires (Sarmiento and Monasterio 1983; Zedler et al. 1983). Humphrey (1958) and McPherson (this volume) reviewed fire's role in desert grasslands, and Bahre (this volume) documents its occurrence. It would be wrong, however, to assume that fire is a universal feature of desert grasslands. In the more arid desert grassland communities, the combination of patchy vegetation and low fuel loads inhibits the propagation of extensive fires (Dick-Peddie 1993).

Grasses differ in their ability to tolerate fire (McPherson, this volume;

Wright 1974). Tobosa recovers rapidly after burning (Neuenschwander 1976), whereas black grama comes back slowly (Cable 1965; Reynolds and Bohning 1956), perhaps because of its suffrutescent growth form. By inference, then, vegetation that features many succulents or a high cover of black grama has seldom burned.

Herbivory

Consumption of plants by animals, or herbivory, is a much more complicated form of disturbance than fire or drought (Ellison 1960; Walker et al. 1981). Herbivory by wild animals and domesticated livestock influences the dynamics among different species and growth forms. Apparently, the removal of kangaroo rats (*Dipodomys* spp.) can convert desertscrub to shrub savanna (J. H. Brown and Heske 1990). In general, herbivory promotes diversity of growth forms by suppressing the more vigorous species, especially grasses, and accelerating vegetation changes. Ecologists are unable to agree whether grass growth is stimulated or suppressed following grazing in semiarid landscapes (Dyer et al. 1993; Painter and Belsky 1993).

The extinction of most large herbivores throughout North America at the end of the Pleistocene probably caused major changes in vegetation (P. S. Martin and Klein 1984). We cannot determine what the desert grasslands looked like with their full complement of large native mammal herbivores, but they may have been quite different from the vegetation that has developed during the last 11,000 years.

Domestic livestock were a major tool for European settlement in the desert grasslands of North America. Although many histories focus on cattle as the primary economic product, sheep were also important consumers on the range before the First World War. Heavy stocking of goats by small homestead operations before the Second World War is often overlooked as well (Hadley et al. 1991).

Moderate livestock grazing can favor the growth of less palatable grass species (Bock and Bock 1993; Brady et al. 1989), and continuous grazing without recovery periods can reduce all grasses, especially during droughts (S. C. Martin 1975). Weakened grasses are less able to suppress the seedlings of extensive exploiters, water storers, and less palatable subshrubs, which may multiply beyond the ability of grasses to constrain them (Walker et al. 1981). Grasses may not recover to their original state after a grazing disturbance because other growth forms are favored by their longer life spans or by particular weather patterns or soil erosion (Westoby et al. 1989).

The desert grasslands of North America have been invaded by nonnative plants associated with livestock use since the nineteenth century (Bahre, this volume). Cool-season annuals originating in the Middle East and Mediterranean lands spread east from California (Heady 1977; Jackson 1985). Filaree (*Erodium cicutarium*) and perhaps red brome (*Bromus rubens*) were intentionally introduced into Arizona (Burgess et al. 1991; Roundy and Biedenbender, this volume). Several species of African grasses were introduced during the mid-twentieth century to improve revegetation success with livestock forage (Roundy and Biedenbender, this volume). Lehmann lovegrass (*Eragrostis lehmanniana*) has been spectacularly successful in some desert grasslands (Cox et al. 1988) and appears to be still expanding its range (Anable et al. 1992). Buffelgrass (*Pennisetum ciliare*) now covers extensive areas in northern Mexico and southern Texas, but its northward spread seems limited by freezing temperatures (Cox et al. 1988).

Invasions by nonnative plants are likely to become even more frequent in the twenty-first century. Decades of acclimation following the initial introduction of a plant may foster a population outbreak (Baker 1986; Kruger et al. 1986), hence plants introduced during the 1980s may not express their invasive potential until after the year 2025. Some desert grassland sites may shift toward grass dominance as more competitive grasses are introduced. As the nonnative species currently used for home landscaping adapt and spread, the proportions of nonnative shrubs, water storers, and subshrubs present may also increase.

I offer a phase diagram that describes the environmental interactions that favor different growth forms (fig. 2.3; Shmida and Burgess 1988). The duration of soil moisture segregates growth forms by determining the length of growth episodes. Duration is not the same as total rainfall. If the rainfall occurs as light showers separated by long dry intervals, the plants experience only short episodes of adequate soil moisture, each allowing only limited growth. If the same amount of rain comes in fewer storms and significant infiltration occurs, a longer period of growth follows. Soil properties and the season when the rain falls both affect the duration of available moisture.

In the more arid contexts described on the left side of the phase diagram (fig. 2.3), disturbance is caused most often by drought; mesic situations, on the right side, are more likely to experience disturbance in the form of fire or grazing.

Trees dominate when soil moisture is abundant and disturbance is low. As moisture duration decreases, trees are reduced to short, shrubby stature.

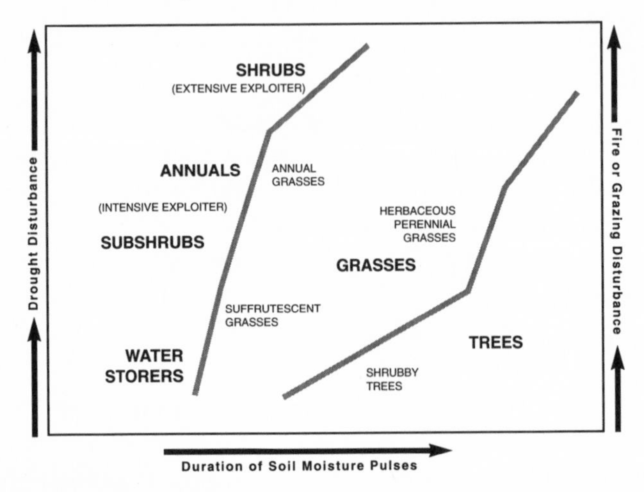

Figure 2.3. Phase diagram of growth form dominance along moisture and disturbance gradients. The horizontal axis represents a gradient of increasing duration of soil moisture pulses. The vertical axis is a gradient of disturbance. The severity of disturbance (the amount of living tissue likely to be lost) increases toward the top. Fire disturbance is more likely when the duration of soil moisture pulses decreases. In the upper left corner, droughts are too severe to support vascular plants.

As disturbance increases, intensive exploiters that regrow rapidly are favored; hence grasses are able to penetrate farther into moist habitats. When the moisture duration is short, most herbaceous perennial grasses are replaced by other growth forms. A climatic regime with short pulses of moisture and relatively short droughts is likely to favor plants that use the water-storer strategy (Burgess and Shmida 1988). As the intensity of drought disturbance increases, subshrubs, then annuals, and finally long-lived, drought-tolerant extensive exploiter shrubs are favored. As drought disturbance increases, continued dominance by grass requires a longer duration of soil moisture.

Within the zone of grass dominance, longer periods of available water favor herbaceous perennial grasses. Annual grasses are favored when there is less moisture and longer drought. Suffrutescent grasses predominate in areas with less severe seasonal droughts and shorter periods of moisture. The area of overlap between annual and suffrutescent grasses is actually greater than the phase diagram (fig. 2.3) indicates; the two forms coexist in many parts of the Southwest (Van Devender et al. 1990).

Stable grasslands occur on sites that are well ensconced near the center of the grass-dominated portion of the phase diagram (fig. 2.3). Most of the time, desert grasslands are found in areas close to the left, dry boundary of the grass dominance area, somewhere in the suffrutescent grass and annual grass zones. When the moisture, drought, grazing, and fire regimes vary from year to year, the crisp conceptual pattern of the diagram becomes blurred, because a mixture of growth forms will be supported. Temporal variation in the duration of soil moisture and frequency of disturbance leads to coexistence of different growth forms.

Similar Areas on Other Continents

There are no exact counterparts of North American desert grasslands in other parts of the world, although there are places with a vegetation of coexisting growth forms that experience semiarid, warm-temperate climates with erratic biseasonal rainfall. These cousin ecosystems are mostly on the eastern sides of deserts, in places where the climate is influenced by north-south-trending mountain ranges.

In South America, the most similar vegetation seems to be the Espinal bordering the pampa grasslands in northwestern Argentina, and perhaps some areas of the Argentine Chaco (Soriano 1979). These areas share many species with North American desert grasslands, including the grasses Arizona cottontop (*Digitaria californica*), burrograss (*Scleropogon brevifolius*), cane beardgrass (*Bothriochloa barbinodis*), cotta grass (*Cottea pappophoroides*), green sprangletop (*Leptochloa dubia*), spike pappusgrass (*Enneapogon desvauxii*), and wolftail (*Lycurus setosus*) (Raven 1963; Soriano 1979).

Southern Africa has climates very similar to the ones that support the North American desert grasslands (Venter et al. 1986). Some of these areas have vegetation that Cole (1986) called "low tree and shrub savanna," described as "widely spaced low-growing perennial grasses (less than 80 cm high) with abundant annuals and studded with widely spaced, low-growing trees and shrubs often less than 2 m high." This structural description would apply equally well to many desert grassland sites. Several veld types defined by Acocks (1988) closely resemble desert grasslands and have a similar recent history. During the past two centuries extensive areas have witnessed substantial declines in grasses and coincidental increases in woody plants and subshrubs, apparently as a result of poor livestock management (Acocks 1988), and nonnative plants from other continents have invaded (C. J. Brown and Gubb 1986). A number of plants are shared by southern Africa and the North American desert grasslands, some spread by Europeans and others apparently dispersed before European settlement, including African sumac (*Rhus lancea*), Bermuda grass (*Cynodon dactylon*), buffelgrass, hop bush (*Dodonaea viscosa* var. *angustifolia*), Lehmann lovegrass, pentzia (*Pentzia incana*), six-weeks threeawn (*Aristida adscensionis*), spike pappusgrass, tanglehead, and Wilman lovegrass (*Eragrostis superba*).

Northern Rajasthan and southern Punjab, in northwestern India, also have climates similar to those of North American desert grasslands (Pandeya et al. 1978). The Indian savannas have undoubtedly been altered by millennia of human exploitation, and their pristine state is unknown. The present conditions in India may indicate possible future states of desert grassland landscapes under sustained intensive use. Buffelgrass, Kleberg bluestem (*Dichanthium annulatum*), and tanglehead are common grasses found there that also occur in North American desert grasslands. Dominant woody plants related to species that grow in North American grasslands include ghaf (*Prosopis spicigera*), sidr (*Ziziphus nummularia*), and thumor (*Acacia senegal*) (Cole 1986; Misra 1983; Pandeya et al. 1978).

The Australian climates most similar to those that support North American desert grasslands lie between the southeastern forests and the interior deserts (Fitzpatrick and Nix 1970). Comparable vegetation includes acacia shrublands, semiarid low woodlands, mallee scrub, arid margins of bimble box woodland (*Eucalyptus populnea*), and grasslands dominated by Mitchell grasses (*Astrebla* spp.) (Beadle 1981; Moore et al. 1970; Perry 1970).

Species from some of these cousin ecosystems have become naturalized in the North American desert grassland, and comparative studies can help us understand their potential effects and how to better integrate these nonnative organisms into local food webs. For example, Cox et al. (1988) described the native habitats of African grasses successfully introduced into the desert grassland. Grasses that have persisted under intense human exploitation in northwestern India might easily invade and dominate areas in the desert grassland.

What's in a Name?

The terms *grassland*, *savanna*, *scrub*, and *steppe* carry unique connotations. Each describes a habitat with a characteristic composition of plant growth forms. The future of the desert grasslands will be influenced by how we perceive them and what we expect of them.

The communities that have been called desert grasslands do not conform with the concepts of grassland developed on the prairies and plains of the Midwest. Nor are they the same as the deserts defined by Shreve (1951) and others. The instability of the growth-form mixtures that constitute desert grasslands has been a barrier to the recognition of these mixtures as a community. Moreover, to label vegetation of coexisting growth forms "desert grassland" focuses on a single growth form and leads to the conception that nongrasses are somehow unnatural and undesirable.

One approach to vegetation classification is based entirely on vegetation structure, ignoring species composition (Mueller-Dombois and Ellenberg 1974). This type of classification system usually restricts the use of the term *grassland* to vegetation composed almost entirely of grasses and other herbaceous plants.

Savanna is a term used to describe grassy vegetation punctuated by taller woody plants. Savannas are often associated with tropical and subtropical climates, however, and the term may seem less appropriate for temperate areas (Bourliere and Hadley 1983; Walter 1971). *Steppe* refers to open herbaceous vegetation (which may include woody plants; Menaut 1983) in dry, temperate climates that produces too little herbage for widespread fires. Woody plants, subshrubs, and succulents can be accepted as normal components of savanna and steppe, although grasses are still emphasized. *Scrub,* also called shrubland, is relatively low woody vegetation in which herbaceous perennials are usually scarce.

We tend to think of vegetation structure as a relatively stable aspect of the landscape, and the instability of its growth-form composition has confused people trying to classify desert grassland vegetation. How many subshrubs or shrubs can appear in a grassland without causing it to be perceived as scrub, steppe, or savanna?

History of the Desert Grassland Concept

The United States was not well prepared to cope with its acquisition of the great American desert during the nineteenth century (Malin 1956). How were these foreign landscapes and cultures to be incorporated into national economic and social frameworks without threatening established institutions? Similarly, United States ecologists were confronted by a bewildering array of novel western landscapes shaped by aridity. Our modern vegetation classification systems might have been very different if these ecologists had been trained in more arid landscapes like those of Mexico instead of in eastern North American universities.

The Carnegie Desert Laboratory was established to provide a scientific basis to help American culture assimilate the deserts (Bowers 1990). Its founder, Daniel MacDougal, wrote (1908), "Of the various separable regions in the transition from the humid to the arid areas of the West, that bearing the sotol bears a vegetation of marked xerophytic type and is true desert." Do we still think of sotol as a desert plant? Sotol usually grows on rocky soils in places too moist for open desertscrub and too dry for oak or juniper woodland. Such places later came to be considered desert grassland (Dick-Peddie 1993; Schmutz et al. 1991), but to a botanist trained in the East, landscapes with sotol were desert.

As settlers and scientists became familiar with western North America they made finer distinctions. Ranching was far more profitable in the plains of southern New Mexico than in the truly arid lower Colorado River valley. Aldous and Shantz (1924) constructed a system of vegetation types to assess the economic potential of western drylands. Their focus was more pragmatic than the simplified schemes of academic ecologists, and they recognized that sotol country was not desert. The 27 vegetation types they defined for the semiarid Southwest provide an early impression of communities that were later to be called desert grassland. Black grama was important in 7 types, tobosa in 4, creosotebush (*Larrea divaricata*) in 4, mesquite in 3, and tarbush in 2. The authors' emphasis on grass was logical from an economic perspective, but that emphasis led to devaluation of the habitat's shrub components.

During the early twentieth century, Frederick Clements developed a theoretical scheme of vegetation classification. Each climate, he believed, would produce a single stable climax vegetation and soil given enough time without disturbance. Vegetation types were organized according to their position along a progression toward the climax community determined by the climate (Allen and Hoekstra 1992; Clements 1963).

Building on his work in the Nebraska grasslands, Clements conducted a reconnaissance of the West. His theoretical grassland climax was consolidated by postulating a former connection between the grasslands of California and those of New Mexico and the Great Plains. His evidence was the relict populations of grasses on both sides of his desertscrub climax, which was centered on the lower Colorado River (Clements 1922, 1924). One report (Clements and Clements 1924) stated that "the extensive areas of Larrea and Prosopis that reach from Texas to Arizona are actually a savannah." To Clements and Weaver (1924) this vegetation resembled a short-grass phase of mixed prairie. Grasses were considered to be unifying elements, and the presence of woody species was thought to be a recent, less relevant phenomenon.

Later, Clements (1963) applied the term *scrub savannah* to vegetation in areas where alternating wet and dry climatic phases favored establishment of scrub and grass growth forms. Although he recognized that variable rainfall allows different growth forms to coexist, he wasn't comfortable with the instability of this vegetation. Therefore he considered scrub savanna an ecotone between desertscrub and grassland climaxes, and he connected them with the transitional term *desert grassland.*

Clements's emphasis on grass was continued by United States Soil Conservation Service scientists (Whitfield and Anderson 1938; Whitfield and Beutner 1938). Their desert plains grassland extended from around Lukeville and Casa Grande, Arizona, east to Sanderson and Big Spring, Texas, including most areas lower than about 1,500 m elevation. An average annual precipitation of 5 inches was considered the demarcation between desertscrub and desert plains grassland. Two faciations were recognized: one dominated by black grama and tobosa at lower elevations, and a second dominated by curly mesquite grass and blue grama on foothills and higher mesas. Communities not dominated by grasses were interpreted as degraded or immature. The authors proposed that the lower desert grassland had been converted by overgrazing into desertscrub communities that included saguaro (*Carnegiea gigantea*), foothills paloverde (*Cercidium microphyllum*), and triangle leaf bursage (*Ambrosia deltoidea*). Shreve (1951) identified this vegetation as part of the Sonoran Desert, but from Whitfield and Anderson's perspective, Shreve's eastern Sonoran Desert was really degraded grassland in need of restoration. In their view (1938), "The reestablishment of the former climax types should be the ultimate goal of conservation activities," and the Southwest should be restored, "from a sadly depleted condition to its high potential value as a range resource." In this case, economic utility was guiding ecological interpretation.

Carpenter (1940) excluded the mesquite–desert grass savanna of Texas from his Grassland Biome of North America. His dismissal challenged the unified grassland climax view established by Clements and expanded by Whitfield and Beutner.

Shreve (1942) viewed the open, arid grasslands with rosette plants and succulents as a transition area between grassland and desert created by species infiltrating from both formations. He considered the grasslands at higher elevations along the eastern Sierra Madre Occidental and at scattered places on the eastern Mexican Plateau to resemble those of the Great

Plains and to be climatically determined. In contrast, he believed the tobosa grasslands on clay soils in the intermountain basins were a desert association rather than part of a climatic grassland formation.

Gentry's dissertation (1957) on the grasslands of Durango and northwestern Zacatecas, Mexico, provided a much-needed description of southern desert grasslands. Between the Sierra Madre Occidental forests and the Chihuahuan Desert, Gentry recognized a grama-grassland formation, which he subdivided into five principal associations. His oak-juniper grassland association was transitional to montane pine-oak woodland. Grama grassland was described as a mosaic of mixed grasses and herbs with scattered subshrubs and sometimes widely scattered shrubs or small trees. A related acacia-cactus grassland had similar dominant grasses with scattered huisache (*Acacia schaffneri*), tuna blanca (*Opuntia megacantha*), and nopal cenizo (*O. durangensis*) up to 7 m tall. Vegetation in which shrubs 1–2 m tall comprised up to 70 percent of the plant cover was termed grassland with shrubs and was considered transitional to desertscrub. The grass layer of this association was shorter, less dense, and had a higher proportion of annual species than grama grassland. Gentry's grassland with shrubs was most often found on rocky slopes or eroded soils with caliche near the surface, in contrast with the deeper, more mature soils of the grama grassland. Chaparillo, dominated by Chihuahuan Desert species, was defined as the arid end of the grassland-desertscrub transition.

Most of the dominant grass species of Durango, Mexico, extend into the United States, forming a nearly continuous grassland along the eastern base of the Sierra Madre Occidental. Many of the woody and succulent species listed by Gentry are widespread around the edge of the Chihuahuan Desert in Texas, but the acacia-cactus grassland has no close counterpart in the United States.

Humphrey (1958) concluded that desert grassland was a subclimax maintained by fire. This was a departure from earlier views that grasses prevented establishment of woody plant seedlings by competition, and that overgrazing was the major cause of the observed increase in brush (Clements 1963). Humphrey's challenge marked a shift in focus from the inferred dynamics of the Clementsian climate-driven systems to the importance of disturbance.

Daubenmire's classification of North American vegetation (1978) places desert grassland in the southwest section of the *Bouteloua gracilis* Province of the Steppe Region. *Steppe* is defined as "the extratropical grassland of areas where the zonal soils are too dry for trees, and herbaceous perennial grasses are well represented" (Daubenmire 1978). The vegetation is a shrub steppe if shrubs form a discontinuous upper layer. Despite its inclusion in the Blue Grama Province, Daubenmire considered black grama to be the characteristic dominant of the area.

D. E. Brown's classification (1982a,b,c) defines vegetation units that can be mapped at a regional scale to provide a context for vertebrate ecology and resource management. He split the broadly conceived desert plains grassland into four biotic communities. The basic concept originally embodied in desert plains grassland—vegetation with black grama, tobosa, and species shared with the Sonoran, Mohave, and Chihuahuan Deserts—he termed semidesert grassland. These landscapes, which experience warm-temperate climates, are dominated by perennial bunchgrass or mixtures of grass with shrubs, subshrubs, cacti, and forbs. The concept includes places where the other growth forms have mostly replaced grasses to form a semidesert scrubland. Brown distinguished as plains grassland the vegetation with more cold-tolerant succulent and woody plants and taller grass species that also occur on the Great Plains. Drier grasslands in colder areas adjoining Great Basin desertscrub were called Great Basin grassland, but they were not clearly demarcated from plains grassland.

Brown's fourth community (1982c) was the almost vanished Sonoran savanna grassland. He believed this vegetation formerly occupied deep, fine-textured soils in the southeastern part of the Sonoran Desert, but livestock grazing had almost entirely transformed it into desertscrub. Brown's recognition of Sonoran savanna grassland has been questioned, however, because its characteristic grasses are mostly short-lived and only transiently dominant (Van Devender et al. 1990).

Schmutz et al. (1991) presented a broad concept of desert grasslands that groups communities extending from the arid edges of oak and juniper woodlands together with desert shrublands that formerly supported grasses. Their four subdivisions are defined by differences in species that correspond with soils, rainfall, and winter temperature extremes. These subregions are similar to those of Whitfield and Beutner (1938) and have more predictive utility for vegetation management than Brown's (1982a,b,c) communities. Their high desert sod grass subregion has stands of sod-forming grasses on mature, heavy-textured soils at higher elevations. Brown (1982a) considered these communities to be plains grassland, but to Schmutz and his colleagues the presence of curly mesquite grass, tobosa, and mesquite, and the near absence of buffalograss (*Buchloë dactyloides*) and galleta (*Hilaria jamesii*), signified a closer relationship with desert grassland. Higher-elevation sites with various bunchgrasses beneath a scattered overstory of oaks or junipers are grouped with floodplains of blue grama, big sacaton (*Sporobolus wrightii*), and mesquite into a high desert bunchgrass subregion. Chihuahuan and Sonoran Desert grasslands are recognized as shrublands that are known or thought to have formerly supported more grasses. This classification retains the tradition of applying the term *desert grassland* to sites with shrubs and few or no grasses if the site has any potential to support grass.

Dick-Peddie's (1993) classification of New Mexico vegetation defines desert grassland as homogeneous stands of grasses and shrubs mixed together, or patchy mosaics of grassland and scrubland. He recognized savanna vegetation as scattered low trees, typically junipers, in a matrix of grass with few shrubs. Savanna in this system is the transition from grassland to more mesic woodland, whereas desert grassland represents a transition from grassland to more xeric desert scrubland or montane scrub. Dick-Peddie separated vegetation clearly dominated by shrubs as scrubland. Desert grassland can form where any scrub community, whether montane or arid, meets grassland. Many of Dick-Peddie's recognized communities are recently derived, but they cannot be easily distinguished from more stable desert grassland vegetation.

Because Dick-Peddie's classification relies heavily on structural dominance and ecotones, communities with similar species composition that have similar vegetation dynamics and growth forms can be separated in very different major vegetation types. His restriction of savanna to open stands of juniper or oak connecting grassland with woodland creates inconsistencies. Velvet mesquite, for example, forms woodlands on floodplains or the wetter margins of semiarid grasslands. To be logically consistent with Dick-Peddie's classification, we should recognize savannas of velvet mesquite as distinct from desert grassland, because velvet mesquite is not typically dominant in desert scrubland.

Several themes recur over the history of desert grassland studies: (1) vegetation is unstable; (2) grass is more important than woody plants, which are to some extent unnatural and undesirable; (3) despite repeated references to its savanna or scrub savanna structure, the vegetation has been called grassland even if grasses are scarce; and (4) Clements's views have generally prevailed.

A Paradigm Shift

Currently the term *desert grassland* is used to describe diverse landscapes in which grass is only one of several common growth forms. On any other continent much of this vegetation would be called savanna, or perhaps steppe. The perception of the North American landscapes as grassland is largely derived from their use for livestock production. The instability of the growth-form mixtures has led to the idea that these communities are ecotones transitional between grassland and desert. Recognizing the integrity of unstable vegetation would have refuted the concept of stable climax formations. The continuing influence of Clements's concepts stems from the widespread use of climax models to assess range condition (Laycock 1991; Westoby et al. 1989). These models are now being eclipsed by chaos theory (Allen and Hoekstra 1992; Worster 1993).

Similar plant communities extend from western Texas to Zacatecas and southern Arizona, where grasses coexist with other growth forms. These areas feature vegetation different from that found on temperate grassland and arid desertscrub. If we continue to perceive these communities as degraded grassland or invading scrub, we unconsciously limit our options for their management. In the most humid areas they can change from savanna to thorn woodland (Archer et al. 1988), and on the arid end they may shift from savanna to desertscrub. To the extent that their dynamics and structure are more influenced by drought than freezing, they resemble tropical savannas. Given their structure, dynamics, and constraints, the term *mixed shrub savanna* seems to be the most appropriate unifying term for these communities.

Mixed shrub savannas have floristic relatives. A few species are shared with Great Plains grasslands, and some dominants can also be found in warm deserts. Many characteristic species—for example, black grama, tobosa, burroweed, mesquite, and sotol—are centered in these savannas. Clements's derivation of desert grassland from Great Plains grassland is questionable. The available evidence suggests that the flora of the Great Plains is "recent and adventive in origin" (Great Plains Flora Association 1986). Most plant communities considered desert grassland have closer ties with floras of subtropical, semiarid Mexico.

Mixed shrub savanna describes the structure of the vegetation, but this designation is not connected to a particular region, as the terms *Sonoran Desert* and *Great Plains grassland* are. It is difficult to find a suitable regional term for the area occupied by these savannas because they are not aggregated in physiographically or politically distinct areas. Mixed shrub savannas are inserted between low-lying desertscrub and montane forests, and they extend well beyond the Basin and Range physiography. In the nineteenth century much of the mixed shrub savanna area was occupied by Apaches, ranging from the Lipan on the east to the Chiricahua on the west (Young 1983). The Spanish vaguely referred to these Apache lands as Apachería. I propose that the best inclusive term for these communities is *Apacherian mixed shrub savanna*, or simply *Apacherian savanna.*

Summary

The vegetation structure of Apacherian mixed shrub savannas typically consists of coexisting growth forms that include grasses, subshrubs, stem succulents, and shrubs. Each growth form represents a different strategy for coping with variable soil moisture and drought. Three basic water-use strategies can be recognized: intensive exploiters, extensive exploiters, and water storers. The relative success of each strategy depends on the constraints imposed by soil, associated species, weather events, and disturbance regimes. Feedback loops and differences in response times among the species create dynamic behaviors that can shift the vegetation structure among several possible states.

Investigators attempting to apply classification schemes to these landscapes and communities have been frustrated by the unstable mixtures of the growth forms. Professional and cultural biases led to a focus on grass components. I suggest that the desert grassland be called Apacherian mixed shrub savanna to acknowledge the dynamic coexistence of many growth forms in a geographic region that corresponds to the former range of the Apache.

Acknowledgments

I am very grateful to Joseph McAuliffe and Peter Warshall for their help in developing and refining the ideas presented here. My appreciation also goes to Carol Cochran, Mitch McClaran, and Tom Van Devender, who put the Desert Grassland Symposium and this volume together. Mitch McClaran deserves special praise for his patient and meticulous editing. I am indebted to the Biosphere 2 Project for forcing me to study tropical savannas. My sincere thanks go to colleagues at the University of Arizona Desert Laboratory for encouragement and for creating an environment that fosters this kind of work.

Literature Cited

Acocks, J. P. H. 1988. Veld types of South Africa. Pp. 1–146 *in* O. A. Leistner (ed.), Memoirs of the Botanical Survey of South Africa 57, 3d ed. Botanical Research Institute, Department of Agriculture and Water Supply, Pretoria.

Aldous, A. E., and H. L. Shantz. 1924. Types of vegetation in the semiarid portion of the United States and their economic significance. Journal of Agricultural Research 28:99–128.

Allen, T. F. H., and T. W. Hoekstra. 1992. Toward a Unified Ecology. Columbia University Press, New York.

Anable, M. E., M. P. McClaran, and G. B. Ruyle. 1992. Spread of introduced Lehmann lovegrass (*Eragrostis lehmanniana* Nees.) in southern Arizona, USA. Biological Conservation 61:181–188.

Archer, S., C. Scifres, C. R. Bassham, and R. Maggio. 1988. Autogenic succession in a subtropical savanna: conversion of grassland to thorn woodland. Ecological Monographs 58:111–127.

Baker, H. G. 1986. Patterns of plant invasion in North America. Pp. 44–57 *in* H. G. Mooney and J. A. Drake (eds.), Ecology of biological invasions of North America and Hawaii. Springer-Verlag, New York.

Beadle, N. C. W. 1981. The Vegetation of Australia. Gustav Fischer Verlag, New York.

Beutner, E. L., and D. Anderson. 1943. The effect of surface mulches on water conservation and forage production in some semidesert grassland soils. Journal of the American Society of Agronomists 35:393–400.

Blydenstein, J. 1966. Root systems of four desert grassland species on grazed and protected sites. Journal of Range Management 19:93–95.

Bock, C. E., and J. H. Bock. 1993. Cover of perennial grasses in southeastern Arizona in relation to livestock grazing. Conservation Biology 7:371–377.

Bourliere, F., and M. Hadley. 1983. Present-day savannas: an overview. Pp. 4–19 *in* F. Bourliere (ed.), Tropical savannas. Ecosystems of the World 13. Elsevier, New York.

Bowers, J. E. 1990. A debt to the future: scientific achievements of the Desert Laboratory, Tumamoc Hill, Tucson, Arizona. Desert Plants 10:9–12, 35–47.

Brady, W. W., M. R. Stromberg, E. F. Aldon, C. D. Bonham, and S. H. Henry. 1989. Response of a semidesert grassland to 16 years of rest from grazing. Journal of Range Management 42:284–288.

Brown, C. J., and A. A. Gubb. 1986. Invasive alien organisms in the Namib Desert, Upper Karroo and the arid and semi-arid savannas of western southern Africa. Pp. 93–108 *in* I. A. W. Macdonald, F. J. Kruger, and A. A. Ferrar (eds.), The ecology and management of biological invasions in southern Africa. Oxford University Press, New York.

Brown, D. E. 1982a. Plains and Great Basin grasslands. Desert Plants 4:115–121.

———. 1982b. Semidesert grassland. Desert Plants 4:123–131.

———. 1982c. Sonoran savanna grassland. Desert Plants 4:137–141.

Brown, D. E., and C. H. Lowe. 1980. Biotic communities of the Southwest. U.S. Department of Agriculture Forest Service General Technical Report RM-78.

Brown, J. H., and E. J. Heske. 1990. Control of a desert-grassland transition by a keystone rodent guild. Science 250:1705–1707.

Buffington, L. C., and C. H. Herbel. 1965. Vegetational changes on a semidesert grassland range from 1858 to 1963. Ecological Monographs 35:139–164.

Burgess, T. L., J. E. Bowers, and R. M. Turner. 1991. Exotic plants at the Desert Laboratory, Tucson, Arizona. Madrono 38:96–114.

Burgess, T. L., and A. Shmida. 1988. Succulent growth-forms in arid environments. Pp. 383–395 *in* E. E. Whitehead, C. F. Hutchinson, B. N. Timmermann, and R. G. Verity (eds.), Arid lands today and tomorrow. Westview Press, Boulder.

Cable, D. R. 1965. Damage to mesquite, Lehmann lovegrass, and black grama by a hot June fire. Journal of Range Management 18:326–329.

———. 1967. Fire effects on semidesert grasses and shrubs. Journal of Range Management 20:170–176.

———. 1969. Competition in the semidesert grass-shrub type as influenced by root systems, growth habits, and soil moisture extraction. Ecology 50:27–38.

———. 1975. Influence of precipitation on perennial grass production in the semidesert Southwest. Ecology 56:981–986.

———. 1977. Seasonal use of soil water by mature velvet mesquite. Journal of Range Management 30:4–11.

Campbell, R. S., and E. H. Bomberger. 1934. The occurrence of *Gutierrezia sarothrae* on *Bouteloua eriopoda* ranges in southern New Mexico. Ecology 15:49–61.

Cannon, W. A. 1911. The root habits of desert plants. Carnegie Institution of Washington Publication no. 131. Washington, D.C.

Carpenter, J. R. 1940. The grassland biome. Ecological Monographs 10:617–684.

Clements, F. E. 1922. The original grassland of Mohave and Colorado Deserts. Carnegie Institution of Washington Year Book 21:350–351.

———. 1924. The original vegetation of Death Valley. Carnegie Institution of Washington Year Book 22:317.

———. 1963. Plant Succession and Indicators. Hafner Press, New York.

Clements, F. E., and E. S. Clements. 1924. Climax formations. Carnegie Institution of Washington Year Book 22:315–316.

Clements, F. E., and J. E. Weaver. 1924. Experimental vegetation; the relations of climaxes to climates. Carnegie Institution of Washington Publication no. 355. Washington, D.C.

Cole, M. M. 1986. The Savannas: Biogeography and Geobotany. Academic Press, New York.

Cox, J. R., M. H. Martin R., F. A. Ibarra F., J. H. Fourie, N. F. G. Rethman, and D. G. Wilcox. 1988. The influence of climate and soils on the distribution of four African grasses. Journal of Range Management 42:127–139.

Daubenmire, R. 1978. Plant Geography. Academic Press, New York.

Dick-Peddie, W. A. 1993. New Mexico Vegetation: Past, Present, and Future. University of New Mexico Press, Albuquerque.

Dyer, M. I., C. L. Turner, and T. R. Seastedt. 1993. Herbivory and its consequences. Ecological Applications 3:10–16.

Dyksterhuis, E. J., and E. M. Schmutz. 1947. Natural mulches or "litter" of grasslands: with kinds and amounts on a southern prairie. Ecology 28:163–179.

Ellison, L. 1960. Influence of grazing on plant succession of rangelands. Botanical Review 26:1–78.

Evenari, M., L. Shanan, and N. Tadmor. 1971. The Negev: The Challenge of a Desert. Harvard University Press, Cambridge, Mass.

Fitzpatrick, E. A., and H. A. Nix. 1970. The climatic factor in Australian grassland ecology. Pp. 3–26 *in* R. M. Moore (ed.), Australian grasslands. Australian National University Press, Canberra.

Gentry, H. S. 1957. Los Pastizales de Durango, Estudio Ecológico, Fisiográfico y Florístico. Ediciones del Instituto Mexicano de Recursos Naturales Renovables, Mexico City.

Gile, L. H. 1966. Coppice dunes and the Rotura soil. Soil Science Society of America Proceedings 30:657–660.

———. 1975. Causes of soil boundaries in an arid region. II. Dissection, moisture, and faunal activity. Soil Science Society of America Proceedings 39:324–330.

Gile, L. H., F. F. Peterson, and R. R. Grossman. 1966. Morphological and genetic sequences of carbonate accumulation in desert soils. Soil Science 101:347–360.

Glendening, G. E., and H. A. Paulson, Jr. 1955. Reproduction and establishment of velvet mesquite as related to invasion of semidesert grasslands. U.S. Department of Agriculture Technical Bulletin 1127.

Glinski, R. L., and D. E. Brown. 1982. Mesquite (*Prosopis juliflora*) response to severe freezing in southeastern Arizona. Journal of the Arizona-Nevada Academy of Science 17:15–18.

Great Plains Flora Association. 1986. Flora of the Great Plains. University Press of Kansas, Lawrence.

Grime, J. P. 1977. Evidence for the existence of three primary strategies in plants and its relevance to ecological and evolutionary theory. American Naturalist 111:1169–1194.

Gurevitch, J. 1986. Competition and the local distribution of the grass *Stipa neomexicana*. Ecology 67:46–57.

Haas, R. H., and J. D. Dodd. 1972. Water-stress patterns in honey mesquite. Ecology 53:674–680.

Hadley, D., P. Warshall, and D. Bufkin. 1991. Environmental change in Aravaipa, 1870–1970; an ethnoecological survey. U.S. Department of Interior, Bureau of Land Management, Arizona State Office, Phoenix.

Harris, G. A. 1967. Some competitive relationships between *Agropyron spicatum* and *Bromus tectorum*. Ecological Monographs 37:89–111.

Heady, H. F. 1977. Valley grassland. Pp. 491–514 *in* M. G. Barbour and J. Major (eds.), Terrestrial vegetation of California. John Wiley & Sons, New York.

Heirman, A. L., and H. A. Wright. 1973. Fire in medium fuels of west Texas. Journal of Range Management 26:331–335.

Hennessy, J. T., R. P. Gibbens, J. M. Tromble, and M. Cardenas. 1983. Vegetation changes from 1935 to 1980 in mesquite dunelands and former grasslands of southern New Mexico. Journal of Range Management 36:370–374.

Herbel, C. H., F. N. Ares, and R. A. Wright. 1972. Drought effects on a semidesert grassland range. Ecology 53:1084–1093.

Humphrey, R. R. 1958. The desert grassland, a history of vegetational change and an analysis of causes. Botanical Review 24:193–252.

Jackson, L. E. 1985. Ecological origins of California's Mediterranean grasses. Journal of Biogeography 12:349–361.

Jameson, D. A. 1966. Competition in a blue grama–broom snakeweed community and responses to selective herbicides. Journal of Range Management 19:121–124.

Kemp, P. R. 1983. Phenological patterns of Chihuahuan Desert (New Mexico, USA) plants in relation to the timing of water availability. Journal of Ecology 71:427–436.

Kemp, P. R., and P. E. Gardetto. 1982. Photosynthetic pathway types of evergreen rosette plants (Liliaceae) of the Chihuahuan Desert. Oecologia 55:149–156.

Kemp, P. R., and G. J. Williams III. 1980. A physiological basis for niche separation between *Agropyron smithii* (C_3) and *Bouteloua gracilis* (C_4). Ecology 61:846–858.

Knoop, W. T., and B. H. Walker. 1985. Interactions of woody and herbaceous vegetation in a southern African savanna. Journal of Ecology 73:235–253.

Kruger, F. J., D. M. Richardson, and B. W. van Wilger. 1986. Processes of invasion by alien plants. Pp. 145–155 *in* I. A. W. Macdonald, F. J. Kruger, and A. A. Ferrar (eds.), The ecology and management of biological invasions in southern Africa. Oxford University Press, New York.

Laycock, W. A. 1991. Stable states and thresholds of range condition on North American rangelands: a viewpoint. Journal of Range Management 44:427–433.

Ludwig, J. A., P. R. Kemp, and P. E. Gardetto. 1980. A simulation model of the productivity of the Spanish bayonet, *Yucca baccata* Torr. Pp. 51–68 *in* W. Guzman G. (ed.), Yucca. El Desierto, vol. 3. Centro de Investigación en Química Aplicada, Saltillo, Coahuila, Mexico.

MacDougal, D. T. 1908. Botanical features of North American deserts. Carnegie Institution of Washington Publication no. 99. Washington, D.C.

Machette, M. N. 1985. Calcic soils of the southwestern United States. Pp. 1–21 *in* D. L. Weide (ed.), Soils and Quaternary geology of the southwestern United States. Geological Society of America Special Paper 203. Geological Society of America, Boulder.

Mack, R. N., and D. A. Pyke. 1984. The demography of *Bromus tectorum*: the role of microclimate, grazing and disease. Journal of Ecology 72:731–748.

Malin, J. C. 1956. The Grassland of North America, Prolegomena to Its History with Addenda. James C. Malin, Lawrence, Kansas.

Martin, S. C. 1975. Ecology and management of semidesert grass-shrub ranges: the status of our knowledge. USDA Forest Service Research Paper RM-156.

Martin, P. S., and R. G. Klein (eds.). 1984. Quaternary Extinctions. University of Arizona Press, Tucson.

Menaut, J. C. 1983. The vegetation of African savannas. Pp. 109–149 *in* F. Bourlier (ed.), Tropical savannas. Ecosystems of the World 13. Elsevier, New York.

Misra, R. 1983. Indian savannas. Pp. 151–166 *in* F. Bourlire (ed.), Tropical savannas. Ecosystems of the World 13. Elsevier, New York.

Mitchell, V. L. 1976. The regionalization of climate in the western United States. Journal of Applied Meteorology 15:920–927.

Moore, R. M., R. W. Condon, and J. H. Leigh. 1970. Semi-arid woodlands. Pp. 228–245 *in* R. M. Moore (ed.), Australian grasslands. Australian National University Press, Canberra.

Mueller-Dombois, D., and H. Ellenberg. 1974. Aims and Methods of Vegetation Ecology. John Wiley & Sons, New York.

Neilson, R. P. 1986. High-resolution climatic analysis and Southwest biogeography. Science 232:27–34.

———. 1987. Biotic regionalization and climatic controls in western North America. Vegetatio 70:135–147.

Neuenschwander, L. F. 1976. The effect of fire in a sprayed tobosa grass-mesquite community on Stamford clay soils. Ph.D. dissertation, Texas Tech University, Lubbock.

Nilsen, E. T., M. R. Sharifi, P. W. Rundel, W. M. Jarrell, and R. A. Virginia. 1983. Diurnal and seasonal water relations of the desert phreatophyte *Prosopis glandulosa* (honey mesquite) in the Sonoran Desert of California. Ecology 64:1381–1393.

Nobel, P. S. 1974. Introduction to Biophysical Plant Physiology. W. H. Freeman, San Francisco.

———. 1980. Water vapor conductance and CO_2 uptake for leaves of a C_4 desert grass, *Hilaria rigida*. Ecology 61:252–258.

———. 1988. Environmental Biology of Agaves and Cacti. Cambridge University Press, New York.

Noy-Meir, I. 1973. Desert ecosystems: environment and producers. Annual Review of Ecology and Systematics 4:25–51.

Painter, E. L., and A. J. Belsky. 1993. Applications of herbivore optimization theory to rangelands of the western United States. Ecological Applications 3:2–9.

Pandeya, S. C., S. C. Sharma, H. K. Jain, S. J. Pathak, K. C. Paliwal, and V. M. Bhanot. 1978. The Environment & Cenchrus Grazing Lands in Western India. Department of Biosciences, Saurashra University, Rajkot, India.

Perry, R. A. 1970. Arid shrublands and grasslands. Pp. 246–259 *in* R. M. Moore (ed.), Australian grasslands. Australian National University Press, Canberra.

Phillips, W. S. 1963. Depths of roots in soil. Ecology 44:424.

Pressland, A. J. 1973. Rainfall partitioning by an arid woodland (*Acacia aneura* F. Muell.) in south-western Queensland. Australian Journal of Botany 21:235–245.

Raven, P. H. 1963. Amphitropical relationships in the floras of North and South America. Quarterly Review of Biology 38:151–167.

Reynolds, H. G., and J. W. Bohning. 1956. Effects of burning on a desert grass-shrub range in southern Arizona. Ecology 37:769–777.

Sala, O. E., R. A. Golluscio, W. K. Lauenroth, and A. Soriano. 1989. Resource partitioning between shrubs and grasses in the Patagonian steppe. Oecologia 81:501–505.

Sala, O. E., and W. K. Lauenroth. 1982. Small rainfall events: an ecological role in semiarid regions. Oecologia 53:301–304.

———. 1985. Root profiles and the ecological effect of light rainshowers in arid and semiarid regions. American Midland Naturalist 114:406–408.

Sala, O. E., W. K. Lauenroth, W. J. Parton, and M. J. Trlica. 1981. Water status of soil and vegetation in a shortgrass steppe. Oecologia 48:327–331.

Sarmiento, G., and M. Monasterio. 1983. Life forms and phenology. Pp. 79–108 *in* F. Bourlire (ed.), Tropical savannas. Ecosystems of the World 13. Elsevier, New York.

Schmidt, R. H., Jr. 1989. The arid zones of Mexico: climatic extremes and conceptualization of the Sonoran Desert. Journal of Arid Environments 16:241–256.

Schmutz, E. M., E. L. Smith, P. R. Ogden, M. L. Cox, J. O. Klemmedson, J. J. Norris, and L. C. Fierro. 1991. Desert grassland. Pp. 337–362 *in* R. T. Coupland (ed.), Natural grasslands: introduction and Western Hemisphere. Elsevier, New York.

Shmida, A., and T. L. Burgess. 1988. Plant growth-form strategies and vegetation types in arid environments. Pp. 211–241 *in* N. J. A. Werger, P. J. M. van der Aart, H. J. During, and J. T. A. Verhoeven (eds.), Plant form and vegetation structure. SPB Academic Publishing, The Hague.

Shreve, F. 1942. Grassland and related vegetation in northern Mexico. Madrono 6:190–198.

———. 1951. Vegetation of the Sonoran Desert. Carnegie Institution of Washington Publication no. 591. Washington, D.C.

Soriano, A. 1979. Distribution of grasses and grasslands in South America. Pp. 84–91 *in* M. Numata (ed.), Ecology of grasslands and bamboolands in the world. Dr. W. Junk, Boston.

Stephenson, N. L. 1990. Climatic control of vegetation distribution: the role of water balance. American Naturalist 135:649–670.

Szarek, S. R., H. B. Johnson, and I. P. Ting. 1973. Drought adaptation of *Opuntia basilaris*. Significance of recycling carbon through crassulacean acid metabolism. Plant Physiology 52:539–541.

Szarek, S. R., and I. P. Ting. 1975. Physiological responses to rainfall in *Opuntia basilaris* (Cactaceae). American Journal of Botany 62:602–609.

Tiedemann, A. R., and J. O. Klemmedson. 1973. Effect of mesquite on physical and chemical properties of the soil. Journal of Range Management 26:27–29.

Tinley, K. L. 1982. The influence of soil moisture balance on ecosystem patterns in southern Africa. Pp. 175–192 *in* B. J. Huntley and B. H. Walker (eds.), Ecology of tropical savannas. Ecological Studies, vol. 42. Springer-Verlag, New York.

Tschirley, F. H., and R. F. Wagle. 1964. Growth rate and population dynamics of jumping cholla (*Opuntia fulgida* Engelm.). Journal of the Arizona Academy of Science 3:67–71.

Van Devender, T. R., L. J. Toolin, and T. L. Burgess. 1990. The ecology and paleoecology of grasses in selected Sonoran Desert plant communities. Pp. 326–349 *in* J. L. Betancourt, T. R. Van Devender, and P. S. Martin (eds.), Packrat middens: the last 40,000 years of biotic change. University of Arizona Press, Tucson.

Venter, J. M., C. Mocke, and J. M. de Jager. 1986. Climate. Pp. 39–52 *in* R. M. Cowling, P. W. Roux, and A. J. H. Pieterse (eds.), The Karroo Biome: a preliminary synthesis. Part 1: Physical environment. South African National Scientific Programmes Report no. 124. Foundation for Research Development, Council for Scientific and Industrial Research, Pretoria.

Walker, B. H., D. Ludwig, C. S. Holling, and R. M. Peterman. 1981. Stability of semi-arid savanna grazing systems. Journal of Ecology 69:473–498.

Walker, B. H., and I. Noy-Meir. 1982. Aspects of the stability and resilience of savanna ecosystems. Pp. 556–590 *in* B. J. Huntley and B. H. Walker (eds.), Ecology of tropical savannas. Ecological Studies, vol. 42. Springer-Verlag, New York.

Waller, S. S., and J. K. Lewis. 1979. Occurrence of C_3 and C_4 photosynthetic pathways in North American grasses. Journal of Range Management 32:12–28.

Walter, H. 1971. Ecology of Tropical and Subtropical Vegetation. Oliver & Boyd, Edinburgh.

Walter, H., and E. Stadelmann. 1974. A new approach to the water relations of desert plants. Pp. 213–310 *in* G. W. Brown, Jr. (ed.), Desert biology, vol. 2. Academic Press, New York.

Weaver, J. E. 1920. Root development in the grassland formation. Carnegie Institution of Washington Publication no. 292. Washington, D.C.

———. 1954. North American Prairie. Johnson Publishing, Lincoln, Neb.

Webb, R. H., and J. L. Betancourt. 1992. Climatic variability and flood frequency of the Santa Cruz River, Pima County, Arizona. U.S. Department of Interior, Geological Survey, Water-Supply Paper 2379.

Westman, W.E. 1981. Seasonal dimorphism of foliage in Californian coastal sage scrub. Oecologia 51:385–388.

Westoby, M., B. Walker, and I. Noy-Meir. 1989. Opportunistic management for rangelands not at equilibrium. Journal of Range Management 42:266–274.

Whitfield, C. J., and H. L. Anderson. 1938. Secondary succession in the desert plains grassland. Ecology 19:171–180.

Whitfield, C. J., and E. L. Beutner. 1938. Natural vegetation in the desert plains grassland. Ecology 19:26–37.

Woodward, F. I. 1987. Climate and Plant Distribution. Cambridge University Press, New York.

Worster, D. 1993. The Wealth of Nature. Oxford University Press, New York.

Wright, H. A. 1974. Effect of fire on southern mixed prairie grasses. Journal of Range Management 27:417–419.

Young, R. W. 1983. Apachean languages. Pp. 393–400 *in* A. Ortiz (ed.), Handbook of North American Indians. Vol. 10: Southwest. Smithsonian Institution, Washington, D.C.

Zedler, P. H., C. R. Gautier, and G. S. McMaster. 1983. Vegetation change in response to extreme events: the effect of a short interval between fires in California chaparral and coastal scrub. Ecology 64:809–818.

3

Desert Grassland History

Changing Climates, Evolution, Biogeography, and Community Dynamics

Thomas R. Van Devender

The grasses include more individuals and encompass a wider environmental range than any other plants on this planet. They reach the limits of vegetation in polar regions and on mountaintops; endure extremes of cold, heat, and drought; and dominate numerous landscapes worldwide. Grasses are highly beneficial for the earth's human inhabitants, especially as nutritious foods and soil anchors that prevent erosion (Thomasson 1987).

In North America, the grasslands of the Great Plains extend from central Canada to Texas between the forests and woodlands of the eastern United States and the Rocky Mountains to the west (McClaran, this volume). In the southwestern United States and on the Mexican Plateau from Chihuahua and Durango south nearly to Mexico City, desert grassland merges into forest, woodland, and desert (Brown and Lowe 1980; Shreve 1951). The vegetation of the northern Chihuahuan Desert in western Texas, southern New Mexico, and southeastern Arizona is a complex mosaic of desert grassland, often invaded by shrubs and succulents, in valleys adjacent to sparse desertscrub on limestone slopes. In northeastern Sonora, desert grassland even contacts the northern edge of subtropical Sinaloan thornscrub.

The modern biotic provinces of North America formed millions of years ago in response to geological events, climatic changes, and evolutionary radiations. Their histories are tightly interwoven and not easily unraveled. In this chapter I present a patchwork discussion of the origins and history of the grasses and grasslands of North America, with emphasis on desert grassland, pieced together from a fragmentary and recalcitrant fossil record.

Grass Evolution

Their importance and taxonomic diversity notwithstanding, grasses have a relatively poor fossil record. Fossil pollen grains, stems, and leaves can often be identified as possibly or probably, but not definitely, grasses. Typical grass pollen has been tentatively recorded as early as the late Cretaceous, 70 mya (millions of years ago), and more definitely from the Paleocene (the first epoch of the Tertiary period, which also includes the Eocene, Oligocene, Miocene, and Pliocene). Fossil pollen indicates that during the Paleocene, archaic grasses (*Graminidites*, *Graminocarpon*, *Monoporites*) were widely distributed in the Southern Hemisphere, including Australia, Brazil, Cameroon, India, and Nigeria (Thomasson 1987). Fossil genera often bear the same names as their modern counterparts, with the addition of prefixes such as *Archeo-* or *Paleo-*, or the suffix *-ites.*

The earliest undoubted fossil record of grasses is based on the identifications of spikelets and inflorescences from the late Paleocene (58 mya) Wilcox formation of western Tennessee (Crepet and Feldman 1991). Although the fossils cannot be aligned with living grasses, their morphological details reflect the early evolution of wind pollination in a dry tropical environment.

Early grasses were likely forest understory herbs with broad leaves and the C_3 photosynthetic pathway, and were ancestral to the living bamboos (fig. 3.1). Broad-leaved grasses are still common in the understory of tropical deciduous forests near Alamos, Sonora, but they are members of the modern subfamily Panicoideae (negrito, *Lasiacis ruscifolia*; panizo caricillo, *Panicum trichoides*; and zacate barbón, *Oplismenus hirtellus*). Most desert grassland grasses have greatly reduced leaf blades and the C_4 photosynthetic pathway (fig. 3.2; McClaran, this volume; Waller and Lewis 1979).

More diverse grasses have been found in late Oligocene (about 24–30 mya) deposits in central North America, including extinct forms

Figure 3.1. A relatively primitive broad-leaved, C_3 grass (*Pharus cornutus*, tribe Phareae, subfamily Bambusoideae) from Coclé, Panama. It was found on a steep bank along a small stream in a tropical forest.

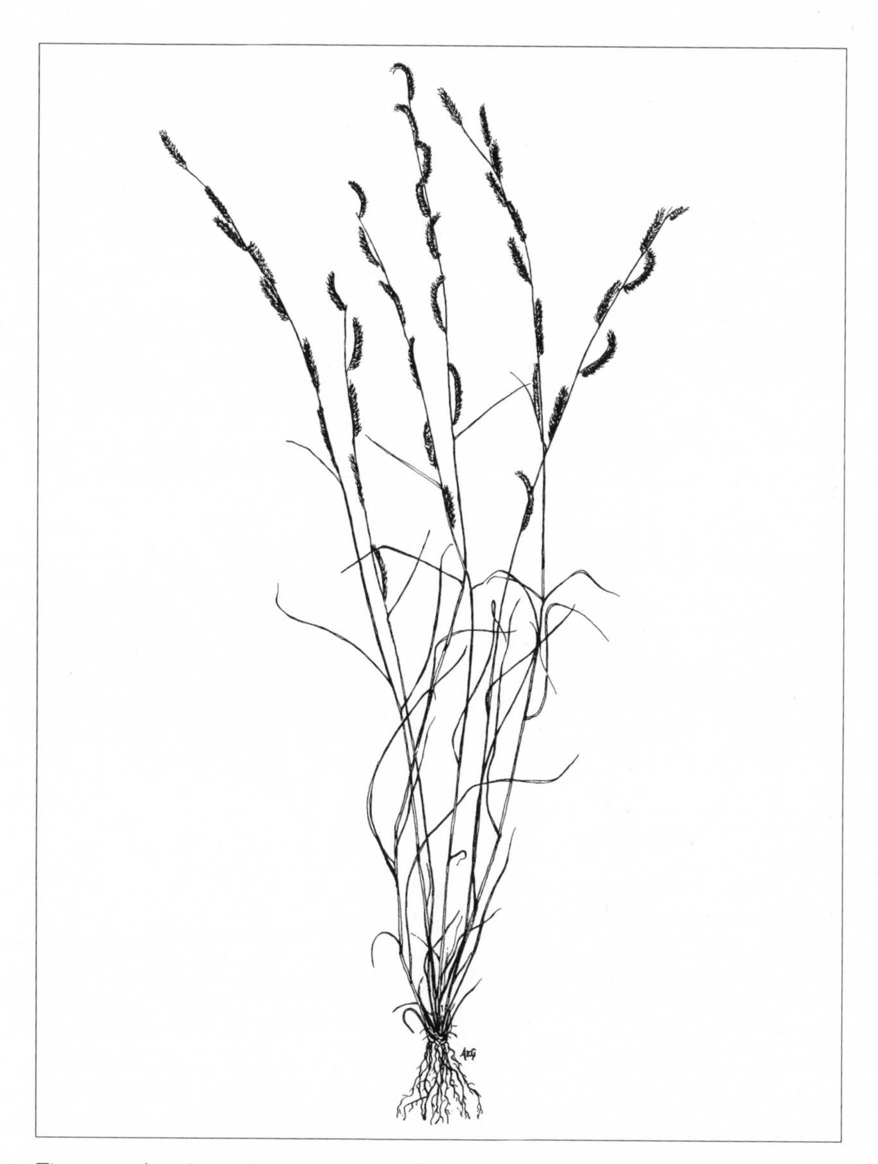

Figure 3.2. An advanced reduced-leaved, C_4 grass (crowfoot grama, *Bouteloua rothrockii*, tribe Chlorideae, subfamily Eragrostoideae) from Tucson, Pima County, Arizona. It is common in desert grassland and occasional in the Sonoran Desert.

(*Graminites*, *Poacites*, *Stipideum*) and the living bentgrass (*Agrostis*), carrizo (*Phragmites*), giant cane (*Arundo*), melicgrass (*Melica*), and pinyon rice-grass (*Piptochaetium*) (Thomasson 1987), although a great deal of the evolutionary history of grasses is clearly missing from the fossil record. The early Miocene (24 mya) fossil record of grasses is much richer and includes representatives all of the modern subfamilies; that is, modern taxonomic and physiological diversity was already well established by then (Thomasson 1987). Many modern genera, including bristlegrass (*Setaria*), fescue (*Festuca*), needlegrass (*Stipa*), and panicgrass (*Panicum*), appeared at this time. The anatomy of a fossil leaf blade from Kansas reflects the evolution of the warm-season C_4 photosynthetic pathway by the Miocene (Thomasson et al. 1986). The leaf is of a species related to fingergrass (*Chloris* spp.), grama grass (*Bouteloua* spp.; fig. 3.2), or tobosa (*Hilaria mutica*). Interestingly, the richest fossil grass deposits are from the Ogallala formation in Colorado, Kansas, Nebraska, New Mexico, and Texas in the heart of the present Great Plains grasslands (Elias 1942). No grass-bearing fossil deposits of early Tertiary age have been found in the areas that now support desert grassland.

Formation of Grasslands

Any reconstruction of the development of grasslands has to be speculative because the fossil record of grasses is so poor. Other plants and animals with richer fossil records can help us understand the environments where grasses evolved.

The rain forests of the late Cretaceous, the first of many evolutionary arenas for grasses, were very different environments from today's grass-dominated landscapes. Temperature and rainfall were relatively constant throughout the year. Archaic tree ferns, cycads, and conifers were dominant; there were relatively few seed-bearing deciduous angiosperms (Wolfe 1987). Broadleaf mixed-mesophytic evergreen forests extended across North America with little regional differentiation. At the end of the Cretaceous, marked by the extinction of the dinosaurs 65 mya, deciduous trees increased dramatically in importance in the rain forests of southeastern Colorado and northeastern New Mexico (Wolfe and Upchurch 1986). Paleocene climates in North America were warm and humid; the forests had strong Asian affinities, and primitive ferns (*Anemia*), cycads (*Dion*, *Zamia*), and palms were found as far north as Alaska (Wolfe 1977). Palms grew at latitude 70° N in Greenland. Studies of fossil leaves from the western United

States indicate the development of a dry season in tropical forests of the Eocene (54–35 mya) as climates continued to be very warm and wet (Axelrod and Bailey 1969; Wolfe and Hopkins 1967).

Today, grass-dominated communities typically occur in areas of harsh environmental regimes—extreme heat, cold, aridity, or salinity. Grasses undoubtedly evolved and spread because of their ability to adapt to the seasonally dry habitats created as tropical deciduous forests developed. Deciduous trees became increasingly more important through the Eocene as many plant and animal groups began adaptive radiations in response to more open, diverse, and variable habitats. The first grass-dominated communities probably appeared in the early Tertiary and were likely seasonally flooded lowlands in dry tropical forests or salt marshes along the coast. The view that many plants and animals can trace their origins to early Tertiary tropical communities has gained general acceptance as the paleobotanical record has unfolded (Axelrod 1979; Leopold and MacGinitie 1972). In sharp contrast to this view is Clements and Weaver's (1924) proposal that the mixed prairie of the temperate central Great Plains was the ancestral North American grassland formation, and that the presence there of woody plants is a more recent, less relevant development (see Burgess, this volume).

Savanna is the term generally used for tropical grasslands with some trees present. Unfortunately, however, the term has come to be used for virtually any community that contains grasses without a closed tree-shrub canopy or a continuous grass cover with trees or shrubs. Broadly defined, *savanna* includes various open woodlands, thornscrub, shrublands, desert grassland, and desertscrub in tropical or temperate climates with or without warm-season rainfall (Burgess, this volume). Nowhere is this bias toward grasslands more apparent than in discussions of the coevolution of grazing animals, grasses, and grass biomes. Grazing animals can be roughly divided into browsers, which feed mostly on trees and shrubs, and grazers, which are specialist feeders on grasses and forbs. The low-crowned (bunodont) teeth of browsing mammals are easily worn down by silica-rich grass stems. From the Eocene through the Miocene, the teeth of many mammals, especially horses, became increasingly high-crowned (hypsodont), apparently as an adaptation to a grass diet. Webb (1977) interpreted the increase of mammals with hypsodont teeth and longer legs in the fossil record as indicating the presence of "savanna" from the Eocene onward. Interestingly, Paleocene fossils from northern Canada suggest that horses may have evolved in a warm Arctic environment rather than in the subtropics (Hickey et al. 1983). Considering the poor fossil record and ubiquitous distribution of grasses, and the ability of hypsodont mammals to feed on trees, shrubs, and forbs as well as on grasses, reconstructions of past grassland communities based on vertebrate fossils alone should be viewed with caution.

Miocene Revolution

Geologic factors have been extremely important in the evolution of plants and animals throughout the earth's history (Tiffany 1985). The modern biogeographic provinces and climatic regimes of western North America were established during the middle Tertiary by the uplifts of the Rocky Mountains in midcontinent and the Sierra Madres Oriental and Occidental bounding the Mexican Plateau. The Rocky Mountains rose continuously from the late Oligocene (about 30 mya) to the middle Miocene (about 15 mya), disrupting the circulation of the upper atmosphere (Axelrod 1979; Leopold and MacGinitie 1972). Climates became continental, with great extremes in temperatures, for the first time in the Tertiary. Moist air masses from the Pacific Ocean were blocked by the mountains, allowing only dry air to descend the lee slopes, and creating the climatic conditions responsible for the development of grasslands in the Great Plains.

Several studies of the Rocky Mountains have provided evidence of the magnitude of the increase in altitude. An Oligocene (35 mya) flora from Florissant, Colorado, was closely allied with plants living today in the highlands of northeastern Mexico (MacGinitie 1953). The flora was a rich mixture that included Bombacaceae (a tropical family containing *Bombax* and *Ceiba*), the ancestor of bristlecone pine (*Pinus crossii*), and palms. The site was then at about 915 m elevation; today Florissant is above 2,440 m—an uplift of at least 1,525 m since the middle Miocene. Over 2,135 m of volcanic rocks were deposited in the Jackson Hole area of west-central Wyoming in the Miocene (Leopold and MacGinitie 1972).

The Sierra Madres Occidental and Oriental were formed when the Mexican Plateau was elevated through volcanic eruptions and regional uplift. The interior of the continent dried out when the mountains blocked tropical moisture from both oceans, and the tropical deciduous forests that had been so widespread in the early Tertiary were restricted to lowland strips along the coast of Mexico.

Harsher extremes in the Miocene created new biomes, including tundra, conifer forests, and grasslands, along latitudinal and elevational environmental gradients. The new climatic regimes segregated drought- and cold-tolerant species into new associations and initiated major evolutionary radiations in many successful groups, including composites and grasses. Vast grasslands likely extended from the northern Great Plains to the southern Mexican Plateau. The plateau was probably as important an evolutionary arena for desert grassland species as the Great Plains, considering that most desert grassland grasses have extensive ranges in the *pastizales* (grasslands) of Chihuahua, Durango, and Zacatecas (Gentry 1957; Shreve 1951).

The evidence for Miocene grasslands in North America is more convincing than biotic reconstructions based on the evolution of vertebrates alone. Grassland pollen assemblages dominated by grasses, ragweed (*Ambrosia*) and other composites, and the pigweed (*Amaranthus*)-goosefoot family (Chenopodiaceae) appeared for the first time (Martin 1975). The number of families of mammals with high-crowned teeth and the diversity of ungulates peaked as the midcontinent became more arid (Webb 1977). The array of Miocene mammals in North America was fully comparable to modern African savanna faunas. The diverse mixture of grazing and browsing mammals suggests that the Miocene grasslands were true savanna, with trees and shrubs, unlike the virtually treeless landscapes of today's Great Plains (Webb 1977). Remains of three species of the extinct C_4 grass genus *Berriochloa* found inside skeletons of rhinoceros (*Teleoceras major*) in late Miocene (9 mya) deposits from Nebraska were the first direct evidence of a large herbivore eating grass (Voorhies and Thomasson 1979).

The deserts formed sometime after the North American biota assumed its modern form in the middle Miocene. Fossil floras in California record a drying trend beginning about 15 mya and culminating in the late Miocene (5–8 mya) in the formation of the Sonoran Desert and the evolution of its biota (Axelrod 1979). On the Mexican Plateau, the Chihuahuan Desert presumably developed at the same time, restricting desert grassland to the south and the eastern flanks of the Sierra Madre Occidental in Chihuahua, Durango, and Zacatecas. Thus, the desert grassland biome may be older than the Chihuahuan and Sonoran Deserts, but the latters' relative areal distributions were established at the same time in the late Miocene.

Elusive Ice Age Grasslands

Although the evolution of grasses and the formation of grasslands were essentially complete by the end of the Miocene, 5 mya, climate fluctuations during the Pleistocene had dramatic impacts on the grasslands of North

America. About 2 mya, massive ice sheets formed at high latitudes and on mountaintops as the earth entered a new climatic era far colder than the middle Miocene conditions. Traditionally, 4 ice ages, or glacial periods, have been recognized, based on terrestrial sedimentary deposits, and widely correlated between Europe, North America, and South America. Recent studies of oxygen isotopes from ocean floor sediments that provide indirect evidence of the amount of ice accumulated in glaciers have forced a revision of this view, however (Imbrie and Imbrie 1979). Instead of 4 glaciations, the isotopic curves indicate that as many as 15 to 20 glacial periods occurred over the last 2.4 million years. Moreover, the ice ages were about 10 times as long as the interglacials, which lasted 10,000–20,000 years. In a sediment core from the Panama Basin near the equator, Porter (1989) found oxygen isotope values indicating environments similar to modern conditions for only 6 percent of the last 340,000 years. The glacial climatic conditions of about 12,000 years ago near the end of the last ice age (the Wisconsin glacial) appear to have been about average for the entire 2.4 million years of the Pleistocene.

During the last glacial, the massive Laurentide ice sheet covered most of Canada and extended as far south as New York and Ohio. Boreal forest with spruce (*Picea*) and jack pine (*Pinus banksiana*) moved southward, displacing mixed deciduous forest in much of the eastern United States (Delcourt and Delcourt 1981). Elsewhere in North America, glaciers covered the tops of the Rocky Mountains, the Sierra Nevada, and the Sierra Madre del Sur. Now-dry playa lakes in the Great Basin were full. Enough water was tied up in ice on land to lower the sea level about 100 m.

A number of Ice Age pollen records are available from sites that are grasslands today. In the central Great Plains, sediments in Muscotah Marsh, northeastern Kansas, record parkland with pine, spruce, and birch (*Betula* spp.) 25,000–23,000 years ago and spruce forest in the full glacial 23,000–15,000 years ago (Grüger 1973). Tall-grass prairie did not develop on the site until 9,930 years ago. Twelve-thousand-year-old sediments from Rich Lake, Texas, on the Llano Estacado in the southern Great Plains yielded abundant pine pollen and low levels of spruce pollen, indicating an open pine parkland instead of the contemporary short-grass prairie (fig. 3.3; Halfsten 1961). A shift from an Ice Age spruce forest to modern oneseed juniper (*Juniperus monosperma*) grassland occurred about 10,000 years ago, as indicated by a pollen core from the San Agustin Plains of west-central New Mexico (Sears and Clisby 1956). A 20,000-year-old pollen profile from the shores of pluvial Lake Cochise in southeastern Ari-

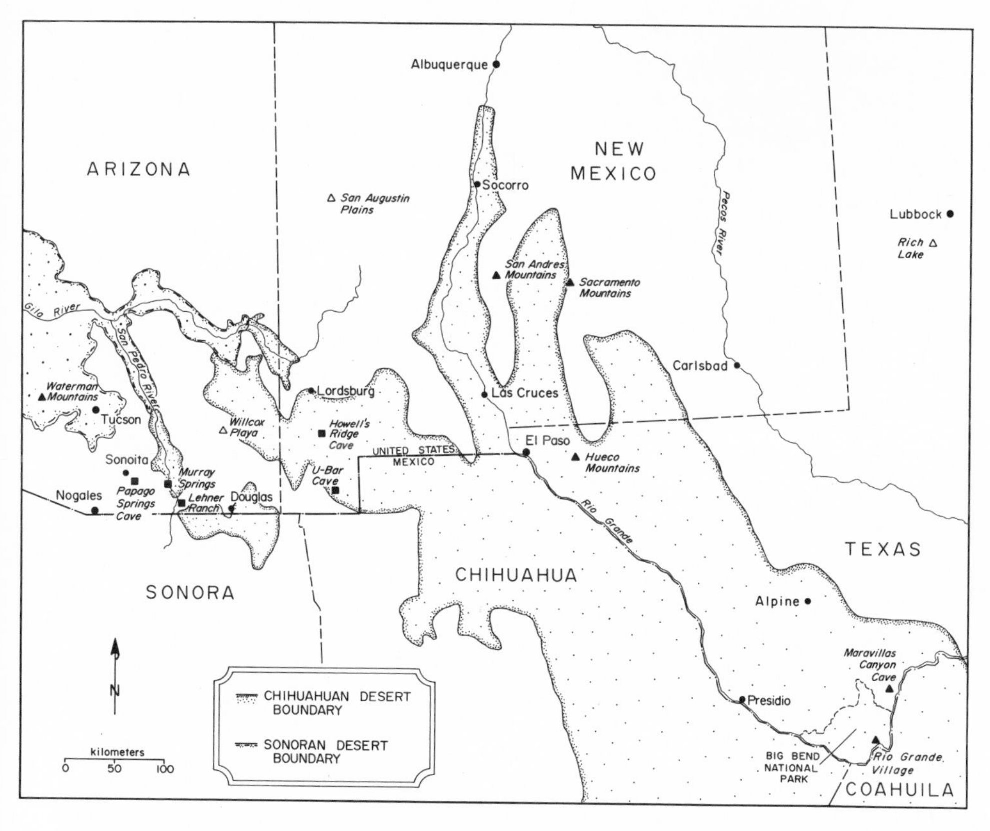

Figure 3.3. Map of study areas mentioned in the text. Solid triangles = packrat midden sites; open triangles = pollen sites; solid squares = vertebrate paleontological sites. Boundary of Chihuahuan Desert after Schmidt 1979; boundary of Sonoran Desert after Brown and Lowe 1980.

zona contained little but pine pollen (Hevly and Martin 1961); today the dry Willcox Playa is surrounded by desert grassland dominated by velvet mesquite (*Prosopis velutina*).

In contrast with pollen records, which indicate domination by forest trees in the Wisconsin glacial period, rich bone beds from the Great Plains to southeastern Arizona reveal the existence of a fauna rivaled today only in such African savanna parks as the Serengeti Plains of Tanzania (Martin 1975). Like the Oligocene records, the fossil record from the Wisconsin indicates the presence of a savanna vertebrate fauna in an area that the plant fossil record shows was covered by trees. Sediments in U-Bar Cave, currently a desert grassland area in southwesternmost New Mexico (fig. 3.3), yielded the bones and teeth of a very diverse fauna. The large mammals present then included the extinct dire wolf (*Canis dirus*), giant short-faced bear (*Arctodus simus*), horses (*Equus* cf. *conversidens*, *E.* cf. *niobrarensis*, *E.* cf. *occidentalis*), mountain deer (*Navahoceros fricki*), pronghorn antelope (*Capromeryx* sp., *Stockoceros onusroagris*), and shrub ox (*Euceratherium collinum*).

A rich late Pleistocene fauna was recovered from Murray Springs, a desert grassland site in the San Pedro River valley on the east base of the Huachuca Mountains of southeastern Arizona (Lindsay 1978). Murray Springs is best known for the elegant flint Clovis spear points found among mammoth (*Mammuthus columbi*) bones that provide glimpses of ancient hunters. Other large mammals recovered from the Ice Age levels include American lion (*Felis atrox*), bison (*Bison* sp.), camel (*Camelops hesternus*), horses, llama (*Hemiauchenia* sp.), and tapir (*Tapirus* sp.). The nearby Lehner Ranch site also yielded mastodon (*Mammut americanum*) teeth. Stock's pronghorn (*Stockoceros onusrosagris*) and flat-headed peccary (*Platygonus compressus*) were found in late Pleistocene sediments at Papago Springs Cave, currently an oak woodland–desert grassland ecotone area in the Canelo Hills east of the Huachuca Mountains (Skinner 1942). These were not the ancestors of the pronghorn (*Antilocapra americana*) or javelina (*Tayassu tajacu*) that live in the area today.

Eleven thousand years ago, about two-thirds of the large mammals of North America became extinct (Martin 1973). Common and widespread grazers such as horses and mammoths disappeared at the very time spruce and pine retreated and grasslands expanded from Canada to Arizona. Martin (1973, 1984) forcefully presented a case that it was hunters, not climate, that caused widespread extinctions within a few hundred years after humans entered North America from Siberia via the Bering Strait. Whether or not one accepts the "overkill" model, there is no evidence in the paleobotanical record of climatic changes severe enough to cause extinction in biotic regimes ranging from the boreal forests of northern latitudes, to conifer parklands and emerging grasslands in the Great Plains, to the deserts of the Southwest, to the chaparral of southern California and throughout the New World Tropics.

Changes in climate and vegetation at the time of the large-mammal extinctions in the southwestern United States were of no greater magnitude than similar fluctuations in many previous interglacials (Imbrie and Imbrie 1979). Moreover, plant remains in ancient packrat (*Neotoma* spp.) middens record the survival of woodland plants in desert lowlands for several thousand years after the extinctions and before the desert grassland formed in the northern Chihuahuan Desert (Van Devender 1990a). It is clear that the last North American grassland fully stocked by large herbivores existed 80,000 years ago at the end of the last interglacial, the Sangamon (Martin 1975). The desert grasslands of the last 10,000 years have had the poorest fauna of large herbivores than at any other time in the last 20 million years.

The vast numbers of bison (*Bison bison*) present on the Great Plains when Europeans arrived give a misleading impression of the abundance of grassland mammals (Martin 1975). Bison immigrated to North America from Eurasia only in the late Pleistocene, perhaps 150,000 years ago. Large bison (*B. latifrons*) with wide-spreading horns were common over the western United States during the late Pleistocene, including in desert grassland sites in southwestern New Mexico (Harris 1985) and southeastern Arizona (Lindsay 1978). The modern species is simply the "extinct" Pleistocene species scaled down to a smaller size. During most of the 11,000 years of the Holocene, bison were far less abundant on the Great Plains and were only sporadically present in the desert grasslands to the southwest (Parmenter and Van Devender, this volume). The ecological roles and impacts of the vanished large grazers and browsers need to be considered if we are to understand and manage modern grassland ecosystems.

In sum, the treeless grasslands of historic times in North America can hardly be traced beyond 11,000 years ago in the fossil record (Martin 1975). Evidence of Ice Age desert grasslands may yet be discovered somewhere in the Southwest or on the Mexican Plateau, but so far such evidence has been elusive. Although grasses themselves were common in Ice Age woodlands, it is clear that during most of the Pleistocene, desert grasslands were extensive only during relatively short periods in interglacials (Van Devender et al. 1987a, 1990a).

Ice Age Woodlands in the Deserts

Although the American Southwest and northern Mexico were distant from the continental glaciers, profound biotic responses to the glacial climates occurred there nevertheless. Abundant, well-preserved plant remains from radiocarbon-dated packrat middens have allowed detailed reconstructions of the history of local desert communities on rocky slopes (Betancourt et al. 1990; Van Devender et al. 1987a). Midden assemblages are excellent for documenting the flora and reconstructing the vegetation within about 30 m of the dry rockshelters on rocky slopes where the middens tend to be found, but only rarely do they permit inferences about nearby riparian communities or the valley-bottom habitats of most desert grasslands.

The midden record extends back far enough to cover several tens of thousands of years of the Wisconsin glacial and the Holocene. These fossils give us an idea of the directions and magnitudes of changes in biotic distributions and communities during the many glacial-interglacial cycles. Plant remains in the middens document widespread expansions of woodland trees and shrubs 45,000–11,000 years ago down to elevations that now support deserts (Betancourt et al. 1990; Van Devender et al. 1987b). Warm desertscrub communities dominated by creosotebush (*Larrea divaricata*) were restricted to the southern Chihuahuan Desert (Van Devender 1990a) and to elevations below 300 m in the lower Colorado River valley in the Sonoran Desert (Van Devender 1990b).

The widespread expansion of woodland into desert elevations in the Ice Age reflects changes in the distribution and abundance of long-lived trees, shrubs, and succulents. Glacial climates had far less impact on other plants, small vertebrates, and invertebrates. Grasses in the Sonoran Desert were especially resilient, and the records show relatively few changes in their distributions (Van Devender et al. 1990). Midden records from the northern Chihuahuan Desert where desertscrub on limestone slopes contacts desert grassland in valley bottoms provide some insight into the Pleistocene history of desert grassland plants and communities.

Rockshelters in the Hueco Mountains just west of El Paso, Texas, in the northern Chihuahuan Desert yielded a 42,000-year series of packrat middens (Van Devender 1990a; Van Devender et al. 1987b). Today the slopes support desertscrub dominated by Chihuahuan whitethorn (*Acacia neovernicosa*), creosotebush, lechuguilla (*Agave lechuguilla*), mariola (*Parthenium incanum*), ocotillo (*Fouquieria splendens*), and Torrey yucca (*Yucca torreyi*; fig. 3.4). Sandy flats in the adjacent Hueco Bolson support a desert grassland that was historically dominated by grasses, especially mesa dropseed (*Sporobolus flexuosus*) (York and Dick-Peddie 1969). Today, the landscape is covered by honey mesquite (*Prosopis glandulosa*) dunes, while broom dalea (*Dalea scoparia*), sand sage (*Artemisia filifolia*), and soaptree yucca (*Yucca elata*) are common in sandier areas.

Samples from 42,000–11,000 years ago were dominated by woodland species, including pinyon pines (*Pinus edulis* and *P. remota*), juniper (*Juniperus* sp.), and shrub oak (*Quercus pungens*) (Van Devender et al. 1990a). The middens also yielded many Ice Age records of important C_4 desert

Figure 3.4. Chihuahuan desertscrub in the Hueco Mountains, El Paso County, Texas. Creosotebush (*Larrea divaricata*) is dominant with Torrey yucca (*Yucca torreyi*) on the lower slopes. Chihuahuan whitethorn (*Acacia neovernicosa*), lechuguilla (*Agave lechuguilla*), mariola (*Parthenium incanum*), and ocotillo (*Fouquieria splendens*) are common on the limestone cliffs. Plant remains in packrat middens from the caves in the cliff base record a pinyon-juniper-oak woodland on the slopes from 42,000 years ago to 11,000 years ago (Van Devender 1990a). Desert grassland plants in the samples were fourwing saltbush (*Atriplex canescens*), variable prickly pear (*Opuntia phaeacantha*), and 13 grasses.

grassland grasses, including black grama (*Bouteloua eriopoda*), blue grama (*B. gracilis*), green sprangletop (*Leptochloa dubia*), sand dropseed (*Sporobolus* cf. *cryptandrus*), sideoats grama (*B. curtipendula*), slim tridens (*Tridens muticus*), and vine mesquite (*Panicum* cf. *obtusum*) (table 3.1). The only important desert grassland shrubs or succulents in the samples were fourwing saltbush (*Atriplex canescens*) and variable prickly pear (*Opuntia phaeacantha*). The teeth of several rodents in the samples—including bannertail kangaroo rat (*Dipodomys spectabilis*), black-tailed prairie dog (*Cynomys* cf. *ludovicianus*), and spotted ground squirrel (*Spermophilus* cf. *spilosoma*)—suggest that open habitats with grasses have been present in the Hueco Bolson for

Table 3.1. Records of desert grassland grasses in Ice Age woodland assemblages from packrat middens greater than 11,000 years old from the Chihuahuan Desert.

Common name	RGV		MC		HM		SM	SAM
(*Scientific name*)	MW	LW	MW	LW	MW	LW	LW	LW
Arizona cottontop (*Digitaria californica*)	x	x		x				
Big bluestem (*Andropogon gerardii*)				x				
Black grama (*Bouteloua eriopoda*)				x	x			
Blue grama (*Bouteloua gracilis*)					x	x	x	x
Cane beardgrass (*Bothriochloa barbinodis*)			x				x	
Chino grass (*Bouteloua ramosa*)		x		x				
Fluffgrass (*Erioneuron pulchellum*)	x	x		x	x	x		
Green sprangletop (*Leptochloa dubia*)	x	x		x	x	x		x
Hall panicgrass (*Panicum hallii*)	x			x		x		
Little bluestem (*Schizachyrium scoparium*)				x				
Mesa muhly (*Muhlenbergia monicola*)		x						x
New Mexico muhly (*Muhlenbergia pauciflora*)							x	
Plains bristlegrass (*Setaria macrostachya/leucopila*)	x	x	x	x	x	x	x	x
Purple threeawn (*Aristida* cf. *purpurea*)			x					
Sand dropseed (*Sporobolus cryptandrus*)					x	x	x	

Table 3.1. *Continued*

Common name (*Scientific name*)	RGV		MC		HM		SM	SAM
	MW	LW	MW	LW	MW	LW	LW	LW
Shortleaf tridens (*Erioneuron grandiflorum*)								x
Sideoats grama (*Bouteloua curtipendula*)	x	x		x	x	x	x	x
Slim tridens (*Tridens muticus*)	x	x	x	x	x		x	
Spike pappusgrass (*Enneapogon desvauxii*)					x	x		
Squirreltail grass (*Elymus elymoides*)		x			x	x	x	x
Tanglehead grass (*Heteropogon contortus*)				x	x	x		
Vine mesquite (*Panicum obtusum*)					x	x		

NOTE: RGV = Rio Grande Village, Big Bend National Park, Tex. (29°15′ N); MC = Maravillas Canyon, Tex. (29°33′ N); HM = Hueco Mountains, Tex. (31°45′ N); SM = Sacramento Mountains, N.M. (32°50′ N); SAM = San Andres Mountains, N.M. (33°10′ N). MW = middle Wisconsin, 22,000–45,600 years ago; LW = late Wisconsin, 11,000–22,000 years ago.

at least 42,000 years (Van Devender et al. 1987a). However, Ice Age grassland in the Bolson without creosotebush, honey mesquite, and ocotillo would have been plains grassland rather than desert grassland. Fourwing saltbush, sand sage, and soaptree yucca were likely present but may have occurred in association with big sagebrushes (*Artemisia tridentata* and relatives), Indian ricegrass (*Oryzopsis hymenoides*), rabbit brushes (*Chrysothamnus* spp.), and winterfat (*Ceratoides lanata*).

Although no Ice Age packrat midden assemblages dominated by grasses have been found, grasses are reasonably common in woodland assemblages from the Chihuahuan Desert (table 3.1). These middens were found at sites ranging from Maravillas Canyon and Rio Grande Village in the Big Bend, north to the Hueco Mountains in Texas (fig. 3.5), to two sites near the

Figure 3.5. Chihuahuan desertscrub near Rio Grande Village in Big Bend National Park, Brewster County, Texas. Important plants include blind prickly pear (*Opuntia rufida*), chino grass (*Bouteloua ramosa*), creosotebush (*Larrea divaricata*), lechuguilla (*Agave lechuguilla*), and ocotillo (*Fouquieria splendens*). Plant remains in packrat middens document Ice Age juniper-oak woodland from 45,600 to 22,000 years ago and pinyon-juniper oak woodland on the rocky limestone slope in front of the rockshelters (Van Devender 1990a). Desert grassland plants including allthorn (*Koeberlinia spinosa*), honey mesquite (*Prosopis glandulosa*), sotol (*Dasylirion wheeleri*), and 12 species of grasses were identified in the Ice Age woodland assemblages.

northern limit of the Chihuahuan Desert in New Mexico (fig. 3.3). Big and little bluestem (*Andropogon gerardii*, *Schizachyrium scoparium*) from the Maravillas Canyon samples reflect southern expansions of typical Great Plains grasses in Ice Age woodlands. The fossil records are somewhat difficult to interpret because few grasses are limited to desert grassland; most also grow between and under trees and shrubs in adjacent woodland and desertscrub. The ranges of typical desert grassland species such as cane beardgrass (*Bothriochloa barbinodis*), feather fingergrass (*Chloris virgata*), plains lovegrass (*Eragrostis intermedia*), sideoats grama, slender grama (*B. repens*), sprucetop grama (*B. chondrosioides*), and tanglehead (*Hetero-*

pogon contortus) extend at least to latitude 27° N in southern Sonora, where they occur in tropical deciduous forest along the Río Cuchujaqui.

Moreover, the Ice Age climates inferred from the more common fossils in packrat middens would not have favored desert grasslands. Summer monsoon rainfall from the tropical oceans was completely absent in the Mohave Desert, and absent or greatly reduced in the Chihuahuan and Sonoran Deserts. The reduction in warm-season moisture likely inhibited C_4 grasses and favored woody cool-season C_3 shrubs, especially big sagebrushes, junipers, rabbit brushes, and snakeweeds (*Gutierrezia* spp.) (Neilson 1986).

Holocene Community Dynamics

In the early part of the twentieth century, plant ecologists proposed and debated two very different concepts of plant communities (see Burgess, this volume; Johnson and Mayeux 1992). The Clementsian climax theory maintains that communities are highly integrated and regulated groups of species that progress through a series of stages to a relatively stable climax. Members of climax communities presumably coevolved in response to similar climatic histories, and thus have similar responses to climatic perturbations. Clements's views on climax and succession continue to be an integral part of management philosophies and policies for range and wildlife in the southwestern United States (Johnson and Mayeux 1992).

According to the opposing individualistic theory, each species evolved unique adaptations and responses to climatic events (Gleason 1939; Shreve 1937). Communities are loosely knit, dynamic assemblages of plants with different physiological tolerances to environmental factors, united by regional and local climatic regimes. Support for the individualistic theory of communities was found in pollen from lake sediments that recorded continuous variation in the trees present in Minnesota forests for thousands of years (Davis 1986). Differential responses of plant species to climatic changes and continuous variation in climate on several time scales mean that plant communities rarely, if ever, reach equilibrium (Davis 1986). Plant remains preserved in packrat middens demonstrate that woodlands, desert grassland, and desertscrub on rocky slopes in the Chihuahuan and Sonoran Deserts appeared to be structurally stable for thousands of years, even though the species composition varied continuously (Van Devender 1986, 1990a,b). Johnson and Mayeux (1992) proposed the concept that communities are relatively stable in physiognomic structure and functional processes even though species composition varies individualistically.

Since 1880, the increasing disturbance of desert grasslands from western Texas to southeastern Arizona has favored the growth of shrubs and succulents. On the deep, fine-grained valley soils, the same species—creosotebush, fourwing saltbush, honey mesquite, joint fir (*Ephedra trifurca*), littleleaf sumac (*Rhus microphylla*), snakeweeds, soaptree yucca, and tarbush (*Flourensia cernua*)—increased throughout the area (Bahre, this volume; Buffington and Herbel 1965; Hastings and Turner 1965; Humphrey 1958; York and Dick-Peddie 1969). Allthorn (*Koeberlinia spinosa*) is becoming increasingly common in desert grasslands in Texas. Important increasers in desert grassland on rocky slopes have been Chihuahuan whitethorn, creosotebush, mariola, ocotillo, sotol (*Dasylirion* spp.), and variable prickly pear. Fossils from Chihuahuan and Sonoran packrat middens provide an opportunity to determine whether shrubs in desert grassland communities have been cohesive or whether species composition has continuously shifted in response to changing climates.

A 45,600-year series of middens from the Big Bend of the Rio Grande along the Coahuila-Texas (Mexico–United States) border yielded records of several desert grassland plants (Van Devender 1990a). Along the Rio Grande, the Chihuahuan Desert reaches its lowest elevation (600 m) and hottest, driest climate. The middens record a juniper-oak woodland in the middle Wisconsin prior to 22,000 years ago and a pinyon-juniper-oak woodland in the late Wisconsin until 11,470 years ago. Chihuahuan desertscrub had formed by 8,980 years ago. None of the assemblages was interpreted as desert grassland. Allthorn and sotol were present in Ice Age woodlands but retreated to higher elevations as Chihuahuan desertscrub developed after 10,400 years ago. A radiocarbon date on honey mesquite verified its presence 21,300 years ago. Honey mesquite was present in four woodland samples but did not become important until after 10,390 years ago. The oldest record of creosotebush is 10,000 years old.

In the Hueco Mountains at 1,385 m elevation in the northern Chihuahuan Desert (fig. 3.4), middens between 8,320 and 4,200 years old yielded desert grassland assemblages dominated by many grasses, fourwing saltbush, honey mesquite, sotol, and variable prickly pear (Van Devender 1990a). Although fourwing saltbush and variable prickly pear were continuously present at low levels in woodlands at the site from 42,000 to about 11,000 years ago, they increased dramatically about 8,100 years ago. Honey mesquite and sotol, which appeared in the record during the last 11,000 years, increased in importance simultaneously. Other common desert grassland shrubs in the northern Chihuahuan Desert, including creosotebush and ocotillo, did not reach the Hueco Mountains until relatively modern Chihuahuan desertscrub developed about 4,500 years ago (Van Devender 1990a). A radiocarbon-dated age of 4,200 years provides a minimum date for the arrival of ocotillo. The oldest creosotebush was found in a midden 3,650 years old.

Additional midden records of desert grassland plants have been obtained from desert grassland sites just above the Chihuahuan Desert in southern New Mexico (Van Devender 1990a). The San Andres Mountains are located on White Sands Missile Range just east of the Jornada del Muerto, about 130 km north of El Paso (fig. 3.3). Fourwing saltbush was associated with Douglas fir (*Pseudotsuga menziesii*), ponderosa pine (*Pinus ponderosa*), and Rocky Mountain juniper (*Juniperus scopulorum*) in a 14,920-year-old sample from an elevation of 1,705 m. Big sagebrush, oneseed juniper, shrub live oak (*Quercus* cf. *turbinella*), and wild rose (*Rosa stellata*) were in an 11,010-year-old sample at 1,465 m. An age of 11,670 years was obtained for fourwing saltbush leaves and fruit. Fourwing saltbush has been in desert grassland in the San Andres Mountains for at least 9,100 years, while creosotebush had reached the area 4,340 years ago. In the Sacramento Mountains about 60 km to the east, on the opposite side of the Tularosa Basin (fig. 3.3), allthorn, creosotebush, mariola, and ocotillo had arrived by 3,300 years ago.

Packrat middens from the northeastern Sonoran Desert have yielded few records of desert grassland plants. Late Wisconsin assemblages from 795 m in the Waterman Mountains, west of Tucson, Arizona (fig. 3.3), included Arizona rosewood (*Vauquelinia californica*), big sagebrush, singleleaf pinyon (*Pinus monophylla*), shrub live oak, tanglehead, and Utah juniper (*Juniperus osteosperma*). Velvet mesquite grew on shady limestone cliff bases from 11,740 years ago until 2,600 years ago but was most common about 6,060 years ago, in the middle Holocene (Anderson and Van Devender 1991). Today, it is restricted to the edges of riparian washes about a kilometer away from the midden sites.

During the late Wisconsin, creosotebush was restricted to elevations below 330 m in the lower Colorado River valley (Van Devender 1990b). Creosotebush twigs from a California juniper (*J. californica*)–Joshua tree (*Yucca brevifolia*) midden assemblage from the Tinajas Altas Mountains just southeast of Yuma, Arizona, were dated at 18,700 years. Creosotebush was first recorded in the Waterman Mountains in a 6,195-year-old assemblage along with blue paloverde (*Cercidium floridum*), brittlebush (*Encelia farinosa*), catclaw (*Acacia greggii*), ironwood, saguaro, and velvet mesquite. Apparently, creosotebush took 5,000 years to move 260 km east and 425 m up in elevation. The midden records suggest that creosotebush reached the northeastern Sonoran Desert several thousand years before it reached the northern Chihuahuan Desert (Van Devender 1990a).

Thus the midden records provide evidence that the shrubs and succulents in the widespread, common desert grassland "community" responded to changing climates as individual as individual species. Some generalists such as fourwing saltbush and variable prickly pear were present in Chihuahuan Desert woodlands and forests throughout the Wisconsin. Others such as allthorn, honey mesquite, and sotol were present in Wisconsin woodlands at lower elevations in the Big Bend but were displaced out of the northern Chihuahuan Desert. Honey mesquite dispersed into the northern Chihuahuan Desert in the early Holocene and then increased in abundance as middle Holocene desert grasslands formed about 9,000 years ago. Finally, increasers that are also widespread in desertscrub, notably creosotebush and ocotillo, reached the northern Chihuahuan Desert less than 4,500 years ago. The physiognomic structure of communities was more stable than species composition, just as Johnson and Mayeux (1992) proposed, but the physiognomic structure shifted with each new bioclimatic regime: late Wisconsin pinyon-juniper-oak woodland, to early Holocene oak-juniper woodland, to middle Holocene desert grassland, to late Holocene Chihuahuan desertscrub. Structural stability was not maintained between bioclimatic regimes because the vegetation changes that took place in the Holocene were fundamental shifts from woodland to desert dominated first by grasses and then by shrubs and succulents.

Holocene Desert Grasslands

Antevs (1955) proposed a series of climatic episodes over the last 10,000 years, based mostly on geological studies of dry lake beds in a winter rainfall area in Oregon in the Great Basin. His proposal of a hot, dry Altithermal period between 4,000 and 7,500 years ago was widely accepted in the southwestern United States and was used to infer drought and desert-like environments during that period. In the desert grassland, which responds primarily to summer monsoon rainfall, increased drought frequency in winter would have reduced the number of C_3 shrubs and favored C_4 grasses (Neilson 1986); shrubs and subtropical succulents would have increased with more frequent summer droughts. Martin (1963) challenged Antev's views, however, concluding that pollen in alluvium from southeastern Arizona indicated summer monsoons 4,000 years ago and earlier that were even stronger than those of today.

In the northern Chihuahuan Desert, the vegetation on limestone slopes shifted from Ice Age pinyon-juniper-oak woodland to oak-juniper woodland about 11,000 years ago, and then to desert grassland about 8,500 years ago (Van Devender 1990a; Van Devender et al. 1987a). The middle Holocene desert grasslands were likely more mesic than they are today, with modest expansions into the upper edges of Arizona Upland desertscrub in the northeastern Sonoran Desert in Arizona and Sinaloan thornscrub in northeastern Sonora (Van Devender 1990b). Instead of an Altithermal drought from 8,500 to 4,000 years ago, the summer monsoons and the grasslands of North America were better developed than they are today, and in fact attained their greatest area.

Thus, the fossil record indicates that desert grasslands were most widespread for perhaps 5,000 years in the middle Holocene and have been more or less similar to today's desert grasslands for the last 4,000 years. Grassland environments may have been present for less than 10 percent of the last 2.4 million years of the Pleistocene (Van Devender 1986). If climatic conditions similar to those of today occurred during only 6 percent of the last 340,000 years, and the average climates were like those 12,000 years ago (Porter 1989), then Ice Age woodlands, not desert grasslands, were the predominant vegetation of the Pleistocene in the Southwest.

Late Holocene Environmental Fluctuations

The historic increase of shrubs in desert grasslands in Texas, New Mexico, and Arizona has triggered heated discussions about environmental changes in the Southwest (Bahre, this volume; Buffington and Herbel 1965; Hastings and Turner 1965; Humphrey 1958; Shreve 1951). Neilson (1986) demonstrated that climatic fluctuations on time scales of years and decades can have profound impacts on the composition and structure of desert grassland, especially in shifting the balance between perennial C_4 grasses and C_3 shrubs.

A 12,000-year-old stratified vertebrate fauna from Howell's Ridge Cave in the Little Hatchet Mountains of southwestern New Mexico (fig. 3.3) helps place the historical increase of shrubs into perspective (Van Devender and Worthington 1977). The cave is a vertical chimney in a steep limestone

Figure 3.6. View of playa in Playas Valley from Howell's Ridge, Little Hatchet Mountains, Grant County, New Mexico. Chihuahuan whitethorn (*Acacia neovernicosa*), creosotebush (*Larrea divaricata*), mariola (*Parthenium incanum*), and ocotillo (*Fouquieria splendens*) are important in desertscrub on limestone slopes. Heavily disturbed desert grassland in the valley is dominated by creosotebush, fourwing saltbush (*Atriplex canescens*), honey mesquite, joint fir (*Ephedra trifurca*), soaptree yucca (*Yucca elata*), and tarbush (*Flourensia cernua*). Shrubs and succulents were probably restricted to drainages in a grassy landscape prior to 1890. Quantitative analyses of reptile bones from Howell's Ridge Cave indicate that similar increases in shrubs and succulents occurred 3,900, 2,500, and 990 years ago (Van Devender and Worthington 1977).

ridge on the edge of the dry lake bed in the Playas Valley (fig. 3.6). Today, Chihuahuan whitethorn, creosotebush, mariola, and ocotillo are important plants in desertscrub on the limestone slopes. The heavily disturbed desert grassland in the valley is dominated by creosotebush, fourwing saltbush, honey mesquite, joint fir, soaptree yucca, and tarbush. Before 1890, shrubs and succulents were probably restricted to drainages in the grassy landscape.

Thousands of bones and teeth of small vertebrates carried to the ridge by owls and hawks were deposited into the cave as regurgitated pellets, providing faunal samples from all the nearby habitats. A total of 2,567 speci-

Figure 3.7. Temporary water in the Lordsburg Playa, Hidalgo County, New Mexico, after a massive Pacific storm in October 1983. The grass is alkali sacaton (*Sporobolus airoides*). The fossil record indicates that this playa and other dry lake beds were permanent lakes throughout the Wisconsin glacial. Tiger salamander (*Ambystoma tigrinum*) and Colorado chub (*Gila* cf. *robusta*) bones in sediments from Howell's Ridge Cave indicate that the playa in the Playas Valley 60 km to the south-southeast contained water from at least 12,000 to 4,000 years ago, and refilled about 3,000 and 1,000 years ago (Van Devender and Worthington 1977).

mens from 45 amphibian and reptile taxa were identified in samples collected at 10-cm intervals through 1.8 m of cave fill. Percentages of the minimum numbers of individuals with specific habitat requirements allowed inferences to be drawn about environmental changes through time.

Bones of an extinct horse (*Equus* cf. *conversidens*) were found in the lower late Wisconsin levels. A California condor (*Gymnogyps californianus*) bone previously collected from the lower levels was radiocarbon dated at 13,460 years (Emslie 1987). A pinyon-juniper-oak woodland was probably growing on the ridge in the late Wisconsin. Tiger salamander (*Ambystoma tigrinum*) and Colorado chub (*Gila* cf. *robusta*) bones along with teeth of voles (*Microtus* spp.) in the deposit indicate that the playa contained peren-

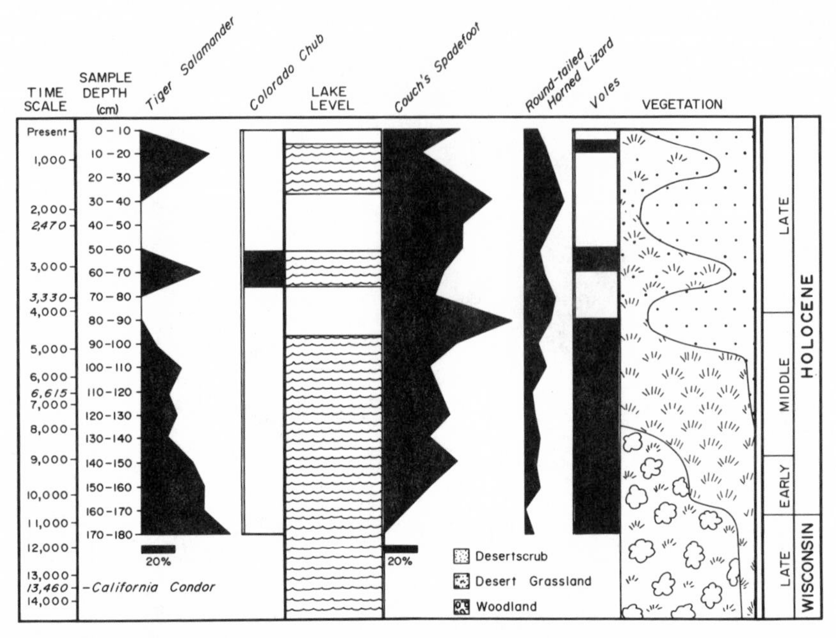

Figure 3.8. Chronological summary of faunal analyses of small vertebrates in 10-cm samples in a 1.8-m-deep cave-fill deposit spanning the last 11,500 years in Howell's Ridge Cave, Grant County, New Mexico (Van Devender and Worthington 1977). The time scale was estimated from sample depth and radiocarbon dates (in italics). Note the California condor (*Gymnogyps californianus*) bone radiocarbon dated as 13,460 years old. Values for tiger salamander (*Ambystoma tigrinum*), Couch's spadefoot (*Scaphiopus couchi*), and round-tailed horned lizard (*Phrynosoma modestum*) are percentages of minimum numbers of individuals. Presence or absence is recorded for Colorado chubs (*Gila* cf. *robusta*) and voles (*Microtus* spp.). Histories of water levels in the playa in the nearby Playas Valley and desert grassland vegetation were inferred from the abundances and temporal distributions of aquatic, grassland, and desertscrub animals.

nial water from 12,000 to 4,000 years ago, and again about 3,000 and 1,000 years ago (figs. 3.7 and 3.8).

The decline of typical grassland animals about 3,900 years ago reflects the first increase of shrubs in the Holocene desert grassland. The increase of typical desert species such as Couch's spadefoot toad (*Scaphiopus couchi*) and round-tailed horned lizard (*Phrynosoma modestum*) reflects a shift to habitats with more exposed soil between shrubs (fig. 3.8). Similar increases in xeric fauna and shrubs were seen 2,500 and 990 years ago.

Thus, the well-documented shrub increase in desert grasslands late in the nineteenth century (Bahre, this volume) was a natural response to drought, differing from earlier invasions only by the unprecedented intensity of disturbances related to ranching and agricultural activities. Unfortunately, the intensity, magnitude, and duration of this recent shrub increase may well preclude reestablishment of prehistoric desert grasslands when the climatic pendulum swings again.

Summary

The history of grasses and grasslands in North America begins with the appearance of grasslike pollen in the fossil record from the late Cretaceous (70 mya); the earliest confirmed grass fossils are from the late Paleocene (58 mya) of Tennessee. The development of the tropical dry season in tropical forests in the Eocene (35–54 mya) precipitated evolutionary radiations in many modern plant groups, including grasses, to adapt to new light and aridity regimes. Late Oligocene (24–30 mya) fossils from the present Great Plains document the presence of diverse, advanced grasses at that time. Major evolutionary radiations from broadleaved, C_3 forest understory herbs ancestral to the living bamboos into the modern tribes apparently were not captured in the fossil record.

The first grass-dominated communities, which probably appeared during the early Tertiary, were tropical savannas in seasonally flooded lowlands in tropical deciduous forests or in salt marshes along the coast. Environmental reconstructions of "savanna" in the Eocene and Oligocene based on fossil evidence of evolutionary trends toward high-crowned teeth and longer limbs in mammals have generally overemphasized the importance and distribution of grass-dominated communities because (1) high-crowned teeth can be used to chew leaves and twigs as well as silica-rich grasses, (2) grasses are ubiquitous in most vegetation types, and (3) the term *savanna* has such a broad definition that it is not useful.

The uplifts of the Rocky Mountains and the Sierra Madres Occidental and Oriental between the late Oligocene and the middle Miocene (15 mya) established the modern climatic regimes and biogeographic provinces, segregated plants into communities limited by environmental extremes (tundra, conifer forests, and grasslands), and stimulated evolutionary radiations in many successful modern plant groups. The first reliable evidence of grasslands is in the fossil record from the Miocene. Warm deserts and, likely, desert grassland developed in the late Miocene (5–8 mya) as climates became more arid. The speciation of modern dominants in these biomes may well have occurred at this time.

Studies of isotope climatic indicators in ocean floor sediment cores indicate that there were between 15 and 20 glacial-interglacial cycles during the 2.4 million years of the Pleistocene. A core from the Panama Basin near the equator indicates that modern climatic conditions were present during only 6 percent of the last 340,000 years. Average Pleistocene climates were similar to those of 12,000 years ago, near the end of the Wisconsin glacial period.

Pollen in sediments from bogs and lakes in areas that now support grasslands and desert grasslands have not yielded evidence of Ice Age grasslands. Instead, spruce forest or parkland was present in the central Great Plains during that period, and pine forest in the southern Great Plains, western New Mexico, and southeastern Arizona. Plant remains preserved in ancient packrat (*Neotoma* spp.) middens record widespread woodlands at elevations that now support the modern Chihuahuan and Sonoran Deserts. No paleobotanical evidence of Ice Age grasslands has been found, even though a rich vertebrate fauna similar to those on modern African savannas was present there.

Desert grasslands formed, as they probably did in each of the 15 to 20 Pleistocene interglacials, about 9,000 years ago in the Holocene as woodland plants retreated to high elevations and desert grassland plants increased in abundance. Fossil records of shrubs and succulents that increased when desert grasslands were present demonstrate that species responded to Holocene climatic changes as individuals rather than as communities. Species such as fourwing saltbush, variable prickly pear, and many grasses were present in Ice Age woodlands. Others dispersed northward into the present northern Chihuahuan Desert or northeastward from the Sonoran Desert at various times during the Holocene.

The fossil plant record indicates that summer monsoon rainfall and desert grasslands peaked between 9,000 and 4,000 years ago in the middle Holocene. Quantitative analyses of fish, amphibian, reptile, and small mammal bones and teeth in Howell's Ridge Cave, southwestern New Mexico, provide evidence that the nearby playa held water until about 4,000 years ago, and again 3,000 and 1,000 years ago. Increases in the number of desert species 3,900, 2,500, and 990 years ago suggest more open habitats with more shrubs and succulents. Thus, increases in shrubs and succulents in desert grasslands are natural processes with precedents in the paleoenvironmental record. The shrub expansion that has occurred in recent history differed from earlier events in that desert grasslands were stressed by intense grazing and agricultural practices before the beginning of a drought. Unfortunately, the severity of the impacts and continued use of the land may well maintain the shrubby phase of desert grassland and prevent reversion to grassier landscapes.

Acknowledgments

Tony Burgess shared his knowledge of desert and desert grassland ecology, ecophysiology, biogeography, and climates, and served as a sounding board for my ideas. Paul S. Martin helped me formulate many of the ideas and conclusions I present here through his careful paleoecological research, standards of scholarship, stimulating teaching and writings, and unflagging encouragement. Laurence J. Toolin developed the skills to identify grass fragments in ancient packrat middens, the basis for understanding the paleoecology of desert grasslands; the identifications in table 3.1 are his. John R. Reeder and Charlotte G. Reeder identified my many grass specimens. The careful and thoughtful editing of Rebecca K. Van Devender, Paul S. Martin, Tony Burgess, Carol Cochran, and Mitchel P. McClaran improved this chapter in many ways. The drawings in figures 3.1 and 3.2, the map in figure 3.3, and the diagram in figure 3.8 were done by Ann Gondor.

Literature Cited

Anderson, R. S., and T. R. Van Devender. 1991. Comparison of pollen and macrofossils in packrat (*Neotoma*) middens: a chronological sequence from the Waterman Mountains of southern Arizona, U.S.A. Review of Paleobotany and Palynology 68:1–28.

Antevs, E. 1955. Geological-climatic dating in the West. American Antiquity 20: 317–355.

Axelrod, D. I. 1979. Age and origin of the Sonoran Desert. California Academy of Sciences Occasional Paper 132:1–74.

Axelrod, D. I., and H. P. Bailey. 1969. Paleotemperature analysis of Tertiary floras. Paleogeography, Paleoclimatology, Paleoecology 6:163–195.

Betancourt, J. L., T. R. Van Devender, and P. S. Martin (eds.). 1990. Packrat middens: the last 40,000 years of biotic change. University of Arizona Press, Tucson.

Brown, D. E., and C. H. Lowe. 1980. Biotic communities of the Southwest. U.S. Department of Agriculture Forest Service General Technical Report RM-78 (map).

Buffington, L. C., and C. H. Herbel. 1965. Vegetational changes on a semidesert grassland. Ecological Monographs 35:139–164.

Clements, F. E., and J. E. Weaver. 1924. Experimental vegetation: the relations of climaxes to climates. Carnegie Institution of Washington Publication no. 355. Washington, D.C.

Crepet, W. L., and G. D. Feldman. 1991. The earliest remains of grasses in the fossil record. American Journal of Botany 78:1010–1014.

Davis, M. B. 1986. Climatic instability, time lags, and community disequilibrium. Pp. 269–284 *in* J. Diamond and T. J. Case (eds.), Community ecology. Harper & Row, New York.

Delcourt, P. A., and H. R. Delcourt. 1981. Vegetation maps for the eastern United States: 40,000 yr B.P. to present. Pp. 123–165 *in* R. C. Romans (ed.), Geobotany II. Plenum Press, New York.

Elias, M. K. 1942. Tertiary prairie grasses and other herbs from the high plains. Geological Society of America Special Paper 41. Geological Society of America, Boulder.

Emslie, S. D. 1987. Extinction of condors in Grand Canyon, Arizona. Science 237: 768–770.

Gentry, H. S. 1957. Los pastizales de Durango. Ediciones del Instituto Mexicano de Recursos Naturales Renovables, Mexico City.

Gleason, H. A. 1939. The individualistic concept of the plant association. American Midland Naturalist 21:92–110.

Grüger, J. 1973. Studies on the vegetation of northeastern Kansas. Geological Society of America Bulletin 84:239–250.

Halfsten, U. 1961. Pleistocene development of vegetation and climate in the southern high plains as evidenced by pollen. Pp. 59–91 *in* F. Wendorf (ed.), Paleoecology of the Llano Estacado. Fort Burgwin Research Center Report 1. University of New Mexico Press, Albuquerque.

Harris, A. H. 1985. Preliminary report on the vertebrate fauna of U-Bar Cave, Hidalgo County, New Mexico. New Mexico Geology 7 (November): 74–77, 84.

Hastings, J. R., and R. M. Turner. 1965. The Changing Mile: An Ecological Study of Vegetation Change with Time in the Lower Mile of an Arid and Semiarid Region. University of Arizona Press, Tucson.

Hevly, R. H., and P. S. Martin. 1961. Geochronology of pluvial Lake Cochise, southern Arizona. I. Pollen analysis of shore deposits. Journal of the Arizona Academy of Science 2:24–31.

Hickey, L. J., R. M. West, M. R. Dawson, and D. K. Choi. 1983. Arctic terrestrial biota: paleomagnetic evidence of age disparity with mid-northern latitudes during the late Cretaceous and early Tertiary. Science 221: 1153–1156.

Humphrey, R. R. 1958. The desert grassland: a history of vegetational change and an analysis of causes. Botanical Review 24:193–252.

Imbrie, J., and K. P. Imbrie. 1979. Ice Ages, Solving the Mystery. Enslow, Short Hills, N.J.

Johnson, H. B., and H. S. Mayeux. 1992. Viewpoint: A view on species additions and deletions and the balance of nature. Journal of Range Management 45:322–333.

Leopold, E. B., and H. D. MacGinitie. 1972. Development and affinities of Tertiary floras in the Rocky Mountains. Pp. 147–200 *in* A. Graham (ed.), Floristics and paleofloristics of Asia and eastern North America. Elsevier, Amsterdam.

Lindsay, E. H. 1978. Late Cenozoic vertebrate faunas, southeastern Arizona. Pp. 269–275 *in* J. F. Callender et al. (eds.), Land of Cochise—southeastern Arizona. New Mexico Geological Society Guidebook, 29th Field Conference, Socorro, N.Mex.

MacGinitie, H. D. 1953. Fossil plants of the Florissant beds, Colorado. Carnegie Institution of Washington Publication no. 599. Washington, D.C.

Martin, P. S. 1963. The Last 10,000 Years. University of Arizona Press, Tucson.

———. 1973. The discovery of America. Science 179:969–974.

———. 1975. Vanishings, and future of the prairie. Geoscience and Man 10: 39–49.

———. 1984. Pleistocene overkill: the global model. Pp. 354–403 *in* P. S. Martin and R. G. Klein (eds.), Quaternary extinctions. University of Arizona Press, Tucson.

Neilson, R. P. 1986. High-resolution climatic analysis and Southwest biogeography. Science 232:27–34.

Porter, S. C. 1989. Some geological implications of average Quaternary glacial conditions. Quaternary Research 32:245–261.

Schmidt, R. H., Jr. 1979. A climatic delineation of the "real" Chihuahuan Desert. Journal of Arid Environments 2:243–250.

Sears, P. B., and K. H. Clisby. 1956. San Agustin Plains—Pleistocene climatic changes. Science 124:537–539.

Shreve, F. 1937. Thirty years of change in desert vegetation. Ecology 18:463–478.

———. 1951. The role of grasses in the vegetation of Arizona. Pp. 10–15 *in* F. W. Gould, Grasses of the southwestern United States. University of Arizona Press, Tucson.

Skinner, M. F. 1942. The fauna of Papago Springs Cave, Arizona, and a study of *Stockoceros* with three new antilocaprines from Nebraska and Arizona. Bulletin of the American Museum of Natural History 80:143–220.

Thomasson, J. R. 1987. Fossil grasses: 1820–1986 and beyond. Pp. 159–167 *in* T. R. Soderstrom, K. W. Hilu, C. S. Campbell, and M. E. Barkworth (eds.),

Grass systematics and evolution. Smithsonian Institution Press, Washington D.C.

Thomasson, J. R., M. E. Nelson, and R. J. Zakrzewski. 1986. A fossil grass (Gramineae: Chloridoideae) from the Miocene with Kranz anatomy. Science 233:876–878.

Tiffany, B. H. 1985. Geological factors and the evolution of plants. Pp. 1–10 *in* B. H. Tiffany (ed.), Geological factors and the evolution of plants. Yale University Press, New Haven.

Van Devender, T. R. 1986. Climatic cadences and the composition of Chihuahuan Desert communities: the late Pleistocene packrat midden record. Pp. 285–299 *in* J. Diamond and T. J. Case (eds.), Community ecology. Harper & Row, New York.

———. 1990a. Late Quaternary vegetation and climate of the Chihuahuan Desert, United States and Mexico. Pp. 104–133 *in* J. L. Betancourt, T. R. Van Devender, and P. S. Martin (eds.), Packrat middens: the last 40,000 years of biotic change. University of Arizona Press, Tucson.

———. 1990b. Late Quaternary vegetation and climate of the Sonoran Desert, United States and Mexico. Pp. 134–165 *in* J. L. Betancourt, T. R. Van Devender, and P. S. Martin (eds.), Packrat middens: the last 40,000 years of biotic change. University of Arizona Press, Tucson.

Van Devender, T. R., G. L. Bradley, and A. H. Harris. 1987a. Late Quaternary mammals from the Hueco Mountains, El Paso and Hudspeth Counties, Texas. Southwestern Naturalist 32:179–195.

Van Devender, T. R., R. S. Thompson, and J. L. Betancourt. 1987b. Vegetation history of the deserts of southwestern North America: the nature and timing of the late Wisconsin–Holocene transition. Pp. 323–352 *in* W. F. Ruddiman and H. E. Wright, Jr. (eds.), North America and adjacent oceans during the last deglaciation. Geological Society of America, Boulder.

Van Devender, T. R., L. J. Toolin, and T. L. Burgess. 1990. The ecology and paleoecology of grasses in selected Sonoran Desert plant communities. Pp. 326–349 *in* J. L. Betancourt, T. R. Van Devender, and P. S. Martin (eds.), Packrat middens: the last 40,000 years of biotic change. University of Arizona Press, Tucson.

Van Devender, T. R., and R. D. Worthington. 1977. The herpetofauna of Howell's Ridge Cave and the paleoecology of the northwestern Chihuahuan Desert. Pp. 85–106 *in* R. H. Wauer and D. H. Riskind (eds.), Transactions of the Symposium on the Biological Resources of the Chihuahuan Desert, United States and Mexico. U.S. National Park Service, Washington, D.C.

Voorhies, M. R., and J. R. Thomasson. 1979. Fossil grass anthoecia within Miocene rhinoceros skeletons: diet in an extinct species. Science 206:331–333.

Waller, S. S., and J. K. Lewis. 1979. Occurrence of C_3 and C_4 photosynthetic pathways in North American grasses. Journal of Range Management 32: 12–28.

Webb, S. D. 1977. A history of savanna vertebrates in the New World. I. North America. Annual Review of Ecology and Systematics 8:355–380.

Wolfe, J. A. 1977. Paleogene floras from the Gulf of Alaska region. U.S. Geological Survey Professional Paper 997.

———. 1987. Late Cretaceous–Cenozoic history of deciduousness and the terminal Cretaceous event. Paleobiology 13:215–226.

Wolfe, J. A., and D. Hopkins. 1967. Climatic changes recorded by Tertiary land floras in northwestern North America. Pp. 67–76 *in* K. Hatai (ed.), Tertiary correlations and climatic changes in the Pacific. Symposium of the 11th Pacific Scientific Congress, Tokyo.

Wolfe, J. A., and G. R. Upchurch, Jr. 1986. Vegetation, climatic and floral changes at the Cretaceous-Tertiary boundary. Nature 324:148–152.

York, J. C., and W. A. Dick-Peddie. 1969. Vegetation changes in southern New Mexico during the past hundred years. Pp. 155–166 *in* W. G. McGinnies and B. J. Goldman (eds.), Arid lands in perspective. University of Arizona Press, Tucson.

4

Landscape Evolution, Soil Formation, and Arizona's Desert Grasslands

Joseph R. McAuliffe

The dominant plant species in the desert grasslands of southern Arizona vary greatly from one place to another. The causes of many of the most striking vegetation and soil patterns in these desert grasslands have not yet been worked out, let alone the causes of the myriad more subtle patterns that exist (Martin 1975). The characteristics and distributions of the soils present today are the product of landscape-forming processes of the immediate geological past. An understanding of how the geological landscapes of the desert grasslands evolved will add significantly to our understanding of vegetation and soil patterns.

A considerable amount of the compositional variation in today's desert grasslands is a consequence of their past use by humans. Widespread vegetation changes in the semiarid Southwest during the last 100 years have been clearly documented (Bahre 1991, this volume; Buffington and Herbel 1965; Hastings and Turner 1965; Humphrey 1958), including the decline of valuable forage grass species and detrimental landscape alteration by erosion. These changes have not occurred uniformly, however; some areas are apparently more resistant to the increase of undesirable species or soil erosion than others. A better understanding of these variable responses can be found through a detailed examination of landscape evolution. This understanding, in turn, can benefit the management of the diverse terrains that make up the desert grasslands.

The first topics covered in this chapter deal with three questions about relationships between landscape evolution, soil formation, and plants: How are soil characteristics a function of landscape development? How do soil characteristics in turn affect the distribution and duration of water stored in the soil? How does the distribution and duration of stored soil water consequently affect plant distributions? Next, I discuss the relationships between geological features and erosion leading to landscape changes. Finally, I address the importance of incorporating landscape features and erosion potential into plans for revegetation.

Landscape Evolution and Soil Characteristics

Most of southern Arizona lies in the Basin and Range Physiographic Province, a region characterized by long, relatively narrow mountain ranges separated by broad alluvial basins (Morrison 1985). The upper elevations of the ranges typically contain evergreen woodlands and coniferous forests. The alluvium-filled basins generally lie below 1,500 m elevation and are the principal domain of the most expansive desert grasslands.

The gentle slopes of the alluvial piedmont flanking the east side of the Santa Rita Mountains, the site of the Santa Rita Experimental Range (Martin 1966), provide an excellent example of landscape, soil, and vegetation heterogeneity within Arizona's desert grasslands. Some areas of the piedmont are dominated by perennial grasses; others are dominated by woody species such as velvet mesquite (*Prosopis velutina*) and burroweed (*Isocoma tenuisecta*).

One of the most distinctive features of the Santa Rita piedmont is a 6-km-wide alluvial fan surface (fig. 4.1) composed of alluvium disgorged from the mouth of Madera Canyon during the Pleistocene (the epoch of geological time containing multiple ice ages that began about 2 million years ago and ended about 11,000 years ago). The deeply dissected central part of this fan (surface Q2a in fig. 4.1) is the highly eroded remnant of a fan estimated to have been deposited approximately 2 million years ago (Pearthree and Calvo 1987). After the deposition of the original fan in earliest Pleistocene times, Madera Canyon Wash began a degradational phase and cut downward into the northern side of this surface and removed a substantial part of the original fan. Subsequent deposition of additional alluvium transported out of Madera Canyon in mid-Pleistocene times (prob-

Figure 4.1. A. Madera Canyon alluvial fan. The photograph includes much of the Santa Rita Experimental Range, looking southeast. Photograph by Peter Kresan, 2 December 1984. B. Map of the same canyon. Designations of geomorphic surfaces Q2a, Q2b, etc., refer to fan deposits of the Quaternary era (approximately the last 2 million years), numbered sequentially starting with the oldest deposits exposed in the basin (Q1 deposits are not present within the view shown here). Designations of surfaces follow Pearthree and Calvo (1987) and the map of Quaternary deposits in Ely and Baker (1985). The deeply dissected Q2a surface (top center) is all that remains of an extensive fan de-

ably in several episodes ranging from an estimated 200,000 to 750,000 years ago) created another distinct fan surface (Q2b in fig. 4.1). This surface is inset within the topographic confines of the slightly higher surfaces of the early Pleistocene (Q1a) surface to the south and the footslopes of the Santa Rita Mountains to the north and east. Similar Q2b fan surfaces were also deposited by Florida Canyon Wash to the north. The aggradation of the mid-Pleistocene surface emanating from the mouth of Madera Canyon ended with a change in the course of Madera Canyon Wash to the south,

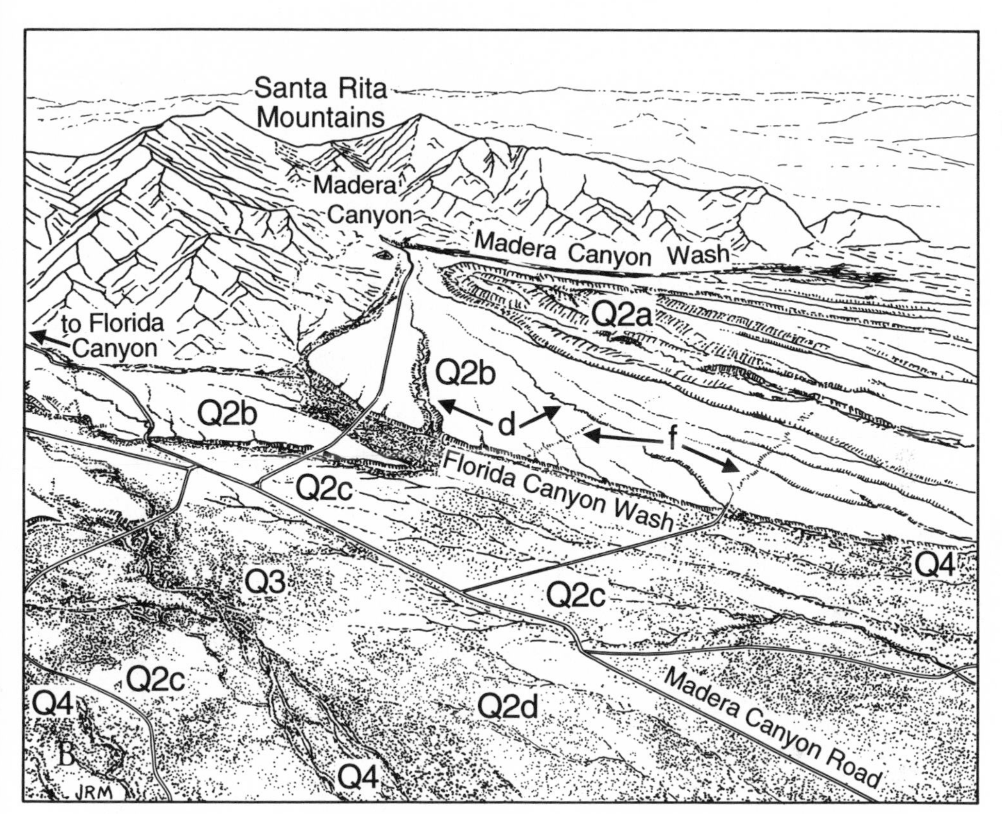

posited in the earliest Pleistocene. The extensive, light-colored Q2b surface contains deposits estimated to be of middle Pleistocene age (several hundred thousand years old). The mesquite-studded (*Prosopis velutina*) surfaces in the foreground (small, dark dots in photo are mesquite trees) are of late Pleistocene (Q2c, Q2d) and Holocene ages (Q3, Q4). On-fan drainages (d) are shown on the extensive Q2b surface. The two linear features (f) are fault scarps produced by earthquakes during Pleistocene times (see Pearthree and Calvo 1987).

where it incised a channel into the original early Pleistocene (Q2a) surface and took the course it follows today (fig. 4.1).

The change in course of Madera Canyon Wash severed the mid-Pleistocene (Q2b) surface from the wash and ended the alluvial aggradation on this surface. Since this fan surface is no longer connected hydrologically to the substantial watersheds of the Santa Rita Mountains, the small drainages across the Pleistocene fan are derived exclusively from precipitation falling directly on the fan. The drainages originating on the fan surface (on-fan drainages) have incised the mid-Pleistocene surface to maximum depths of 10–12 m, only about one-fourth to one-fifth the depth of incisions in the early Pleistocene (Q2a) surface cut by on-fan drainages. This great difference in incision depth between the two surfaces is a result of the much greater age of the fan and the longer period of incision that has occurred on the early Pleistocene (Q2a) surface.

The alluvial surfaces originally deposited by Madera Canyon Wash have been subsequently shaped by another stream system, Florida Wash, which flows onto the piedmont from Florida Canyon several kilometers north of Madera Canyon. In late Pleistocene and Holocene times, Florida Wash cut into and removed northern parts of the mid-Pleistocene (Q2b) Madera fan surface and also contributed new alluvium that created a series of more recent surfaces (Q2c–Q4 in fig. 4.1).

The mosaic of alluvial fan surfaces below Madera Canyon is typical of alluvial piedmonts in basins throughout southern Arizona such as the Santa Cruz, San Pedro, Gila, and San Simon River valleys (McAuliffe 1994; Morrison 1985). In each valley the various geomorphic surfaces represent episodes of aggradation, partially in response to changes in the local base level of the river valley. During times of incision of the main exterior drainage, aggradational surfaces of fans are abandoned and become incised. Younger surfaces are typically inset within the topographic confines of the older surfaces and contribute to a general pattern of successively elevated surfaces of increasing age within the piedmont. In some areas, older surfaces may also be buried by more recent alluvium. This pattern of burial occurs most commonly in lower parts of the piedmont and is frequently encountered in closed basins lacking exterior drainage such as Sulphur Springs Valley in southeastern Arizona or in exteriorly drained basins where the base level of the valley floor is relatively stable. Peterson (1981) provided a comprehensive and well-illustrated description of these alluvial landforms and their formation within the Basin and Range Province.

The ages of alluvial landforms are relevant to an understanding of desert grassland composition because soil formation is in part dependent on the passage of time. The characteristics of the soils on the various fan remnants vary according to the geological ages of those surfaces.

Two distinct soil horizons form and become increasingly well developed over time on noncalcic, gravelly parent materials (derived from rocks relatively devoid of calcium carbonate) in warm arid and semiarid regions: an argillic horizon enriched with clay and a calcic horizon enriched with calcium carbonate (Gile et al. 1981). The formation of the argillic horizon is

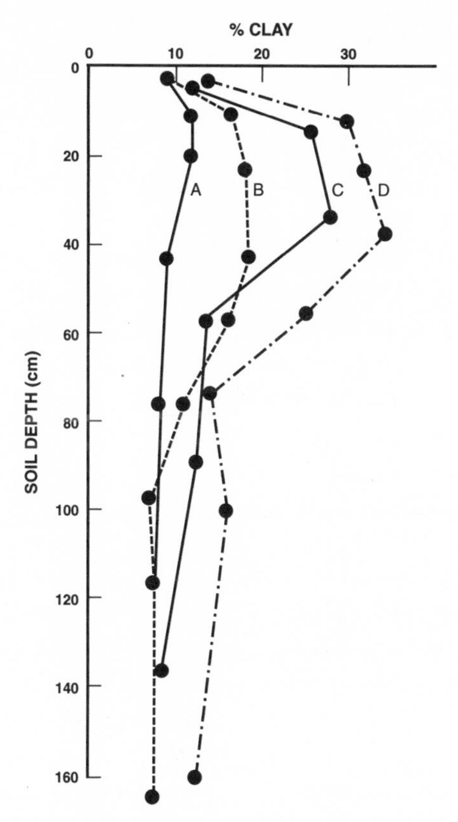

Figure 4.2. Clay accumulation as a function of soil age in a semiarid environment in southern New Mexico. A. Mid-to-late Holocene deposit with estimated age of 1,000–4,000 ybp (years before present). B. Latest Pleistocene, estimated age 8,000–15,000 ybp. C and D. Late Pleistocene, estimated ages 25,000–75,000 ybp. (Data from Gile et al. 1981, table 28).

especially important because it affects the capacity of soil to absorb and store water. Typically, noncalcic, gravelly parent materials of gently sloping fan deposits in the Southwest contain little clay at the time of deposition. Holocene surfaces (less than 11,000 years old) typically exhibit very weak profile development with little or no accumulation of clay in distinct horizons (fig. 4.2). As time passes in semiarid climates, a soil on an undifferentiated deposit of coarse alluvium will undergo increasing development of the argillic and calcic horizons due to a combination of weathering of minerals in the original parent materials and the substantial addition of mate-

Table 4.1. Change in soil characteristics with increasing age of geomorphic surfaces of the Madera Canyon fan.

Geomorphic Surface	Estimated Age (10^3 years)	Maximum % Clay	Increase in % Clay in Lower B Soil Horizon vs. Upper A Horizon
Q3b	<4	7	no B horizon
Q3a	4–8	4	3
Q2d	8–20	6	5
Q2c	75–130	22	15
Q2b-2	200–300	31	19
Q2b-1	400–750	44	33
Q2a	1,000–2,000	67	42

SOURCE: Pearthree and Calvo 1987.

rials from dustfall and precipitation (Gile et al. 1981; Machette 1985; Marion 1989; McFadden and Tinsley 1985). In contrast with the relatively unaltered gravelly alluvium of late Holocene deposits, soils developed on stable, relict Pleistocene surfaces show very strong soil profile development with the clay content of argillic horizons increasing as a function of age (table 4.1, fig. 4.2). The textural, structural, and chemical characteristics of some horizons on these older relict surfaces differ dramatically from those of the original gravelly parent materials.

The various geomorphic surfaces of the Madera Canyon fan system exhibit this progression in development of the argillic horizon as a function of age. Maximum clay content within profiles ranges from 4–6 percent in the youngest Holocene surfaces to 67 percent on the oldest remnant Pleistocene surfaces. Calcic horizons similarly exhibit increasingly strong development as a function of soil age (table 4.1; Pearthree and Calvo 1987).

Soil Characteristics and Water Storage

In semiarid environments, the argillic horizon exerts an extremely important control over the depth of water infiltration and the duration of its availability to plants. Soils from Pleistocene fan remnants in southern Arizona with relatively coarse-textured A, or upper, horizons typically possess water infiltration capacities on the order of 1.5–5.0 cm/hr, a rate comparable to that of coarse-textured soils on Holocene surfaces (Richardson et al. 1979; Vogt 1980). Infiltration capacities typically drop to one-tenth of this rate or less in clay-rich argillic horizons. Although the abundance of extremely fine capillary spaces within argillic horizons limits the rate of infiltration, these spaces permit the storage of large quantities of water. On the other hand, coarser soil textures permit rapid infiltration to great depth, but a comparable volume of such soil is capable of holding substantially less water.

The water from precipitation or runoff from upslope areas penetrates into the soil to a depth at which the capillary tension on the water equals the downward gravitational pull. When this equilibrium is reached, the downward movement of water ceases and the wetted soil is said to be at field capacity with respect to soil moisture. The separation between wetted and unwetted parts of the soil profile, called the wetting front, is typically quite abrupt. The amount of water held at field capacity by a coarse-textured soil is typically a small fraction of the amount capable of being held by an equivalent volume of soil with an argillic horizon. The clay-rich horizons of Pleistocene fan remnants (sandy clay loam, sandy clay, and clay textural classes) may hold between 20 and 50 percent (or more) water by volume at field capacity, whereas coarse-textured soils (loamy sands and sandy loams) reach field capacity at approximately 10 percent water by volume. Because of the considerably greater volumes of water that can be stored in clayey soil horizons compared with coarser soils, the depth of wetting in soils with strongly developed argillic horizons is typically much shallower than the depth of wetting on Holocene surfaces or some Pleistocene surfaces whose argillic horizons have been truncated by erosion. In semiarid lands such as the desert grasslands, small precipitation events are typically far more common than large events (Sala et al. 1982). Consequently, wetting of deeper layers in soils with exceptionally well developed argillic horizons is infrequent in most semiarid grasslands of Arizona, especially during the summer rainy season.

In addition to argillic horizons, other soil conditions such as the soil volume occupied by coarse stony fragments and the depth to shallow petrocalcic horizons (caliche) or bedrock on rocky slopes also affect infiltration and the storage of water within soil profiles. Consequently, semiarid landscapes usually exhibit considerable heterogeneity in the amount and timing of water stored in soils and the plants that occupy different sites.

Vegetation Responses to the Distribution and Duration of Soil Water

The influence of the texture of various soil horizons on the downward movement and storage of water ultimately has a pronounced effect on the types of plants present. Surface soil horizons may be texturally quite similar even in soils that have markedly different B, or lower, horizons (fig. 4.2). Studies that attempt to relate vegetational differences to surface soil characteristics alone (e.g., J. H. Bock and Bock 1986; Nicholson and Bonham 1977) ignore physical characteristics of deeper soil horizons that may in fact exert the strongest influences over the types of plants that dominate.

In addition to warm-season perennial grasses, the desert grasslands contain a diverse array of other kinds of plants, including leaf succulents such as the Spanish daggers and soaptree (*Yucca* spp.) and century plants (*Agave* spp.); stem succulents, most notably the prickly pears and chollas (*Opuntia* spp.); small, shallow-rooted shrubs including fairy duster (*Calliandra eriophylla*) and snakeweed (*Gutierrezia sarothrae*); and deeper-rooted shrubs and small trees including whitethorn acacia (*Acacia constricta*), catclaw (*Acacia greggii*), wait-a-minute bush (*Mimosa biuncifera*), and velvet mesquite. The different life forms of the perennial plants of the desert grasslands permit these plants to acquire moisture from various levels within soil profiles and to use water in different ways.

Shallow-rooted plants—including perennial grasses, succulents, and some small, drought-deciduous shrubs—are subjected to a highly variable water supply in the uppermost soil horizons. This surface layer is the first to reach field capacity and the first to dry. These plants rely on their capacity for drought-induced dormancy or their ability to store water in succulent tissues for continued photosynthetic activity (see Burgess, this volume).

On the other hand, in semiarid environments the deeper roots of most large woody plants occupy a soil environment that exhibits less seasonal variation in the availability of water (Burgess, this volume; Noy-Meir 1973; Schlesinger et al. 1987). Many woody plants have well-developed lateral root systems in addition to deeper tap roots, and thus have access to shallow soil moisture as well. The deeper soil layers are the last to wet but are more constant in the seasonal quantity of available water. Deep-rooted perennial plants maintain photosynthesis or other water-demanding activities (such as flowering) by extensive extraction of deeper soil water reserves even during seasonal drought intervals. For example, velvet mesquite and catclaw rely on stored soil moisture that arrived during the winter season

Table 4.2. Cover classes[a] of dominant species on three geomorphic surfaces (1,060–1,120 m elevation) in the Santa Rita Experimental Range, 19 August 1992.

	Geomorphic Surface and Estimated Age (10^3 years)		
Vegetation	Q2b (200–300)	Q3 (4–8)	Q4 (<4)
GRASSES			
Arizona cottontop (*Digitaria california*)		+	
Bush muhly (*Muhlenbergia porteri*)		+	
Lehmann lovegrass (*Eragrostis lehmanniana*)	6	5	4
WOODY PLANTS			
Burroweed (*Isocoma tenuisecta*)		1	
Catclaw (*Acacia greggii*)		1	
Desert hackberry (*Celtis pallida*)		1	
Fairy duster (*Calliandra eriophylla*)	1		
Joint fir (*Ephedra trifurca*)		1	
Ocotillo (*Fouquieria splendens*)	+		
Range ratany (*Krameria erecta*)	+		
Velvet mesquite (*Prosopis velutina*)	+	3	4
Velvet pod mimosa (*Mimosa dysocarpa*)	+		

[a]Percentage canopy cover is ranked into the following classes of relative cover, based on visual estimates: + = present but 1%; 1 = 1–5%; 2 = 6–12%; 3 = 13–25%; 4 = 26–50%; 5 = 51–75%; 6 = >75%.

for the subsequent production of leaves and flowers during the most arid portion of the year—the presummer drought extending from April through July in Arizona (Mooney et al. 1977; Turner 1963). Similarly, burroweed, a small, deep-rooted shrub, produces most of its stem and leaf growth during winter and spring, even during the spring drought season (Cable 1969; Humphrey 1937; Tschirley and Martin 1961).

This pattern of belowground root distribution and seasonal activity in different kinds of desert grassland plants is analogous to that described for the grasses and woody plants of tropical savannas (Walker and Noy-Meir 1981; Walker et al. 1981; Walter 1973). Any factor that profoundly affects the vertical distribution of water within the soil will greatly affect the relative success of different plant growth forms.

The geomorphically diverse Santa Rita Experimental Range demonstrates the link between soil profile characteristics and the dominance of different kinds of plants. Photographs taken at the turn of the century indicate that woody plants were far less abundant on many parts of the range than they are today (Bahre 1991; Hastings and Turner 1965; Humphrey 1953); however, the extent of the increase in woody plants and the species that have become established are not uniform across the area. Deep-rooted woody perennial plants, especially velvet mesquite and burroweed, increase in close conformity with the boundaries of soils characteristic of the distinct geomorphic surfaces. The wide, flat surfaces of the mid-Pleistocene fan remnants (surface Q2b in fig. 4.1) with strongly developed argillic horizons (table 4.2) have remained essentially free of deep-rooted woody plants, despite widespread increases of velvet mesquite on adjacent surfaces of younger geological age. Although the introduced Lehmann lovegrass (*Eragrostis lehmanniana*) has increased on these remnants (Anable et al. 1992), large, deep-rooted woody plants such as velvet mesquite are very rare. A series of matched photographs taken on this surface beginning in 1936 shows continued dominance by grasses—first native warm-season grasses, then the invading Lehmann lovegrass (fig. 4.3). The woody plants that are found on this surface—such as fairy duster and ocotillo (*Fouquieria splendens*)—are relatively shallow rooted and capable of rapid, ephemeral response to precipitation. The rarity of velvet mesquite cannot be explained by the occurrence of fires. The area shown in the matched photographs (fig. 4.3) has not been burned since at least 1937 (M. McClaran, University of Arizona, pers. comm., January 1993). Although there have been some fires on the Santa Rita Experimental Range during the last half century

Figure 4.3. Matched photographs taken on the Q2b surface (mid-Pleistocene, estimated age 200–400 million years) of the Madera Canyon fan, looking southward toward Elephant Butte. A. Photo taken in 1936. B. Photo taken in 1959.

continued

Figure 4.3 *Continued.* C. Photo taken in 1990. Photographs provided by M. McClaran.

(Cable 1965; Martin 1983; Reynolds and Bohning 1956), their extent has been extremely limited.

The few velvet mesquite plants that do occur on the Q2b surface (fig. 4.3) do not attain large size, and they exhibit periodic mortality of large branches, an indication of severe occasional drought stress. Velvet mesquites in this condition are most obvious on surfaces with strongly developed argillic horizons (M. McClaran, University of Arizona, pers. comm., January 1993). This type of stem mortality indicates how strongly an argillic horizon can reduce the infiltration and storage of moisture in deeper soil layers. In semiarid environments, drought conditions are amplified on soils with strongly developed argillic horizons.

Velvet mesquite and other deep-rooted woody plants increase in abundance on the more recently deposited late Pleistocene to Holocene fan surfaces, where argillic horizons are very weak or absent (fig. 4.1, table 4.1). Similarly, deep-rooted plants increase at the margins of the mid-Pleistocene fan remnants, where argillic horizons have been partially or completely truncated by erosion, thus permitting deeper water infiltration. In addition, areas at the erosional margins of Pleistocene fan remnants may receive considerable runoff from clay-rich soils in the interiors of the remnants. On Holocene and the most recent Pleistocene surfaces, velvet mesquite is dominant regardless of the distance from the nearest wash (fig. 4.1); therefore, the relatively uniform dominance of velvet mesquite and other deep-rooted species on these surfaces is better explained by the infiltration of precipitation to greater depths on surfaces that lack strong argillic horizons.

Some of these relationships between soils and vegetation are recognized and applied in the management of rangelands. Modern soil surveys of western rangelands conducted by the United States Department of Agriculture Soil Conservation Service classify different parts of the landscape into *range sites* based on soil characteristics and typical vegetation that can potentially exist on such soils (e.g., see Vogt 1980). In southeastern Arizona, soils on Pleistocene fan remnants containing well-developed argillic horizons are often classed in range sites with descriptive names such as "clay loam upland" and "loamy upland." Soils on Pleistocene surfaces from which argillic horizons have been erosionally removed, exposing underlying calcic horizons, are typically designated "limy upland" or "limy slopes." Deep, coarse-textured soils of Holocene surfaces often fall within the designations "sandy loam upland" and "sandy bottoms." Land management agencies such as the Bureau of Land Management use range site designations from soil surveys in their formulation of management plans for given areas. However, the ability to effectively use soil maps for this kind of management application depends on the extent and accuracy of knowledge of vegetation-soil relationships, and there is much yet to be learned about these relationships.

Geologic and Vegetation Influences on Erosional Incision and Soil Surface Characteristics

The erosional incision of alluvial piedmonts greatly alters patterns of water distribution and availability to plants. Erosional episodes involving the incision of fan surfaces have operated in these landscapes throughout their geological histories (Bull 1991), but accelerated landscape incision and erosion in some areas is at least partly due to human activities. These kinds of landscape and associated hydrological alterations can reinforce vegetation changes in some kinds of desert grassland communities.

Erosional incision occurs and arroyos form when runoff energy increases as a result of changes in climate and rainfall patterns, removal of vegetation cover by livestock, or the concentration of runoff in roadways and other structures. Vegetation cover protects the soil surface from extreme drop-splash erosion and slows overland flow, which enables greater infiltration. Different landscapes vary widely in their potential to erode when the energy of runoff is increased. Management practices (or lack of them) that drive one terrain into a mode of landscape incision and long-term diminished potential for production may cause little or no long-term landscape changes in other terrains. For example, erosion of alluvial fan deposits on the south flank of the Date Creek Mountains northwest of Wickenburg, Arizona, is markedly different from erosion of fan deposits along Cave Creek north of Phoenix, Arizona, even though tobosa grass (*Hilaria mutica*) has been reduced on both sites as a result of livestock grazing. The sensitivity of these landscapes to pronounced erosional processes depends on their geological contexts. One particularly important geological attribute that exerts a strong control on erosional processes is lithology, or rock type (Bull 1991).

Lithology Controls Erosion

The types of rock and their different rates of weathering (physical and/or chemical alteration and decomposition) greatly influence the nature of the alluvium that forms alluvial fans and piedmonts. As a consequence, alluvial piedmonts flanking mountains of different rock types exhibit varying resistance to erosion and landscape incision.

Weatherability varies among different kinds of rocks in semiarid and arid climates (Bull 1991). Coarse-grained, crystalline, intrusive igneous rocks such as granites, quartz monzonite, and diabase weather rapidly because some common minerals in these rocks (e.g., biotite and other micas) are rapidly decomposed. Many fine-grained, extrusive igneous rocks, including andesites and basalts, are considerably more resistant to weathering in semiarid regions. Weathering in these rocks occurs primarily on exposed surfaces and does not result in rapid disintegration. Fine-grained, dense metamorphic rocks such as quartzite (metamorphosed sandstone) and hornfels (metamorphosed, fine-grained mafic rocks) are extremely resistant to weathering.

Alluvial landforms composed of large, weathering-resistant cobbles are much more resistant to erosion than are surfaces of alluvial deposits composed of finer-grained materials or large particles of more rapidly weathered types of rocks. For example, the alluvial fans associated with Cave Creek north of Phoenix (545–575 m elevation) are composed of coarse, well-rounded basalt cobbles that were transported from the extensive basalt escarpments of Black Mesa to the north. The flat, extensive surfaces of some of these fan deposits are extremely stable and have persisted with minimal incision for a very long time. The extremely well developed argillic

Figure 4.4. Early Pleistocene fan deposit at Cave Creek, Maricopa County, Arizona. A. Tobosa (*Hilaria mutica*) stand at T5N, R3E, Sec. 1, elev. 570 m, view toward southwest, 12 August 1990. Tobosa on this flat fan surface typically exists in discontinuous stands, as shown. Runoff from nonvegetated parts of the landscape (foreground) contributes to the high density of tobosa in other areas (middle and background). B. Fenceline contrast on same fan surface shown above, southwest corner of T11N, R5W, Sec. 1, view north, 4 October 1991. The left side of the fence has been heavily grazed; the right side seems to have been ungrazed for an extended period. Grazing by livestock has apparently completely eliminated tobosa from some areas of the Cave Creek fan and other similar-aged fan deposits of the Agua Fria and New Rivers north of Phoenix.

and calcic horizons formed in soils on one of the most extensive, terracelike surfaces of the Cave Creek fan (Carefree cobbly clay loam; Camp 1986) suggest an age of approximately 2 million years for this surface (Gorey 1990). Dense stands of tobosa are common on some of the extensive planar surfaces of these terracelike fan remnants. The high density of tobosa is due to the concentration of runoff received from adjacent areas with little perennial vegetation cover (fig. 4.4; Cornet et al. 1988). Heavy grazing has all but eliminated tobosa from some sites, however, and little vegetation cover remains (fig. 4.4). Despite the loss of vegetation cover, the armor provided by the large basalt cobbles located on the soil surface and within soil profiles has prevented surface runoff from markedly eroding and incising these landscapes. Consequently, the potential of this landscape to produce tobosa may not be greatly diminished, even in places where tobosa has been nearly or completely eliminated by livestock. The potential for growth has been maintained because surface hydrology characteristics are primarily controlled by the nature of the coarse, stony to bouldery alluvium rather than vegetation, and grazing has not markedly altered these characteristics of the physical substrate.

Erosion occurs on a completely different spatial and temporal scale on landforms composed of alluvium derived from highly weatherable rock types. Fan deposits derived from granite are often composed of fine, gravelly particles. These fine alluvial materials provide little resistance to movement by runoff generated on the fan surfaces. Even if larger cobble- to boulder-sized particles of highly weatherable rock are incorporated in the original fan deposit, they are weathered fairly rapidly to smaller particles that are easily eroded. Consequently, fan deposits derived from rock types that weather rapidly are highly susceptible to rapid incision and altered surface hydrology.

An example of this instability is found on the alluvial fan deposits flanking the south side of the Date Creek Mountains, near Congress, Yavapai County, Arizona. These deposits were emplaced by wash systems originating in a granitic mountain range and are composed almost entirely of fine, gravelly alluvium. The presence of moderately well developed argillic horizons in some of these deposits (Vekol-Mohave soils series complex; Wendt et al. 1976) suggests that some of them are of at least late Pleistocene age. Photographs of the area taken in 1950 show a substantial cover of tobosa grass. Today, tobosa has been nearly eliminated from many upland sites in this area (fig. 4.5), probably because of the increase in year-round use by

Figure 4.5. A. A well-developed tobosa (*Hilaria mutica*) stand southwest of Congress, Yavapai County, Arizona (T9N, R6W, Sec. 4), looking northeast. The photograph was taken by R. R. Humphrey on 2 February 1950. Humphrey regarded the stand as being in fair condition and noted the presence of pancake prickly pear (*Opuntia chlorotica*) and velvet mesquite (*Prosopis velutina*). Even though the tobosa was grazed by stock, the stand was originally relatively dense on this level upland surface. B. The same spot, 20 September 1992. The fenceline visible in the background of the 1950 photograph is hidden by small velvet mesquites. Tobosa has been completely eliminated from the vicinity of the photograph.

livestock after extensive watering places (earthen tanks) were developed in the vicinity. The area contains no natural perennial surface water, and earthen tanks were only rarely constructed in the uplands during most of the first half of this century because of the expense and the lack of heavy equipment. Similarly, the drilling of deep wells was generally unfeasible. Increased development of water in such upland locations led to considerable range deterioration in the decades following the 1930s and 1940s (Cox et al. 1983; Martin 1975).

Changes in the Date Mountains piedmont include not only loss of vegetation cover due to grazing but also extreme hydrological and geomorphic alteration. Numerous roadways and dirt tracks have contributed directly to landscape incision by compacting soil and concentrating runoff. Many incisions can be traced to old roadbeds.

Many of the late Pleistocene surfaces containing the Vekol-series soils that originally supported substantial tobosa cover are experiencing extensive and rapid erosion (fig. 4.6). Individual tobosa plants can occasionally be found in unincised areas, especially beneath the spiny canopies of Christmas cholla (*Opuntia leptocaulis*), where they are inaccessible to livestock. The almost complete restriction of tobosa plants to refugia is an indication that extensive removal of tobosa by livestock generally preceded the rapid erosion of this landscape. With erosional incision, the rapid exit of water, especially summer precipitation, on which tobosa depends, further inhibits growth and survival of tobosa.

In stark contrast to Cave Creek, the recent incision of this piedmont is the result of the inherent instability of a landscape composed of fine, easily eroded alluvium. The deep incisions have greatly changed the surface hydrologic regime, and therefore the loss of many tobosa stands probably cannot be easily reversed. In this landscape, return to a hydrologic system that can once again support tobosa will require a change from the current erosional degradation to one in which alluvial aggradation can back-fill the extensive channel systems. Such a geomorphic change may require centuries to millennia to accomplish.

Vegetation Controls of Erosion

A cover of low grasses growing near the soil surface is more effective at preventing drop-splash erosion and slowing overland flow in desert grasslands than are the canopies of many woody plants. Therefore, a change from a diffuse cover of perennial grasses to a more patchy cover of shrubs can even-

Figure 4.6. Incisions have ramified through much of the late Pleistocene surface of the piedmont of the Date Creek Mountains (T9N, R6W, Sec. 7). This photograph was taken 3.7 km west-southwest of the photograph in figure 4.5, on 13 August 1992. Much of the landscape dissection in this vicinity can be traced to old roadbeds. The soils at this site (Vekol soil series) are the same as those at the site shown in figure 4.5 and likewise probably supported more extensive tobosa cover in the past. In the area shown within the photograph, four tobosa grass (*Hilaria mutica*) plants were found: two in the open and two beneath canopies of prickly pear cactus (*Opuntia* spp.).

tually lead to diminished potential for infiltration of precipitation. In the short term, increased loss of water as runoff reduces the potential for grass development. Over the longer term, this reduced grass cover can lead to a cycle of self-enhancing feedbacks in which erosion of surface soil horizons and exposure of less permeable underlying horizons leads to slower infiltration, additional runoff and erosion, reduced storage of soil moisture, and altered distributions of plant nutrients (Bull 1991; Schlesinger et al. 1990).

Parsons et al. (1992) described moundlike accumulations beneath small shrubs (coppice mounds) in the Walnut Gulch Experimental Watershed that apparently have developed only within the last century as a consequence of a decrease in grass cover and an increase in shrubs. The authors used directional slash collectors to study the origin of the moundlike accumulations beneath individual shrub canopies and found that splash-drop erosion led to net removal of materials from the surface of exposed areas between the canopies. These materials accumulated beneath shrub canopies, and the surface of areas in between became lower and depleted of fine sand-sized particles. The pavementlike, pebbly surface that remained between shrubs after the selective removal of fine particles increased runoff, further diminishing infiltration (Abrahams and Parsons 1991).

Very slight changes in the landscape and soils that diminish infiltration of summer precipitation can have a marked effect on the relative success of warm-season perennial grasses versus woody plants more dependent on the receipt and storage of winter moisture. The violent summer convective storms in southern Arizona generate a disproportionate amount of runoff in comparison with the gentler precipitation of winter frontal storms. In the Walnut Gulch Experimental Watershed, approximately 70 percent of the year's precipitation occurs during the summer months, and 95 percent of the runoff for the year is derived from these summer storms (Osborn 1983). Because shrubs are generally less effective than grasses at diminishing runoff, the shift from grass to shrub cover is likely to further reduce the potential for storage of summer precipitation. The historic vegetation change at this site appears to be self-maintaining because of the change in soil conditions and infiltration. Consequently, any management scheme to reverse this vegetation change must take into account the alteration of surface soil characteristics.

Desert Grasslands Revegetation versus Geomorphic Reality

The loss of vegetation cover is a principal factor contributing to changes in runoff regimes, which in turn can contribute to increased soil erosion, landscape incision, and further reduction of productivity. Concern over these types of landscape changes has prompted various attempts to control runoff and erosion in the arid Southwest. These efforts have involved structural modification of the physical landscape and attempts to reduce runoff with the reestablishment of vegetation cover (see Roundy and Biedenbender, this volume). Some costly attempts to thwart erosion with physical structures may be ineffective. Once set in motion, degradational episodes often proceed until a new quasi equilibrium of erosion and deposition is reached. This may describe the situation in the recently incised Date Creek Mountains piedmont. The common occurrence of headward erosion and arroyo cutting through earthen dams and other soil conservation structures throughout the Southwest is testimony to the land's capacity to continue a course of change regardless of the degree of human intervention.

The reestablishment of vegetation is a means to control or reverse soil erosion and incision because vegetation reduces erosion-causing runoff and increases the capacity of soils to absorb precipitation. Before World War II, many attempts were made to restore lost perennial grass cover on semiarid rangelands with native species such as black grama (*Bouteloua eriopoda*) (Cox et al. 1982). Most of these trials met with very little success, however, because of difficulty in establishing seedlings. Long-term studies of black grama populations at the Jornada Experimental Range in New Mexico have confirmed the extreme rarity of natural seedling establishment (Neilson 1986). Other difficulties with using native species for revegetation included the problem of obtaining sufficient quantities of seed for large-scale seeding efforts (see Roundy and Biedenbender, this volume).

Their lack of success with native grasses led range managers to try nonnative species (Cox et al. 1982). One grass species that exhibited a great capacity for rapid establishment was Lehmann lovegrass from southern Africa. Introduced in Arizona in the 1930s, Lehmann lovegrass has now spread far beyond the areas in which it was originally introduced (Anable et al. 1992), and it may eventually detrimentally affect desert grassland fauna (C. E. Bock et al. 1986). Another southern African grass, buffelgrass (*Pennisetum ciliare*), was introduced and has spread in low-elevation sites in the Sonoran Desert region of Arizona and in Sonora and southern Baja California, Mexico. The spread and increase of buffelgrass has the potential to considerably alter the composition of native plant communities by promoting a regime of recurrent fires to which desertscrub communities are not adapted (Burgess et al. 1991).

Plant introduction trials driven in part by concerns about runoff and soil erosion have involved exotic species other than grasses as well. One case in particular warrants attention because of its potential future impact on desert grasslands in Arizona. The area of immediate concern is on Frye Mesa, located on the north flank of the Pinaleño Mountains overlooking Safford, Arizona (fig. 4.7). Frye Mesa is an ancient alluvial fan remnant whose surface alluvial deposits are estimated to be 2–2.5 million years old (Morrison 1985), approximately the same age as the oldest parts of the Madera Canyon fan (surface Q2a in fig. 4.1). The elevated, mesalike appearance of Frye Mesa is the result of the rapid downcutting of the Gila Valley. The original fan surface (of which Frye Mesa is a small remnant) was graded to a much higher basin floor because the upper Gila Valley in the vicinity of Safford did not have exterior drainage by the Gila River until sometime between the latest Pliocene (2.3 million years ago) and the middle Pleistocene

Figure 4.7. Frye Mesa. A. The mesa as viewed from the cemetery in Thatcher, Arizona. B. The surface of Frye Mesa showing cobbly to bouldery, weathering-resistant gneiss clasts and a dense cover of curly mesquite (*Hilaria belangeri*). C. An area on Frye Mesa

(0.6 million years ago) (Houser 1990; B. B. Houser, U.S. Geological Survey, pers. comm., September 1993). The connection of this valley to an exterior drainage system led to rapid downcutting and lowering of the base level and the isolation of surfaces such as Frye Mesa.

The surface alluvium found on Frye Mesa consists of cobble- to boulder-sized particles of relatively weathering-resistant gneiss (fig. 4.7B). Soils with

has been covered by resinbush (*Euryops multifidus*) to the complete exclusion of grasses and most woody perennials. Posts and fence mark the area where the exotic shrub was introduced in 1938. Photographs taken in April 1993.

very strongly developed argillic horizons have formed within this stony alluvium, and the coarse, stony surface provides an effective armor against erosional incision. Because of this, the principal mode of erosion on Frye Mesa is the retreat of the relatively steep slopes along its margins. Although the mesa is clearly an erosional landscape, and in a view from the basin floor one might initially view the erosional rim with a sense of alarm, the type and scale of the erosion occurring on its margins are far beyond the control of human intervention. Further, the processes of erosion operating at the margins of the mesa do not significantly affect plant productivity on the planar (flat) surfaces atop the mesa.

In 1938, resinbush (*Euryops multifidus*), a southern African shrub, was introduced by the Soil Conservation Service on a small test plot 26 m × 26 m (ca. 0.06 ha) on the upper surface of Frye Mesa (B. Pierson and R. M. Turner, U.S. Geological Survey, Tucson, Arizona, pers. comms., August 1993). Resinbush is one of many species collected in southern Africa and brought to Arizona in the 1930s for erosion control and other economic uses. Apparently it is not eaten by cattle (J. Goff, rancher and former lessee of state lands on Frye Mesa, pers. comm., October 1992). The documentation of the introduction of resinbush on Frye Mesa is extremely sketchy, but erosion control may have been one of the objectives. The shrub has since spread to cover most of the western half of the mesa surface, an area approaching 100 ha. The ripened fruits bear fine, stiff hairs that stick to clothing and fur, enabling the dispersal of seeds by mammals. Seeds also appear to be dispersed by water, and populations of resinbush can be found along washes downslope from the mesa.

The native desert grassland flora has been devastated by the spread of this shrub. The vegetation in areas of Frye Mesa not yet colonized by resinbush consists of an extensive cover of curly mesquite (fig. 4.7B) with lesser amounts of sideoats grama (*Bouteloua curtipendula*), several other species of native warm-season grasses, and scattered woody plants including catclaw, fairy duster, Mormon tea (*Ephedra* spp.), snakeweed, and small velvet mesquite. In the areas where resinbush has spread, grasses and other woody plants have all but disappeared (fig. 4.7C). Velvet mesquites and catclaw acacia present within dense stands of resinbush exhibit considerable stem mortality. Snakeweed, very common throughout the remaining grassland stands, is completely absent from areas dominated by resinbush. Considerable mortality of curly mesquite and sideoats grama often occurs at distances of 1.5–2 m from the advancing edge of the zone completely covered by resinbush. This indicates some sort of strong interference from resinbush other than competition for light or water, since resinbush grows actively only in winter and spring. The dried, tuftlike remains of dead curly mesquite plants found just inside borders of dense resinbush stands demonstrate the exclusion of grasses by this shrub. In terms of its usefulness for controlling soil erosion, resinbush forms a very sparse ground cover that provides far less protection from drop-splash erosion and rapid runoff than the sod-forming curly mesquite (fig. 4.7).

The elevational range on Frye Mesa over which this dramatic and devastating spread of resinbush has occurred includes the zone of some of the most productive desert grassland areas in southern Arizona (1,000–1,500 m). Within this elevational range, resinbush predominates on many kinds of soils, including those with extremely well developed argillic horizons atop the mesa; young, sandy-textured soils on Holocene deposits at the foot of the mesa; and extremely calcic soils derived from the degradation of older surfaces with strongly developed calcic horizons. This range of general soil types includes nearly all of those found within the desert grasslands of southern Arizona.

This exotic shrub has clearly shown its ability to spread and its potential to seriously alter a wide range of desert grasslands. Such a spread could render many desert grasslands in southeastern Arizona (and perhaps a considerably wider area) essentially worthless for ranching and for wildlife species that depend on native vegetation. If resinbush continues to spread unchecked, it may alter desert grasslands in a far more detrimental way than introduced grasses such as Lehmann lovegrass have done. Looking at this problem with perfect hindsight, I believe that resinbush might not have been introduced on Frye Mesa if the erosional processes that shape this landscape had been better understood by land managers.

Summary

An understanding of landscape evolution, geomorphology, and soils adds a great deal to our understanding of plant distributions in North American desert grasslands and the varying responses of these ecosystems to human use over the last century. The distribution of different soils is a direct consequence of geological processes that have shaped Basin and Range landscapes over the past few million years. Soil characteristics that require tens to hundreds of thousands of years to form, such as strongly developed argillic horizons, exert a dominant control over the distribution and duration of soil water, and ultimately over the types of plants that dominate on these surfaces.

The effect and extent of erosion depends on the geological context in which it is occurring, including the lithology of the mountains that contribute alluvium to landforms on which extensive areas of desert grasslands are located. Once we understand the complicated relationships between landforms, soils, vegetation, and the hydrological and erosional behavior of landscapes, we can more efficiently manage, protect, and restore our desert grasslands.

Literature Cited

Abrahams, A. D., and A. J. Parsons. 1991. Relation between infiltration and stone cover on a semiarid hillslope, southern Arizona. Journal of Hydrology 122:49–59.

Anable, M. E., M. P. McClaran, and G. B. Ruyle. 1992. Spread of introduced Lehmann lovegrass (*Eragrostis lehmanniana* Nees.) in southern Arizona, USA. Biological Conservation 61:181–188.

Bahre, C. J. 1991. A Legacy of Change: Historic Impact on Vegetation in the Arizona Borderlands. University of Arizona Press, Tucson.

Bock, C. E., J. H. Bock, K. L. Jepson, and J. C. Ortega. 1986. Ecological effects of planting African lovegrasses in Arizona. National Geographic Research 2:456–463.

Bock, J. H., and C. E. Bock. 1986. Habitat relationships of some native perennial grasses in southeastern Arizona. Desert Plants 8:3–14.

Buffington, L. C., and C. H. Herbel. 1965. Vegetational changes on a semidesert grassland range from 1858 to 1963. Ecological Monographs 35:139–164.

Bull, W. B. 1991. Geomorphic Responses to Climatic Change. Oxford University Press, Oxford.

Burgess, T. L., J. E. Bowers, and R. M. Turner. 1991. Exotic plants at the Desert Laboratory, Tucson, Arizona. Madrono 38:96–114.

Cable, D. R. 1965. Damage to mesquite, Lehmann lovegrass, and black grama by a hot June fire. Journal of Range Management 18:326–329.

———. 1969. Competition in the semidesert grass-shrub type as influenced by root systems, growth habits, and soil moisture extraction. Ecology 50:28–38.

Camp, P. D. 1986. Soil survey of Aguila-Carefree area, parts of Maricopa and Pinal Counties, Arizona. U.S. Department of Agriculture, Soil Conservation Service. Government Printing Office, Washington, D.C.

Cornet, A., J. P. Delhoume, and C. Montana. 1988. Dynamics of striped vegetation patterns and water balance in the Chihuahuan Desert. Pp. 221–231 *in* J. J. Durine, M. A. Weiger, and H. J. Willems (eds.), Diversity and patterns in plant communities. SPB Academic Publishers, Champaign, Ill.

Cox, J. R., H. L. Morton, T. N. Johnson, Jr., G. L. Jordan, S. C. Martin, and L. C. Fierro. 1982. Vegetation restoration in the Chihuahuan and Sonoran Deserts of North America. U.S. Department of Agriculture, Agricultural Research Service, Agricultural Reviews and Manuals, ARM-W-28.

Cox, J. R., H. L. Morton, J. T. LaBaume, and K. G. Renard. 1983. Reviving Arizona's rangelands. Journal of Soil and Water Conservation 38:342–345.

Ely, L. L., and V. R. Baker. 1985. Geomorphic surfaces in the Tucson Basin, Arizona. Field Guidebook for the Workshop on Global Mega-Geomorphology, 14–17 January 1985. Department of Geosciences, University of Arizona, Tucson.

Gile, L. H., J. W. Hawley, and R. B. Grossman. 1981. Soils and geomorphology in the Basin and Range area of southern New Mexico—Guidebook to the Desert Project. New Mexico Bureau of Mines and Mineral Resources Memoir 39. Socorro, N.Mex.

Gorey, T. L. 1990. Stream terraces of middle Cave Creek, Maricopa County, Arizona. M.S. thesis, Arizona State University, Tempe.

Hastings, J. R., and R. M. Turner. 1965. The Changing Mile: An Ecological Study of Vegetation Change with Time in the Lower Mile of an Arid and Semiarid Region. University of Arizona Press, Tucson.

Houser, B. B. 1990. Late Cenozoic stratigraphy and tectonics of the Safford Basin, southeastern Arizona. Pp. 20–24 *in* G. E. Gehrels and J. E. Spencer (eds.), Geological excursions through the Sonoran Desert region, Arizona and Sonora. Arizona Geological Survey Paper 7.

Humphrey, R. R. 1937. Ecology of the burroweed. Ecology 18:1–17.

———. 1953. The desert grassland, past and present. Journal of Range Management 6:159–164.

———. 1958. The desert grassland: a history of vegetation change and an analysis of causes. Botanical Review 28:193–253.

Machette, M. N. 1985. Calcic soils of the southwestern United States. Pp. 1–21 *in* D. L. Weide (ed.), Soils and Quaternary geology of the southwestern United States. Geological Society of America Special Paper 201. Geological Society of America, Boulder.

Marion, G. M. 1989. Correlation between long-term pedogenic $CaCO_3$ formation rate and modern precipitation in deserts of the American Southwest. Quaternary Research 32:291–295.

Martin, S. C. 1966. The Santa Rita Experimental Range. U.S. Department of Agriculture, Forest Service Research Paper RM-22.

———. 1975. Ecology and management of southwestern semidesert grass-shrub ranges: the status of our knowledge. U.S. Department of Agriculture, Forest Service Research Paper RM-156.

———. 1983. Responses of semi-desert grasses and shrubs to fall burning. Journal of Range Management 36:604–610.

McAuliffe, J. R. 1994. Landscape evolution, soil formation and ecological patterns and processes in Sonoran Desert bajadas. Ecological Monographs 64: 111–148.

McFadden, L. D., and J. C. Tinsley. 1985. Rate and depth of pedogenic-carbonate accumulation in soils: formulation and testing of a compartment model. Pp. 23–41 *in* D. L. Weide (ed.), Soils and Quaternary geology of the southwestern United States. Geological Society of America Special Paper 203. Geological Society of America, Boulder.

Mooney, H. A., B. B. Simpson, and O. T. Solbrig. 1977. Phenology, morphology, physiology. Pp. 26–45 *in* B. B. Simpson (ed.), Mesquite: its biology in two desert ecosystems. Dowden, Hutchinson & Ross, Stroudsburg, Pa.

Morrison, R. B. 1985. Pliocene/Quaternary geology, geomorphology, and tectonics of Arizona. Pp. 123–146 *in* D. L. Weide (ed.), Soils and Quaternary geology of the southwestern United States. Geological Society of America Special Paper 203. Geological Society of America, Boulder.

Neilson, R.P. 1986. High-resolution climatic analysis and Southwest biogeography. Science 232:27–34.

Nicholson, R. A., and C. D. Bonham. 1977. Grama (*Bouteloua* Lag.) communities in a southeastern Arizona grassland. Journal of Range Management 30: 427–433.

Noy-Meir, I. 1973. Desert ecosystems: environment and producers. Annual Review of Ecology and Systematics 4:25–51.

Osborn, H. B. 1983. Precipitation characteristics affecting responses of southwestern rangelands. U.S. Department of Agriculture, Agricultural Research Service, Agricultural Reviews and Manuals ARM-W-34.

Parsons, A. J., A. D. Abrahams, and J. R. Simanton. 1992. Microtopography and soil-surface materials on semi-arid piedmont hillslopes, southern Arizona. Journal of Arid Environments 22:107–115.

Pearthree, P. A., and S. S. Calvo. 1987. The Santa Rita fault zone: evidence for large magnitude earthquakes with very long recurrence intervals, Basin and Range Province of southeastern Arizona. Bulletin of the Seismological Society of America 77:97–116.

Peterson, F. F. 1981. Landforms of the Basin and Range Province defined for soil survey. Nevada Agricultural Experiment Station Technical Bulletin 28. University of Nevada, Reno.

Reynolds, H. G., and J. W. Bohning. 1956. Effects of burning on a desert grass-shrub range in southern Arizona. Ecology 37:769–777.

Richardson, M. L., S. D. Clemmons, and J. C. Walker. 1979. Soil survey of Santa Cruz and parts of Cochise and Pima Counties, Arizona. U.S. Department of Agriculture, Soil Conservation Service and Forest Service, in cooperation with Arizona Agricultural Experiment Station. Government Printing Office, Washington, D.C.

Sala, O. E., W. K. Lauenroth, and W. J. Parton. 1982. Long-term soil water dynamics in the shortgrass steppe. Ecology 73:1175–1181.

Schlesinger, W. H., P. J. Fonteyn, and G. M. Marion. 1987. Soil moisture content and plant transpiration in the Chihuahuan Desert of New Mexico. Journal of Arid Environments 12:119–126.

Schlesinger, W. H., J. F. Reynolds, G. L. Cunningham, L. F. Huenneke, W. M. Jarrell, R. A. Virginia, and W. G. Whitford. 1990. Biological feedbacks in global desertification. Science 247:1043–1048.

Tschirley, F. H., and S. C. Martin. 1961. Burroweed on southern Arizona range lands. Arizona Agricultural Experiment Station Technical Bulletin 146. University of Arizona, Tucson.

Turner, R. M. 1963. Growth in four species of Sonoran Desert trees. Ecology 44: 760–765.

Vogt, K. D. 1980. Soil survey of San Simon area, Arizona, parts of Cochise, Graham and Greenlee Counties. U.S. Department of Agriculture, in cooperation with Arizona Agricultural Experiment Station. Government Printing Office, Washington, D.C.

Walker, B. H., D. Ludwig, C. S. Holling, and R. M. Peterman. 1981. Stability of semi-arid savanna grazing systems. Journal of Ecology 69:473–498.

Walker, B. H., and I. Noy-Meir. 1981. Aspects of the stability and resilience of savanna ecosystems. Pp. 556–590 *in* B. J. Huntley and B. H. Walker (eds.), Ecology of tropical savannas. Springer-Verlag, New York.

Walter, H. 1973. Vegetation of the Earth and Ecological Systems of the Geobiosphere. Springer-Verlag, New York.

Wendt, G. E., P. Winkelaar, C. W. Weisner, L. D. Wheeler, R. T. Meurisse, A. Leven, and T. C. Anderson. 1976. Soil survey of Yavapai County, Arizona, western part. U.S. Department of Agriculture, Soil Conservation Service and Forest Service, in cooperation with Arizona Agricultural Experiment Station. Government Printing Office, Washington, D.C.

5

The Role of Fire in the Desert Grasslands

Guy R. McPherson

Fire is a pervasive and powerful force in desert grasslands. Its importance in controlling ecosystem structure and function rivals that of precipitation (Kimmins 1987). As the frequency, season, and behavior of fires have shaped plant communities, communities have in turn shaped the frequency, season, and behavior of fires. The long-term fire regime is probably more a consequence than a cause of vegetation patterns; that is, vegetation probably affects fire regime to a greater extent than fire regime affects vegetation, at least at a coarse level of resolution (Clark 1990). Nonetheless, as an integral component of ecosystems, fire should not be viewed as external to these systems; rather, fire is part and parcel of community organization and development.

Perhaps nowhere is the role of fire more widely acknowledged than in grasslands. In fact, some North American researchers have proposed that treeless grasslands were a product of repeated fires set by aborigines (Sauer 1944; Stewart 1951). A more comprehensive and accurate view is that fire interacts with other factors (including topography, soil, insects, rodents, lagomorphs, and herbaceous plants) to restrict woody plant establishment in grasslands (Grover and Musick 1990; Wright and Bailey 1982). There is general agreement, however, that fire is necessary (though usually not sufficient) to control the abundance of woody plants and maintain most grasslands. In the absence of periodic fires, grasslands usually give way to dominance by woody plants.

Three conditions must be met for fires to spread: an ignition source must be present, there must be adequate fine fuel (0.5 cm in diameter), and the fuel must be dry enough to burn. In the desert grassland, ample ignition sources are provided by the dry lightning storms that signal the beginning of the southwestern monsoon in late June or early July. Native Americans were probably responsible for some ignitions as part of their hunting activities (Bahre 1991; Dobyns 1981). If rainfall is adequate, there is usually sufficient quantity and continuity of fine fuel present to support fire spread in mesic areas of the desert grassland, and the extended hot, dry period that occurs virtually every year in the southwestern United States in late spring produces very dry fine fuels. Thus, all the conditions conducive to fire ignition and spread occur frequently in desert grasslands. After a fire, grasses generally recover to preburn levels within three years (Cable 1967; Wright and Bailey 1982).

In this chapter I discuss historic and recent fire frequencies, outline fire behavior and effects, and discuss likely trends in future fire regimes.

Changes in Fire Frequency

Fire has historically been common in most desert grasslands. Before 1882, fires were extensive, sometimes covering hundreds of square miles (Bahre 1991; Humphrey 1949). Humphrey (1958) reviewed historical accounts of frequent fires dating to 1528 and concluded that fire is critical to the maintenance of desert grasslands. Bahre (1991) agreed with this assessment and concluded that fire size and frequency have diminished greatly since the 1880s.

Prehistoric fire frequencies are more difficult to determine; grassland fires leave behind no fire-scarred trees and the direct evidence of fire dates and extents that these scars provide (Wright and Bailey 1982). Historical accounts and indirect evidence that have been used to estimate prehistoric fire frequency indicate that fires were relatively common, but these are imprecise estimates of frequency and extent. Early settlers and newspaper reports made note of fires only if they were large or particularly destructive; no comprehensive records were maintained until well into the twentieth century. Nevertheless, several lines of indirect evidence suggest that fires occurred at least every 10 years.

Shrubs were inconspicuous in desert grasslands before 1880, which suggests that fires occurred frequently enough to prevent widespread shrub recruitment. Most desert grassland shrubs are susceptible to fire, at least as seedlings (J. H. Bock and Bock 1992; Cable 1967; Cox et al. 1993; Glendening and Paulsen 1955; Humphrey 1949; Reynolds and Bohning 1956; Wright et al. 1976). For example, velvet mesquite (*Prosopis velutina*) plants usually do not resprout following fire until stems were larger than 1 cm in diameter when they burned (Glendening and Paulsen 1955). Furthermore, many woody species do not produce seeds until they are at least 10 years old (Chew and Chew 1965; Humphrey 1958; Martin 1975), and seeds on the soil surface are easily killed by fire (Cox et al. 1993). Thus, a fire frequency of once every 7 (Schmutz et al. 1985) to 10 years (Griffiths 1910; Leopold 1924; Wright and Bailey 1982) would have maintained relatively shrub-free grasslands. Extrapolation from nearby ponderosa pine (*Pinus ponderosa*) forests, where fires are known to have burned through grassy understories every 2 to 10 years (Swetnam 1990), has been cited as further evidence that grassland fires were frequent as well (Wright and Bailey 1982).

Considerable evidence suggests that widespread livestock grazing reduced fine fuel, and therefore fire frequency, in desert grasslands after 1880. Historical accounts (Bahre 1991) and direct evidence of reduced fire frequency in nearby pine forests (Savage and Swetnam 1990; Swetnam 1990) document reduced incidence of fires concomitant with the buildup of the livestock industry. In fact, forest administrators encouraged overgrazing to reduce fire hazard and promote tree growth (Leopold 1924). It is very likely that fires occur less frequently today than they did before Europeans settled the desert grassland (Bahre, this volume). Given the magnitude of the other changes that occurred during this period of settlement, we may never know what, if any, changes in structure and function of desert grassland resulted directly from this presumed decline in fire frequency.

Fire Behavior and Effects

No two fires are identical. Fire behavior (size, intensity, and rate of spread) is influenced by (1) physical factors such as fuel conditions (e.g., moisture, total combustible material, and continuity), weather conditions (e.g., wind, temperature, and relative humidity), and topography (e.g., slope, aspect); and (2) biological factors such as plant morphology and physiology (fig. 5.1). Plants' response to fire varies widely both between and within species and is affected by the fire's behavior and postfire physical and bio-

Figure 5.1. Prescribed burning in winter. The response of vegetation to this fire will be very different from the response of the same vegetation to a fire in the summer.

logical conditions. Understanding how fire affects the physical environment is a useful starting point for understanding and interpreting biotic responses to fire.

Effects on the Physical Environment

Grassland fire temperatures vary from 50°C to 682°C at the soil surface, depending on fuel load—the amount of combustible material available (Britton and Wright 1971; Stinson and Wright 1969). Soil temperatures decrease rapidly with increasing depth, and changes are generally negligible 1 cm below the surface (Stott 1986; Wright and Bailey 1982). Thus, grassland fires have little direct effect on soil organic matter, microbial populations, roots, or buried seeds (Wright and Bailey 1982). However, removal of biomass and the consequent lower albedo contribute to increases in the light striking the soil surface, evaporation, and diurnal temperature extremes.

Fire typically reduces the total amount of nutrients such as nitrogen, phosphorus, and sulfur but increases their cycling rate. For example, total nitrogen usually is reduced by fire because nitrogen volatilizes at 200°C. Although soil nitrogen usually is not affected, up to 90 percent of the aboveground nitrogen is volatilized in grassland fires (Sharrow and Wright 1977). Some chemical forms of nutrients that are readily absorbed by plants increase after fires—for example, the nitrate (NO_3) form of nitrogen (Wright and Bailey 1982). The nitrate level rises as the result of the increased microbial activity stimulated by higher daytime soil temperatures, thereby increasing the rate at which organic matter is converted into nitrate (Wright and Bailey 1982). Nutrients such as calcium, magnesium, potassium, and sodium are not volatilized in grassland fires and thus are retained on-site (Wright and Bailey 1982).

Effects on Plants

Many mechanisms facilitate the survival and persistence of plants after exposure to fire (table 5.1). In some cases, traits that enhance success in fire-prone communities may also enhance a plant's success in the presence of other stressful environmental factors (Lotan et al. 1985). Many fire-adapted species evolved or achieved dominance during the middle Miocene epoch (ca. 15 million years ago), presumably in response to the shift to a much drier climate that took place during that period (Van Devender, this volume). In other words, caution is necessary when interpreting the stimulus for adaptive traits. Although they may not have evolved in response to fire, adaptive traits nonetheless may confer a competitive advantage to species in communities where fire is common.

Fire interacts with other disturbances (Collins 1987) to change the light, temperature, moisture, and nutrient regimes that plants experience (Raison 1979). Fire effects thus range from direct to indirect, and from short term to long term. For example, fire may stress individual plants by destroying aboveground plant parts (direct), and may affect plant communities by reducing soil moisture through increased evapotranspiration or runoff of precipitation (indirect). Fire may virtually eliminate grass cover in the short term but enhance grass cover over a longer period by killing woody plants. A plant's response to fire is affected by physical and biological factors, which vary at different stages of the plant's life cycle. Calling a species "fire tolerant" or "fire adapted" nearly always requires qualification (Steuter and McPherson 1995). For example, many herbs are easily killed by fires but may become established at high densities from seed during the first one or two years after a fire (C. E. Bock and Bock 1990; J. H. Bock et al. 1976). Winter fires typically kill herbs that germinate in autumn (e.g.,

Table 5.1. Mechanisms exhibited by plants at different life stages that enable them to survive fire.

Life Stage	General Response	Mechanisms
Seeds	Avoidance	Burial
	Resistance	Insulative seed coat; protective tissue around fruit
	Stimulus	Increased germination; mortality of established neighbors
Juveniles	Avoidance	Rapid growth to resistant (protected) size
	Resistance	Aboveground buds protected by insulative plant tissue; belowground buds protected by soil
	Stimulus	Rapid growth of resprouts
Adults	Avoidance	Life cycle shorter than fire-return interval; flowering and fruiting phenology out of phase with fire season; suppression of understory fine fuel production
	Resistance	Thick, platy, corky, fissured bark; aboveground buds protected by insulative plant tissue; belowground buds protected by soil
	Stimulus	Rapid growth of resprouts; fire-obligate flowering; increased flowering (?)

SOURCE: Adapted from Steuter and McPherson, in press.

annual broomweed, *Xanthocephalum dracunculoides*) but do not harm, and may even favor, herbs that germinate in early spring (e.g., sunflower, *Helianthus annuus*).

The desert grassland's response to fire depends much more on the season and frequency of the fires than on fire behavior. Species are most susceptible to damage when they are actively growing and are relatively tolerant of fire when dormant. Most perennial plants experience high mortality from early-summer fires, when growth is just beginning and accelerating (Cable 1965, 1967, 1973). Conversely, spring, fall, and winter fires produce significant mortality only among the few herbaceous plants (e.g., plains lovegrass [*Eragrostis intermedia*] and many dicots) that initiate growth during the cool season. The effects of early-summer fires are more obvious and longer lasting than the effects of fires in other seasons (Glendening and Paulsen

1955; Martin 1983; Pase 1971; Tschirley and Martin 1961). For example, November fires have virtually no effect on the density of velvet mesquite, but the impact of June fires can be substantial. Grasses' response to fire is variable and generally less conspicuous than shrubs' response; however, production of big sacaton (*Sporobolus wrightii*) and buffelgrass (*Pennisetum ciliare*) is slightly higher following fires in summer than in fall or winter (Cox et al. 1990).

Desert grassland communities developed under conditions of relatively lower fire frequencies than many grassland types (e.g., tallgrass and mixed-grass prairies; Wright and Bailey 1980). Fire, particularly in combination with other factors, is capable of producing long-term changes in desert grassland structure. Plant production and cover are usually reduced for one to three years following a fire, depending on postfire precipitation (J. H. Bock and Bock 1992; J. H. Bock et al. 1976; Cable 1967; Humphrey 1949; Reynolds and Bohning 1956). Recovery is rapid during wet years, but long-term changes in community structure may result if fire is followed by drought or livestock grazing. For example, black grama (*Bouteloua eriopoda*) may not recover to its preburn basal area within 50 years if the fire is followed by drought and moderate livestock grazing (Canfield 1939; Nelson 1934; Reynolds and Bohning 1956). In fact, black grama is so negatively affected by fire that some authors have questioned whether fires ever occurred in desert grasslands dominated by this species (Buffington and Herbel 1965; Dick-Peddie 1993).

The plant's growth form and the location of its growing points are important factors contributing to its survival and response to fire. For example, the growing points of rhizomatous plants are located below the soil surface, and thus usually escape lethal temperatures (Humphrey 1949; Wright and Bailey 1982). On the other hand, stoloniferous species like buffalo grass (*Buchloë dactyloides*) and black grama, whose growing points are above the soil, are severely damaged by fire, especially if they have low vigor when burned or are grazed soon after the fire (Cable 1975; Reynolds and Bohning 1956). Bunchgrasses are also susceptible to fire damage if they have growing points near the soil surface (e.g., threeawns, *Aristida* spp.) or if they are leafy and large amounts of dead plant material accumulate within the bunch (e.g., Arizona cottontop, *Digitaria californica*). The accumulation of dead matter provides additional fuel during a fire, and this additional fuel maintains a high temperature in the center of the plant for a long time (above 250°C for more than an hour) after a fire passes (Wright

1971). Thus, rhizomatous grasses and small bunchgrasses recover much more quickly after a fire than large, decadent bunchgrasses and stoloniferous species (Wright and Bailey 1982), although many additional factors interact to influence the response of bunchgrasses to fire.

The spatial arrangement of plant species in desert grasslands also affects their survival. For example, Santa Rita threeawn (*Aristida glabrata*) grows between burroweed (*Isocoma tenuisecta*) plants, whereas some other threeawns typically grow within burroweed crowns. The latter species are subjected to greater heat in the smoldering burroweed crowns than Santa Rita threeawn in more open areas. In combination with the shallow subsurface buds of threeawn plants in general, the spatial arrangement of the crown-inhabiting species predisposes them to higher mortality than Santa Rita threeawn (Cable 1967).

Grassland species exhibit considerable variability in rooting habit (Schmutz et al. 1991), but, with the exception of the general difference between rhizomatous and stoloniferous plants, that generally does not directly affect their response to fire. As little as 1 cm of soil is an excellent insulator, usually enough to protect even shallow-rooted species. The postfire environment typically has a greater impact on plant roots than the fire itself. Burning may indirectly increase primary above- and belowground productivity by elevating soil temperatures and increasing nitrate availability for several weeks or months afterward (Raison 1979). However, increased soil temperatures tend to make burned sites more xeric than unburned sites, and shallow-rooted species are thus sometimes susceptible to drought stress on burned sites (Wright and Bailey 1982).

Although fire survivors have access to more resources (e.g., light and nutrients) than they had before the fire, it must be noted that a plant species need not survive a fire to reap its benefits. Species that produce abundant, widely dispersed seeds capable of establishing in the high-light, fluctuating-temperature environment characteristic of recently burned grasslands may also benefit from fire. Examples include annual grasses such as six-weeks needle grama (*Bouteloua aristidoides*) and six-weeks threeawn (*Aristida adscensionis*) (Cable 1967) and the nonnative perennial Lehmann lovegrass (*Eragrostis lehmanniana*) (fig. 5.2; Ruyle et al. 1988; Sumrall et al. 1991).

To some degree, succulents can tolerate fire (i.e., the aboveground growing points survive; Thomas and Goodson 1992), and a few plants are usually left unburned in refugia (low-fuel areas produced by rocky soils). The apical meristem (growing point) of cacti usually is protected by the surrounding

A

B

Figure 5.2. Fire frequency strongly affects the structure of desert grassland communities. A. Woody plants are inconspicuous in areas burned annually; Lehmann lovegrass (*Eragrostis lehmanniana*) is the dominant plant in the foreground. B. Woody plants are uncommon and scattered in areas burned every 5 to 10 years. C. Woody plants dominate many areas that have not been burned for at least 20 years.

tissue (e.g., hairs and spines), and fire-induced mortality of most species in the Sonoran region is similar to that of unburned controls (Humphrey and Everson 1951; Thomas 1991; Thomas and Goodson 1992). However, a second fire within 10 years of the first may be catastrophic to cacti in this region (Brown and Minnich 1986; Cable 1967; Thomas 1991). Fire-induced mortality of succulents is reportedly much higher (up to 70 percent) in Chihuahuan Desert grasslands (Bunting et al. 1980; Wright and Bailey 1982), at least partly as a result of their increased susceptibility to attack by insects such as the cactus bug (*Chelinidea vittiger*) (Sickerman and Wangberg 1983). Fire may also contribute indirectly to cactus mortality by removing the spines (which protect the plant against mammalian herbivores) and waxy cuticle (which protects against insects, diseases, and water loss). The sum total of fire-related mortality may not be evident until several years after a fire. Shorter individuals are generally more subject to mortality than taller ones, presumably because the apical meristem of the former suffers more damage (Bunting et al. 1980; Humphrey 1974; McLaughlin and Bowers 1982). Of course, interspecific variability produces exceptions to this generalization (Humphrey 1949; Humphrey and Everson 1951). An additional survival mechanism employed by most *Opuntia* species is their ability to regenerate vegetatively from small fragments lying on the ground (Cable 1973; Thomas 1991). The apical meristem of leaf succulents (e.g., *Agave* spp., *Dasylirion wheeleri*, and *Yucca* spp.) is usually protected from fires by a cluster of green leaves. These species also commonly resprout from roots or rhizomes after a fire (Benson and Darrow 1981; Freeman 1973; Gentry 1972).

Resprouting, relatively common in woody plants following fire, is initiated at vegetative buds protected by bark, dense leaf bases, or soil (Gill 1977). Protection of buds is probably more important than protection of cambial tissue (formative cells between the wood and bark) for woody plants in desert grasslands. Following fire-induced mortality of aboveground tissue, woody plants with protected buds can rapidly resprout and exploit the postfire environment. In general, individual plants must be several years old before they have the ability to resprout. The recurrence of fires at intervals short enough to prevent woody plants from reaching the age of resprouting potential partially explains the relative absence of woody plants from frequently burned desert grasslands. Resprouting vigor decreases with decreased soil moisture content; resprouting is generally less common following fires that occur during the growing season compared with dormant-season fires (Wright and Bailey 1982). Catclaw (*Acacia greggii*), mesquites (*Prosopis glandulosa*, *P. velutina*), and wait-a-minute bushes (*Mimosa biuncifera*, *M. dysocarpa*) are among the most conspicuous woody plants that resprout following fire in desert grasslands. Blue paloverde (*Cercidium floridum*), burroweed, ocotillo (*Fouquieria splendens*), oneseed juniper (*Juniperus monosperma*), and snakeweed (*Gutierrezia sarothrae*) rarely resprout following fire (Wright and Bailey 1982).

The long-term absence of fire may produce dramatic changes in community structure and function, particularly if soils do not limit shrub establishment (cf. Burgess, this volume; McAuliffe 1994). Shrubs associated with many desert grasslands generally resprout following fire, and some readily reestablish from seed. Resprouters are well adapted to maintain their presence when the interval between fires is less than 20 years (Cable 1965, 1973).

In the absence of fire, desert grasslands may develop dense, woody overstories that significantly reduce herbaceous plant production. The resulting lack of fine fuel reduces fire intensity and frequency, and the community changes from grassland to shrubland (Archer 1989; Brown 1982). After this threshold between grassland and shrubland has been crossed, land management strategies must be reevaluated (Archer 1989; Westoby et al. 1989). Once a site is dominated by woody plants, fire alone cannot return it to the earlier composition: fine fuel is too scarce and discontinuous to produce fires of sufficient intensity to kill woody plants. Thus, woody plants become permanent occupants of the site; the physiognomic change from grassland to shrubland is irreversible without significant cultural inputs (e.g., herbicides or mechanical shrub control) over temporal scales appropriate to management. The net result of the absence of periodic fires is a reduction in herbaceous production.

Effects on Vertebrates

The direct effects of fire on animals are minimal. Most species are able to escape, and fire-induced mortality is rare (Wright and Bailey 1982). Large mammals and birds use speed to get away, and small animals that live in burrows, such as the kangaroo rats (*Dipodomys* spp.), hispid pocket mouse (*Perognathus hispidus*), spotted ground squirrel (*Spermophilus spilosoma*), and white-footed mouse (*Peromyscus leucopus*), are well insulated from the heat. Small mammals that live aboveground, such as packrats (*Neotoma* spp.), may die from the direct effects of heat.

The indirect effects of fire on animal populations, such as changes in vegetation structure and altered quantity and quality of forage resources (i.e., changes in habitat and food sources), are much more important. The response of animals to fire is closely related to the vegetation's response. Since fires usually reduce woody plant cover and herbaceous litter, species that use woody plants or dense litter for cover or homes—for example, Botteri's sparrow (*Aimophila botterii*), Cassin's sparrow (*A. cassini*), grasshopper sparrow (*Ammodramus savannarum*), masked bobwhite quail (*Colinus virginianus ridgewayi*), and most small rodents—are detrimentally affected. On the other hand, species that eat large-seeded herbaceous dicots, including the chipping sparrow (*Spizella passerina*), mourning dove (*Zenaidura macroura*), and vesper sparrow (*Pooecetes gramineus*), typically respond positively to fire-induced increases in their food sources and ease of finding seeds. Species that prefer bare ground, such as the horned lark (*Eremophila alpestris*) and lark sparrow (*Chondestes grammacus*), breed in high numbers on recently burned sites. Animals that rely on sight and speed for escape, like the pronghorn (*Antilocapra americana*), benefit from the unimpeded views.

Despite the examples given above, the effects of fire on most vertebrate species are complex and difficult to predict. Fire may reduce the nesting habitat of some species (e.g., masked bobwhite quail) while increasing the abundance of large-seeded herbaceous dicots they use for food (C. E. Bock and Bock 1990). Or fire may increase the habitat used by some species during one season while reducing the habitat they use during another (e.g., by reducing woody plant cover). Some species (e.g., mourning dove) switch from nesting in trees to nesting on the ground in recently burned areas (Soutiere and Bolen 1972). A relatively detailed knowledge of a species' habitat requirements is necessary to predict how it will respond to fire. Having evolved with periodic fires, most grassland animals probably are well suited to deal with a typical fire (i.e., one that produces patches of burned and unburned areas, and hence many types of vegetation) (C. E. Bock and Bock 1990). The response of an animal to a specific fire depends both on its habitat requirements (which change seasonally and throughout an animal's life) and on the characteristics of the fire itself (e.g., size, season, patchiness). Habitat requirements are not fully known for most animal species, and effects of fire on vegetation are not always predictable, particularly at scales that influence animals. Some fire characteristics (e.g., patchiness, intensity) are difficult to quantify and have seldom been correlated with animal responses. In short, the response of animals to fire is complex, variable, and unpredictable.

An additional indirect effect of fire on animals is its impact on forage quality. Grassland fires remove dead, low-quality forage, and the ensuing new growth is often succulent and of high quality (i.e., has a high concentration of some nutrients). The nutrient quality of resprouting woody plants has not been measured in desert grasslands; however, resprouting woody plants in other systems often are high in moisture (Short et al. 1977) and crude protein (Reynolds 1967). These increases rarely persist beyond one to three years, if they are present at all (Regelin et al. 1974). Herbaceous forage quality increases also are usually short-lived (less than a month), and may be insignificant (Cox et al. 1990), although fires usually increase the availability of herbaceous forage (Wright and Bailey 1982).

Effects on Invertebrates

Relatively little is known about the effects of fire on invertebrates in desert grasslands. C. E. Bock and Bock's (1991) study of grasshoppers (Orthoptera: Acrididae) is exceptional in this regard. The density of adult and nymph grasshoppers on burned areas was 60 percent less than on unburned areas following a July wildfire; however, these differences disappeared by the second year. The same pattern was seen in five of the nine most common species of grasshoppers. The authors concluded that temporary reductions in grasshopper densities resulted from direct mortality and the indirect effect of reduced grass cover. Recolonization of burned areas was relatively rapid, particularly by flying adults. Some species known to prefer habitats with considerable bare ground, herbaceous foods, or both—for example, the bandwinged grasshopper (*Trimerotropis pallidipennis*) and Gladstone spur-throat grasshopper (*Melanoplus gladstoni*)—temporarily increased in numbers following the fire. Bock and Bock (1991) concluded that grasshoppers, like the plants and other animals in desert grasslands, are highly fire tolerant but not fire dependent.

Future Trends

Fire regimes in desert grasslands have changed considerably since the late 1800s (Bahre 1991). Desert grasslands are not stable, unchanging communities at any temporal scale (Burgess, this volume). Moreover, many factors—biological, managerial, political, and climatological—are likely to contribute to altered fire regimes within the next century.

Inadequate fine fuel is the most common constraint on fire spread in desert grasslands—ignition sources are plentiful, and an extended drying period occurs virtually every year. Cattle grazing over the last 150 years has reduced the biomass enough to limit fire spread during most years. The number of cattle grazing southwestern ranges has declined in the last few decades (Allen 1992), however, and this trend is expected to continue for the foreseeable future (McClaran et al. 1992). Nonnative species such as Boer lovegrass (*Eragrostis curvula* var. *conferta*), buffelgrass, and Lehmann lovegrass produce more fine fuel than native species (Cable 1971; Cox et al. 1984), and the recent and continuing spread of these species (e.g., Anable et al. 1992; Cox and Ruyle 1986; Cox et al. 1988) will probably contribute to increased fire spread. Lehmann lovegrass, the most common of the introduced species, increases after fire (Ruyle et al. 1988; Sumrall et al. 1991), suggesting that a positive feedback pattern of lovegrass–fire–increased lovegrass–increased fire may develop (Anable et al. 1992). The net result of decreased cattle stocking rates (a managerial factor) and the introduction of nonnative grasses (a biological factor) likely will be more frequent fires than have occurred in the last century, but less frequent than historic fire return intervals.

Desert grasslands are more fragmented than ever before (e.g., from roads, rights-of-way, and subdivisions) and are often efficiently managed for fire suppression. Grassland fires will probably never regain the frequency or extent of pre-1882 fires.

Prescribed fire became an important land management tool in the 1970s and 1980s, and fires are used today to accomplish several objectives (Wright 1974; Wright and Bailey 1982). Recently imposed air quality regulations constrain the use of prescribed fire in some states, however. For example, smoke management concerns have contributed to a long and complex permitting process in Arizona, making prescribed fire an unattractive management tool for many land managers. Thus the use of prescribed fire, increasingly recognized as biologically and economically appropriate for grassland management, is expected to decrease in many states as a result of the changing politics of rural land use.

Finally, fire frequency is likely to change if increasing greenhouse gas concentrations in the atmosphere produce climatic changes. Long-term fire frequency and intensity are probably more a result than a cause of the vegetation present (Clark 1990), suggesting that changes in community structure resulting from changes in atmospheric composition or climate may have substantial effects on fire regimes. Increased atmospheric CO_2 may favor woody plants (cool-season plants that use the C_3 photosynthetic pathway; see McClaran, this volume) and lead to a decline in warm-season grasses (which employ the C_4 photosynthetic pathway; Idso 1992; Idso and Quinn 1983), resulting in large-scale physiognomic shifts in vegetation. Although woody species have increased in many desert grassland areas in the twentieth century (Van Devender, this volume), the rate of change resulting from recent increases in CO_2 may be unprecedented. One expected effect of this grassland-to-woodland transition is lower fire frequency due to decreased accumulation of fine fuel. The hotter temperatures predicted to occur in the future may contribute to either increased fire frequency (because hotter, drier weather results in drier fuel) or decreased fire frequency (because of decreased fine fuel production resulting from the hotter, drier weather). Climatic variability is expected to increase as a result of increasing greenhouse gas concentrations (Katz and Brown 1992; Mitchell et al. 1990). Again, the impact of this change on fire frequency is difficult to predict, especially in light of the multitude of other changes that are likely to occur simultaneously.

Summary

Fires exert long-term impacts on desert grassland structure and function. Periodic fires maintain grasslands in a relatively shrub-free state: the resulting structure is less vertically complex, nutrients cycle more rapidly between above- and belowground biomass, and plant growth is more readily available to grazers. Historical accounts and indirect evidence indicate that fires were once large and prevalent in desert grasslands, although fire frequency declined considerably after 1882. The response of individual plant species to fire varies widely, although most species possess traits that make them well adapted to periodic fires. Conversely, the long-term absence of fire may produce irreversible changes in structure and function of desert grasslands (e.g., a transition from grassland to shrubland, with resulting shifts in structural complexity, nutrient cycling, and forage availability). Fire regimes are likely to change in the near future as a result of recent and possible future changes (e.g., in atmospheric chemistry, management strategies, or policy) in desert grasslands. For example, the recent and continuing spread of highly productive nonnative grasses, coupled with fewer livestock, may contribute to increased fire frequency. The effects of other biological, managerial, political, and climatological factors are difficult to predict; these changes may either accelerate or counteract the effects of declining livestock numbers and the introduction of nonnative plant species.

Acknowledgments

This chapter is dedicated to the memory of my mentor and friend, Henry A. Wright. In addition to being a pioneer in research on fire ecology and management, Dr. Wright was an outstanding prescribed fire practitioner.

This chapter benefited from reviews by Rena Ann Peck and Jake Weltzin and insightful editing by Mitch McClaran.

Literature Cited

Allen, L. S. 1992. Livestock-wildlife coordination in the encinal oak woodlands: Coronado National Forest. Pp. 109–110 *in* P. F. Ffolliott, G. J. Gottfried, D. A. Bennett, V. M. Hernandez C., A. Ortega-Rubio, and R. H. Hamre (tech. coords.), Ecology and management of oak and associated woodlands: perspectives in the southwestern United States and northern Mexico. U.S. Department of Agriculture, Forest Service General Technical Report RM-218.

Anable, M. E., M. P. McClaran, and G. B. Ruyle. 1992. Spread of introduced Lehmann lovegrass (*Eragrostis lehmanniana* Nees.) in southern Arizona, USA. Biological Conservation 61:181–188.

Archer, S. 1989. Have southern Texas savannas been converted to woodlands in recent history? American Naturalist 134:545–561.

Bahre, C. J. 1991. A Legacy of Change: Historic Impact on Vegetation in the Arizona Borderlands. University of Arizona Press, Tucson.

Benson, L. D., and R. A. Darrow. 1981. Trees and shrubs of the southwestern deserts. 3d ed. University of Arizona Press, Tucson.

Bock, C. E., and J. H. Bock. 1990. Effects of fire on wildlife in southwestern lowland habitats. Pp. 50–64 *in* J. S. Krammes (tech. coord.), Effects of fire management of southwestern natural resources. U.S. Department of Agriculture, Forest Service General Technical Report RM-191.

———. 1991. Response of grasshoppers (Orthoptera: Acrididae) to wildfire in a southeastern Arizona grassland. American Midland Naturalist 125: 162–167.

Bock, J. H., and C. E. Bock. 1992. Short-term reductions in plant densities following prescribed fire in an ungrazed semidesert shrub-grassland. Southwestern Naturalist 37:49–53.

Bock, J. H., C. E. Bock, and J. R. McKnight. 1976. A study of the effects of grassland fires at the Research Ranch in southeastern Arizona. Journal of the Arizona Academy of Science 11:49–57.

Britton, C. M., and H. A. Wright. 1971. Correlation of weather variables to mesquite damage by fire. Journal of Range Management 24:136–141.

Brown, D. E. (ed.). 1982. Biotic Communities of the American Southwest—United States and Mexico. Desert Plants 4:1–342.

Brown, D. E., and R. A. Minnich. 1986. Fire and changes in creosote bush scrub of the western Sonoran Desert, California. American Midland Naturalist 116:411–422.

Bunting, S. C., H. A. Wright, and L. F. Neuenschwander. 1980. Long-term effects of fire on cactus in the southern mixed prairie of Texas. Journal of Range Management 33:85–88.

Cable, D. R. 1965. Damage to mesquite, Lehmann lovegrass, and black grama by a hot June fire. Journal of Range Management 18:326–329.

———. 1967. Fire effects on semi-desert grasses and shrubs. Journal of Range Management 20:170–176.

———. 1971. Lehmann lovegrass on the Santa Rita Experimental Range, 1937–1968. Journal of Range Management 24:17–21.

———. 1973. Fire effects in southwestern semidesert grass-shrub communities. Proceedings of the Tall Timbers Fire Ecology Conference 12:109–127.

———. 1975. Influence of precipitation on perennial grass production in the semidesert Southwest. Ecology 56:981–986.

Canfield, R. H. 1939. The effect of intensity and frequency of clipping on density and yield of black grama and tobosa grass. U.S. Department of Agriculture Technical Bulletin 681.

Chew, R. M., and A. E. Chew. 1965. The primary productivity of a desert-shrub (*Larrea tridentata*) community. Ecological Monographs 35:355–375.

Clark, J. S. 1990. Landscape interactions among nitrogen, species composition, and long-term fire frequency. Biogeochemistry 11:1–22.

Collins, S. L. 1987. Interaction of disturbances in tallgrass prairie: a field experiment. Ecology 68:1243–1250.

Cox, J. R., A. De Alba-Avila, R. W. Rice, and J. N. Cox. 1993. Biological and physical factors influencing *Acacia constricta* and *Prosopis velutina* establishment in the Sonoran Desert. Journal of Range Management 46:43–48.

Cox, J. R., F. A. Ibarra F., and M. H. Martin R. 1990. Fire effects on grasses in semiarid deserts. Pp. 43–49 *in* J. S. Krammes (tech. coord.), Effects of fire management of southwestern natural resources. U.S. Department of Agriculture, Forest Service General Technical Report RM-191.

Cox, J. R., M. H. Martin R., F. A. Ibarra F., J. H. Fourie, N. F. Rethman, and D. G. Wilcox. 1988. The influence of climate and soils on the distribution of four African grasses. Journal of Range Management 41:127–139.

Cox, J. R., H. L. Morton, T. N. Johnsen, Jr., G. L. Jordan, S. C. Martin, and L. C. Fierro. 1984. Vegetation restoration in the Chihuahuan and Sonoran Deserts of North America. Rangelands 6:112–115.

Cox, J. R., and G. B. Ruyle. 1986. Influence of climatic and edaphic factors on the distribution of *Eragrostis lehmanniana* Nees. in Arizona, USA. Journal of the Grassland Society of South Africa 3:25–29.

Dick-Peddie, W. A. 1993. New Mexico Vegetation: Past, Present and Future. University of New Mexico Press, Albuquerque.

Dobyns, A. F. 1981. From Fire to Flood. Ballena Press Anthropology Paper 20. Socorro, N. Mex.

Freeman, C. E. 1973. Some germination responses of lechuguilla (*Agave lechuguilla* Torr.). Southwestern Naturalist 18:125–134.

Gentry, H. S. 1972. The agave family in Sonora. U.S. Department of Agriculture, Agricultural Research Service Handbook no. 399.

Gill, A. M. 1977. Plants' traits adaptive to fires in the Mediterranean land ecosystems. Pp. 17–26 *in* H. A. Mooney and C. E. Conrad (tech. coords.), Symposium on the environmental consequences of fire and fuel management in Mediterranean ecosystems. U.S. Department of Agriculture, Forest Service General Technical Report WO-3.

Glendening, G. E., and H. A. Paulsen, Jr. 1955. Reproduction and establishment of velvet mesquite as related to invasion of semidesert grasslands. U.S. Department of Agriculture Technical Bulletin 1127.

Griffiths, D. A. 1910. A protected stock range in Arizona. U.S. Department of Agriculture, Bureau of Plant Industries Bulletin 177.

Grover, H. D., and H. B. Musick. 1990. Shrubland encroachment in southern New Mexico, U.S.A.: an analysis of desertification processes in the American Southwest. Climatic Change 17:305–330.

Humphrey, R. R. 1949. Fire as a means of controlling velvet mesquite, burroweed, and cholla on southern Arizona ranges. Journal of Range Management 2:175–182.

———. 1958. The desert grassland. Botanical Review 24:193–253.

———. 1974. Fire in the deserts and desert grassland of North America. Pp. 365–400 *in* T. T. Kozlowski and C. E. Ahlgren (eds.), Fire and ecosystems. Academic Press, New York.

Humphrey, R. R., and A. C. Everson. 1951. Effect of fire on a mixed grass-shrub range in southern Arizona. Journal of Range Management 4:264–266.

Idso, S. B. 1992. Shrubland expansion in the American Southwest. Climatic Change 22:85–86.

Idso, S. B., and J. A. Quinn. 1983. Vegetational redistribution in Arizona and New Mexico in response to a doubling of the atmospheric CO_2 concentration. Climatological Publications Science Paper 17. Laboratory of Climatology, Arizona State University, Tempe.

Katz, R. W., and B. G. Brown. 1992. Extreme events in a changing climate: variability is more important than averages. Climatic Change 21:289–302.

Kimmins, J. P. 1987. Forest Ecology. Macmillan, New York.

Leopold, A. 1924. Grass, brush, timber and fire in southern Arizona. Journal of Forestry 22:1–10.

Lotan, J. E., J. K. Brown, and L. F. Neuenschwander. 1985. Role of fire in lodgepole pine forests. Pp. 133–152 *in* D. E. Baumgartner, R. G. Krebill, J. T. Arnott, and G. F. Weetman (eds.), Lodgepole pine: the species and its management, symposium proceedings. Cooperative Extension Service, Washington State University, Pullman.

Martin, S. C. 1975. Ecology and management of southwestern semidesert grass-shrub ranges: the status of our knowledge. U.S. Department of Agriculture, Forest Service General Technical Report RM-156.

———. 1983. Responses of semi-desert grasses and shrubs to fall burning. Journal of Range Management 36:604–610.

McAuliffe, J. R. 1994. Landscape evolution, soil formation and ecological patterns and processes in Sonoran Desert bajadas. Ecological Monographs 64: 111–148.

McClaran, M. P., L. S. Allen, and G. B. Ruyle. 1992. Livestock production and grazing management in the encinal oak woodlands of Arizona. Pp. 57–64 *in* P. F. Ffolliott, G. J. Gottfried, D. A. Bennett, V. M. Hernandez, C. A. Ortega-Rubio, and R. H. Hamre (tech. coords.), Ecology and management of oak and associated woodlands: perspectives in the southwestern United States and northern Mexico. U.S. Department of Agriculture, Forest Service General Technical Report RM-218.

McLaughlin, S. P., and J. E. Bowers. 1982. Effects of wildfire on a Sonoran Desert plant community. Ecology 63:246–248.

Mitchell, J. F. B., S. Manabe, V. Meleshko, and T. Tokioka. 1990. Equilibrium climate change and its implications for the future. Pp. 313–372 *in* J. T. Houghton, G. J. Jenkins, and J. J. Ephraums (eds.), Climate change: the IPCC scientific assessment. Cambridge University Press, Cambridge, England.

Nelson, E. W. 1934. The influence of precipitation and grazing upon black grama grass range. U.S. Department of Agriculture Technical Bulletin 409.

Pase, C. P. 1971. Effect of a February fire on Lehmann lovegrass. Journal of Range Management 24:454–456.

Raison, R. J. 1979. Modification of the soil environment by vegetation fires, with particular reference to nitrogen transformations: a review. Plant and Soil 51:73–108.

Regelin, W. L., O. C. Wallmo, J. G. Nagy, and D. R. Dietz. 1974. Effects of logging on forage values for deer in Colorado. Journal of Forestry 72:4–7.

Reynolds, H. G. 1967. Chemical constituents and deer use of some crown sprouts in Arizona chaparral. Journal of Forestry 65:905–908.

Reynolds, H. G., and J. W. Bohning. 1956. Effects of burning on a desert grass-shrub range in southern Arizona. Ecology 37:769–777.

Ruyle, G. B., B. A. Roundy, and J. R. Cox. 1988. Effects of burning on germinability of Lehmann lovegrass. Journal of Range Management 41:404–406.

Sauer, C. O. 1944. A geographic sketch of early man in America. Geographical Review 34:529–573.

Savage, M., and T. W. Swetnam. 1990. Early 19th-century fire decline following sheep pasturing in a Navajo ponderosa pine forest. Ecology 71: 2374–2378.

Schmutz, E. M., E. L. Smith, P. R. Ogden, M. L. Cox, J. O. Klemmedson, J. J. Norris, and L. C. Fierro. 1991. Desert grassland. Pp. 337–362 *in* R. T. Coupland (ed.), Natural grasslands: introduction and Western Hemisphere. Ecosystems of the World 8A. Elsevier, Amsterdam.

Schmutz, E. M., M. K. Sourabie, and D. A. Smith. 1985. The Page Ranch story: its vegetative history and management implications. Desert Plants 7: 13–21.

Sharrow, S. H., and H. A. Wright. 1977. Effects of fire, ash and litter on soil nitrate, temperature, moisture, and tobosagrass production in the Rolling Plains. Journal of Range Management 30:266–270.

Short, H. L., W. Evans, and E. L. Boeker. 1977. The use of natural and modified pinyon pine–juniper woodlands by deer and elk. Journal of Wildlife Management 41:543–559.

Sickerman, S. L., and J. K. Wangberg. 1983. Behavioral responses of the cactus bug, *Chelinidea vittiger* Uhler, to fire damaged host plants. Southwestern Entomologist 8:263–267.

Soutiere, E. C., and E. G. Bolen. 1972. Role of fire in mourning dove nesting ecology. Proceedings of the Tall Timbers Fire Ecology Conference 12:277–283.

Steuter, A. A., and G. R. McPherson. 1995. Fire as a physical stress. Pp. 550–579 *in* R. E. Sosebee and D. J. Bedunah (eds.), Wildland Plants: Physiological Ecology and Developmental Morphology. Society for Range Management, Denver.

Stewart, O. C. 1951. Burning and natural vegetation in the United States. Geographical Review 41:317–320.

Stinson, K. J., and H. A. Wright. 1969. Temperature of headfires in the southern mixed prairie of Texas. Journal of Range Management 22:169–174.

Stott, P. 1986. The spatial pattern of dry season fires in the savanna forests of Thailand. Journal of Biogeography 13:345–358.

Sumrall, L. B., B. A. Roundy, J. R. Cox, and V. K. Winkel. 1991. Influence of canopy removal by burning or clipping on emergence of *Eragrostis lehmanniana* seedlings. International Journal of Wildland Fire 1:35–40.

Swetnam, T. W. 1990. Fire history and climate in the southwestern United States. Pp. 6–17 *in* J. S. Krammes (tech. coord.), Effects of fire management of southwestern natural resources. U.S. Department of Agriculture, Forest Service General Technical Report RM-191.

Thomas, P. A. 1991. Response of succulents to fire: a review. International Journal of Wildland Fire 1:11–22.

Thomas, P. A., and P. Goodson. 1992. Conservation of succulents in desert grasslands managed by fire. Biological Conservation 60:91–100.

Tschirley, F. H., and S. C. Martin. 1961. Burroweed on southern Arizona range lands. University of Arizona Technical Bulletin 146. University of Arizona, Tucson.

Westoby, M., B. Walker, and I. Noy-Meir. 1989. Opportunistic management for rangelands not at equilibrium. Journal of Range Management 42:266–274.

Wright, H. A. 1971. Why squirreltail is more tolerant to burning than needle-and-thread. Journal of Range Management 24:277–284.

———. 1974. Range burning. Journal of Range Management 27:5–11.

Wright, H. A., and A. W. Bailey. 1980. Fire ecology and prescribed burning in the Great Plains—a research review. U.S. Department of Agriculture, Forest Service General Technical Report INT-77.

———. 1982. Fire Ecology: United States and Southern Canada. John Wiley & Sons, New York.

Wright, H. A., S. C. Bunting, and L. F. Neuenschwander. 1976. Effect of fire on honey mesquite. Journal of Range Management 29:467–471.

6

Diversity, Spatial Variability, and Functional Roles of Invertebrates in Desert Grassland Ecosystems

Walter G. Whitford, Gregory S. Forbes, and Graham I. Kerley

Any analysis of the community structure and functional roles of desert grassland invertebrates is constrained by the fact that there are few, if any, patches of grasslands undisturbed by humans and their domestic livestock. The desert grasslands currently exist as scattered patches surrounded by large expanses of desertscrub or as a grassland matrix with shrubs. Much of the area that was desert grassland in the 1850s is now classified as desertscrub (Buffington and Herbel 1965; York and Dick-Peddie 1969). Further, the remaining desert grassland remnants are diverse in terms of soil characteristics and grass species composition. In some areas, exotic introduced grasses have become dominant; in others, remnant grasslands are dominated by species of grasses that are relatively unpalatable to domestic livestock.

The invertebrate fauna of desert grasslands is incredibly diverse and includes several phyla. While mammal, bird, reptile, and vascular plant species occur in the tens to hundreds, invertebrate species in desert grasslands number in the thousands or tens of thousands, and many of the less conspicuous species have never been described by taxonomists. Much of this fauna is never seen by the casual observer. If you were asked, "What invertebrates did you see on your morning walk in the desert grassland?" your response would probably be "ants." Ants are the most conspicuous and, along with unseen subterranean termites, the most numerous macroinvertebrates in arid environments. However, the fauna of desert grasslands includes many microscopic organisms and many large but cryptic species as well as the obvious and familiar ones (Crawford 1981; Wallwork 1982). Most of what we know about desert grassland invertebrates and their general life history characteristics is based on data from economically important species (Crawford 1981).

The Invertebrate Fauna of Desert Grasslands

Protozoans (single-celled animals that feed on bacteria and single-celled plants) populate the soil of all desert grasslands, but practically nothing is known of their distribution and biology. Naked amoebae predominate, along with smaller numbers of other orders of protozoans. The abundance of protozoans in desert soils is astounding: there are 25,000 naked amoebae, 4,900 flagellates, and 700 ciliates in every gram of dry soil (Parker et al. 1984). These numbers are deceiving, however, because most of the protozoans in dry soil are encysted (in an inactive physiological state; Whitford 1989). After a rain they quickly resume activity and remain active until the soil dries and they once again encyst.

Nematodes represent another group of extremely abundant microscopic soil animals. There are approximately 100,000 bacteria feeders, 1,000 fungus feeders, 10,000 omnivore-predators, and 10–1,000 plant root feeders (root parasites) per square meter of soil (Freckman et al. 1987). Nematodes are aquatic animals confined to single-molecule-thick water films surrounding soil particles, although they are well adapted to deal with dehydration. As the water films disappear in a drying soil, the nematodes enter an inactive state called anhydrobiosis, which is immediately reversible when the soil is wetted. Anhydrobiotic nematodes can survive desiccation for days to years at any stage of their life cycle. Despite the ubiquity of free-living nematodes, however, there have been few studies of their ecology, and many species are undescribed. Nematodes are often classified into feeding groups based on their morphology. The bacteria-feeding forms have a muscular buccal pump (a structure in the throat region) and a sucking mouth. The fungus feeders have oral stylets (hypodermic needle-like mouthparts) with which they pierce the fungal hyphae and suck out the cellular contents (fig. 6.1). The stylets and modified mouths of the omnivore-predators allow them to feed on fungi and other nematodes.

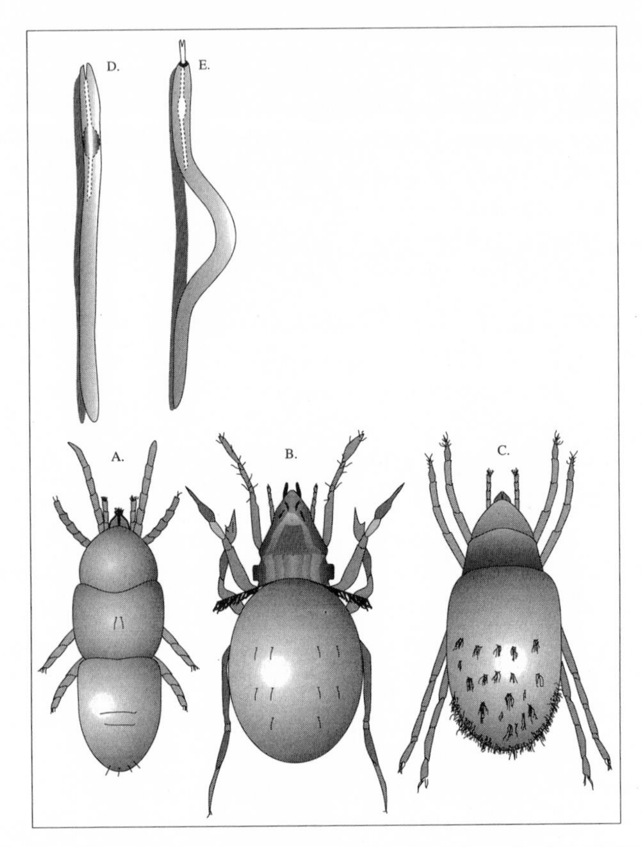

Figure 6.1. The most common types of soil mites in desert grasslands and the two most common nematode feeding types. A. Tydeidae mite. B. Oppiid mite (Cryptostigmata, Oribatida). C. Nanorchestidae mite. D. Bacteria-eating nematode with buccal pump. E. Fungal-eating or predatory nematode with oral stylets.

The soil mites (Acarina) are the smallest of all the arachnids—the group that contains spiders and their relatives. Most mites are large enough to be visible to the unaided eye (about the size of grains of ground pepper) but not large enough to allow determination of their morphological characteristics. The mite fauna in desert grassland soils includes more than 30 families and 100 species (Cepeda-Pizarro and Whitford 1989). The most abundant are the generalist microbe feeders such as the nanorchestid and tydeid mites (fig. 6.1). Some mite species feed on nematodes as well as fungi, yeasts, and bacteria. Included in this fauna are a variety of predators that capture and eat other mites and nematodes. The most abundant mites in desert soils are prostigmatids; the soils in more mesic environments (forest ecosystems) are dominated by cryptostigmatid mites (sclerotized mites about as large as the head of a pin). The ratio of prostigmatid to cryptostigmatid mites in desert grassland soils is between 10:1 and 20:1; in forest soils this ratio is reversed. Unlike the nematodes and protozoans, soil mites do not depend on water films, and they remain active even in very dry soils (MacKay et al. 1988).

Many of the larger arachnids live in burrows or in plant litter and are active on the surface at night, and hence are infrequently seen. Among these are some of the most interesting and feared of all desert animals: sun spiders (Solifugae); vinegaroons, or whip scorpions (Uropygi: Thelyphonidae); pseudoscorpions (Pseudoscorpiones); and the true scorpions (Scorpiones). Although they occur throughout desert grasslands, their population densities are low in most settings ($<1/m^2$). In addition, a number of relatively common short-lived spiders inhabit desert grasslands. Several kinds may become locally abundant during some periods and are associated with certain plant species. For example, crab, jumping, and orb-weaving spiders are abundant on soaptree yucca (*Yucca elata*) when it flowers but are rare at other times. Tarantulas, wolf spiders, and funnel-web spiders are relatively common, occurring at densities in the tens per hectare.

The detritus-feeding millipedes (Diplopoda), represented by the large (>20 cm long, >1 g) desert millipede (*Orthoporus ornatus*: Spirostepidae) and centipedes (Chilopoda) such as the giant desert centipede (*Scolopendra heros*; up to 20 cm long), are among the largest invertebrates in desert grasslands. These animals occur at low densities except in particularly favorable habitats. Their activity periods on the soil surface are largely confined to early morning, evening, and nights during the rainy season (Crawford 1981).

Cockroaches (Blattodea), crickets and grasshoppers (Orthoptera), ant lions (Neuroptera: Myrmeliontidae), beetles (Coleoptera), butterflies and moths (Lepidoptera), flies (Diptera), bees and wasps (Hymenoptera), and true bugs (Hemiptera) make up the desert grassland insect fauna. Their numbers fluctuate over orders of magnitude depending on immediate past climatic conditions. Little is known of the life histories of most of these organisms. The few quantitative data available regarding grassland insects show biomass in the range of less than 1 mg to 16 mg dry mass per square meter; peak biomass and species diversity occur at the end of the summer rainy season (Whitford 1974). The biomass determined by Whitford (1974) was divided among 139 kinds of insects recognized as distinct species based on morphology and included large numbers of species of beetles, flies, and plant-sucking bugs.

Life Spans and Reproduction

Desert grassland invertebrates run the spectrum of life spans, ranging from a few days to months or years. Those with the shortest life spans include protozoans, nematodes, aphids, thrips, and mirid bugs. Several groups of flies produce several generations per year whenever environmental conditions are suitable (Crawford 1981). Populations of some species of aphids increase rapidly by asexual reproduction (parthenogenesis) when high-quality food is abundant. When the quantity and quality of resources decline, they switch back to sexual reproduction. Most desert invertebrates are annual species (one generation per year), including most grasshoppers and crickets, many beetles, butterflies, moths, flies, wasps, bugs, sun spiders, and many spiders. Some species reproduce at fixed times determined by photoperiod, degree-days, or other predictable environmental cues; other species are more flexible and reproduce when environmental conditions (temperature, moisture, and food availability) approach some optimum.

Some desert grassland invertebrates—including millipedes, collembolans, cicadas, some beetles, whip scorpions, tarantulas, some wolf spiders, and scorpions—live for more than one year. They survive stress with long periods of dormancy, and their reproduction is generally coordinated with favorable environmental conditions.

Adaptations and Microhabitats

It is reasonable to classify most desert invertebrates as "avoiders," because their behavior allows them to avoid desiccation and thermal stress (Crawford 1981). Most surface-dwelling desert invertebrates avoid physiological stress by limiting their activity to periods of the day and times of the year when the physical environment is relatively benign. Generally they are nocturnal, sometimes extending their activity to the early morning hours when temperatures are still moderate. During the driest and hottest periods these animals escape the lethal conditions of the soil surface by entering burrows in the soil where relative humidities remain above 90 percent and temperatures rarely exceed 35°C. Invertebrates that live in the vegetation canopy are not exposed to the high temperatures found on the soil surface. The temperature in the canopy half a meter above the soil surface will remain at 40°C when the midday soil surface temperature is above 60°C. Grass canopies do not provide the microclimatic moderation that shrub canopies provide. This may be one reason why arthropods are generally more abundant and diverse in desertscrub than in grassland.

Desert macroinvertebrates exhibit few unique physiological specializations, although some arthropods have an epicuticle that reduces the loss of water through the cuticle, and some can absorb water from moist surfaces (Crawford 1981). The other physiological and morphological features that allow invertebrates to survive in deserts, such as their water-conserving excretory physiology, are not limited to invertebrates that inhabit deserts, and hence do not represent special adaptations.

Jornada del Muerto Basin Study Area

Much of the data presented in this chapter was collected on sites in the Jornada del Muerto Basin (32°30′ N, 106°45′ W) 40 km north-northeast of Las Cruces, Doña Ana County, New Mexico. The basin lies between 1,300 and 1,350 m elevation at the northern edge of the Chihuahuan Desert. The desert grasslands of the Jornada are structurally similar to those in southeastern Arizona and have had a similar history of land use and vegetation change (Bahre, this volume; McClaran, this volume). Data from this area should therefore also apply to desert grasslands in northern Mexico and southeastern Arizona in terms of species composition and functional attributes of faunal assemblages.

The data were collected in a number of different types of desert grasslands. The black grama (*Bouteloua eriopoda*) grasslands remain as remnant patches in the basin floor on sandy to sandy loam soils (fig. 6.2) and on montane piedmonts, or bajadas. The most extensive grasslands are areas dominated by a variety of relatively short-lived perennial grasses such as dropseeds (*Sporobolus* spp.) and threeawns (*Aristida* spp.) on sandy soils. The dominant shrubs in these areas are honey mesquite (*Prosopis glandulosa*) and creosotebush (*Larrea divaricata*). Many of the data on invertebrates in this grassland habitat were collected on the lower slopes of a watershed. A number of desert grassland habitats are located on clayey soils, including playa lake basins dominated by vine mesquite (*Panicum obtusum*) and swales

Figure 6.2. An ungrazed black grama (*Bouteloua eriopoda*) and mesa dropseed (*Sporobolus flexuosus*) grassland viewed from inside a grazing exclosure in the Jornada del Muerto Basin, Doña Ana County, New Mexico.

dominated by tobosa (*Hilaria mutica*), alkali sacaton (*Sporobolus airoides*), and burrograss (*Scleropogon brevifolius*) (fig. 6.3). The dominant shrub on clayey soils is tarbush (*Flourensia cernua*).

Diversity and Biology of Selected Groups

The soil biota. The biota of desert grassland soils is very similar to that of other water-limited subtropical ecosystems; any differences are likely related to the basic energy source for the food web. In desert grasslands, most of the energy-releasing decomposition occurs when the grasses are in the standing-dead state, not as leaf litter on the soil surface, and the basic energy source for soil food webs is thus the dense root systems of desert grasses. Invertebrates that live in litter layer patches under shrubs in desertscrub are not present in the desert grassland.

The biota of the soil food web includes bacteria and fungi, the primary

Figure 6.3. A burrograss (*Scleropogon brevifolius*) and tobosa (*Hilaria mutica*) swale on the Jornada Experimental Range, Doña Ana County, New Mexico. The burrograss is the short grass in the foreground; the tall grass in the background is tobosa.

consumers of dead and dying roots (fig. 6.4). This microflora provides the energy base for the soil microfauna: protozoans, nematodes, mites, collembolans, and insect larvae. The rhizosphere (the cylinder of soil immediately adjacent to a root) microflora is dominated by fungi that serve as food for the microfauna. Stanton (1988) reported that between 16 and 26 species of fungi served as the substrate for the microfauna in a semiarid grassland in Colorado. The rhizosphere microfauna was dominated by protozoans, which were estimated to occur in the hundreds of millions per square meter of soil (Stanton 1988). Microarthropods (microscopic mites and insects) occur at much lower densities than protozoans and nematodes (tables 6.1 and 6.2).

Most of the dominant species of microarthropods in the rhizosphere of desert grasses feed on both fungi and nematodes. Except during the brief periods following rains, most of the soil microfauna exists in an inactive state. As soils dry, the protozoans encyst, nematodes and collembolans

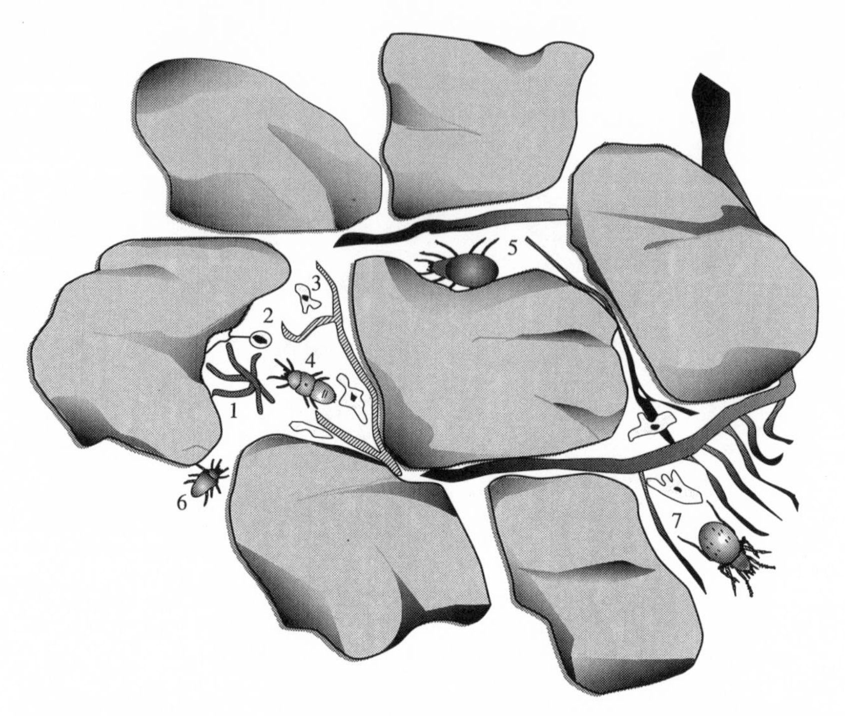

Figure 6.4. Microscopic view of the soil biota in a desert grassland soil: (1) bacteria-eating nematodes, (2) flagellate protozoan, (3) naked amoeba, (4) fungus-eating mite, (5) predatory mite, (6) collembolan, and (7) oribatid mite.

Table 6.1. Soil microarthropods extracted from rhizospheres of native black grama (*Bouteloua eriopoda*), introduced Lehmann lovegrass (*Eragrostis lehmanniana*), and unvegetated soil between grass clumps.

Taxon	*Bouteloua eriopoda*	*Eragrostis lehmanniana*	Unvegetated
Microbe-eating mites	14,109	11,343	7,124
Predatory mites	577	796	398
Collembola	816	1,890	298
Speleorchestes spp.	2,189	1,512	1,612
Tydeidae	4,597	4,995	2,334
Prostigmata/Cryptostigmata	17	23	42

NOTE: Data are numbers per square meter in the upper 10 cm of soil.

Table 6.2. Population densities (no./m^2) of soil microarthropods in areas grazed by domestic livestock in summer or all year versus densities in ungrazed exclosures.

	Summer Grazing Only		Yearlong Grazing	
Taxon	Grazed	Exclosure	Grazed	Exclosure
Collembola	5,975	3,781	716	915
Microbe-eating acari	8,915	9,910	13,452	9,194
Predatory acari	358	398	796	159
Speleorchestes spp.	876	1,552	2,145	2,706
Tydeidae	1,353	3,144	2,985	3,065
Prostigmata/Cryptostigmata	21	33	19	4

become anhydrobiotic, and some species of soil acari enter a cryptobiotic (literally, "hidden life"; inactive but not dead) state (Whitford 1989). Thus, during most of the year the active component of the desert grassland soil biota consists of a few species of mites feeding on fungi (MacKay et al. 1987; Whitford 1989).

The abundance of soil microarthropods depends on the availability of food. Microarthropods are twice as numerous in soil cores collected from the rhizospheres of perennial grasses than in cores from unvegetated patches between grass clumps (table 6.1). There are no obvious differences in the abundance and diversity of microarthropods extracted from the rhizosphere of the native black grama grass and those from introduced Lehmann lovegrass (*Eragrostis lehmanniana*) (table 6.1). Apparently there is little difference between the quality or quantity of organic matter available to the soil microarthropod fauna in the rhizospheres of native grasses and introduced grasses. A similar relationship was apparent in soil core data from grazed and ungrazed grasslands (table 6.2). The absence of differences among sites suggests that vegetation composition has little effect on soil microarthropods.

Ants. Ants are the most conspicuous invertebrates in many ecosystems, and this generally holds true for desert grasslands as well. Most desert grassland ant species are diurnal and frenetically active when food supplies are available. They tend to be either seed harvesters or omnivores, although even the species usually classified as seed harvesters are omnivorous to some extent (Whitford 1978a,b). The nests of desert grassland ants tend to be conspicuous and abundant. The construction and maintenance of the subterranean nests affect the structure and function of desert grassland ecosystems by altering soil nutrient distribution and water infiltration.

A variety of techniques can be used to estimate the density, species composition, and activity patterns of ants. Colony densities can be estimated by recording their presence in randomly placed quadrants, in belt transects, or by plotless techniques (Whitford 1978b). Qualitative information can be obtained by using bait boards, seed trays, and pitfall traps. The data reported in this section were obtained using all of these methods, depending on the question being addressed and the suitability of the technique for use in the particular habitat.

The species composition, species richness, and density of ant colonies in desert grasslands depend on the landscape and soil characteristics. Grasslands at the lowest elevations are generally characterized by clay to clay-loam soils. These heavy soils with their characteristic low infiltration rates frequently have standing water on the surface for several days to several weeks following large rainfalls. Periodic flooding plus the resistance of clayey soils to excavation eliminates many ant species as potential residents of these areas.

Alkali sacaton, burrograss-tobosa, and tobosa swales characteristically have few species and low densities of ants (table 6.3). On the Jornada del Muerto, the lowest species richness was found in alkali sacaton swales, where the clay soils have a high gypsum content. The only ant species we found was the crazy ant (*Conomyrma insana*), a small (worker length 2.5–3.8 mm) ant found in every grassland habitat we examined. The crazy ant colonies appear to be small (<500 workers) and exhibit rapid colony turnover. Crazy ants are dietary generalists that collect honeydew and cadavers of small arthropods, a characteristic that allows them to colonize extremely marginal habitats.

Of the low-elevation, clayey soil habitats we examined, the highest species richness was in the burrograss-tobosa basin, where microtopographic relief allows species to construct nests whose shallow chambers are above ponded floodwater levels (table 6.3). The nests of hairless honeypot ants (*Myrmecocystus depilis*) were in clumps of grass between 5 and 10 cm above the crusted ponding areas. In the low, crusted areas, several crazy ant colonies had constructed cemented mud chimneys 1–3 cm tall around the nest entrance holes to prevent water from inundating the nests. Because rainfalls in this region are usually less than 2.5 cm, a 3-cm-tall chimney will

Table 6.3. Densities of ant colonies per hectare in a variety of grassland habitats in the Jornada del Muerto Basin of the northern Chihuahuan Desert, Doña Ana County, N.M.

	Grassland Habitats					
Ant Species	AS	TS	BG	BT	BGDN	BGDS
Aphaenogaster cockerelli	0	0	6	0	0	0
Conomyrma bicolor	0	0	250	20	0	0
Conomyrma insana	71	50	117	340	1,639	1,282
Crematogaster depilis	0	0	267	0	0	0
Iridomyrmex pruinosum	0	50	150	0	0	0
Myrmecocystus depilis	0	0	2	20	0	0
Pheidole desertorum	0	0	333	0	0	0
Pheidole xerophila	0	0	33	0	0	0
Pogonomyrmex desertorum	0	0	33	0	181	100
Pogonomyrmex rugosus	0	17	2	0	0	0
Pogonomyrmex texanum	0	0	0	20	0	0
Solenopsis krockowi	0	0	0	100	0	0
Solenopsis xyloni	0	0	33	0	28	17

NOTE: AS = alkali sacaton swale, TS = tobosa swale, BG = bunchgrass grassland, BT = burrograss-tobosa basin, BGDN and BGDS = black grama–mesa dropseed grasslands.

prevent most flooding. The chimneys were found in areas surrounded by honey mesquite coppice sand dunes, which have little if any runoff.

The largest number of ant species (11) were in bunchgrass grassland on the lower slopes of a watershed (table 6.3), although the density of nests was not as high as in black grama grassland in the basin, where crazy ants were abundant. Most of the colonies in the bunchgrass grassland were seed-harvesting ants. In this habitat, seeds of a wide range of sizes and qualities from the abundant annual plants add significantly to the small seeds of grasses. Small seed-harvester ants (*Pheidole* spp.) and desert fire ants (*Solenopsis xyloni*) foraged primarily on small grass seeds, while harvester ants (*Pogonomyrmex* spp.) preferred the larger seeds of annual dicots and grasses (Whitford 1978a; Whitford et al. 1981). Species richness was low in the sandy soils of the basin black grama grasslands, where nests of only three species were recorded (table 6.3).

Comparisons of species richness among these habitats must be made with caution because the data were collected only in early summer, not over a full year, and the activities of ant species vary seasonally and within seasons (Whitford 1978b). Small seed-harvester species and desert harvester ants (*Pogonomyrmex desertorum*) are likely to be underrepresented in early-summer samples.

Seed-harvesting ants divide seed resources in a variety of ways. The black desert harvester ant (*Pogonomyrmex rugosus*) constructs large, permanent nests with several thousand workers per colony and forages primarily in groups or columns. Group foragers leave the nest and follow a discrete trail for some distance from the nest to collect seeds. Foraging trails are easily recognized by the stream of ants going toward and away from the nest. Other species, including desert harvester ants, are characterized by relatively small colonies (usually fewer than 500 workers), and individual workers search for seeds, apparently at random (Whitford and Ettershank 1975). Brown et al. (1979) suggested that these behavioral differences constrain species like the black harvester ant to forage in areas with dense seed accumulations while species that forage as individuals can exploit areas with scattered seeds.

Foraging behavior is an important variable that affects the activity time of the species and the impact of the species on seed reserves. Group foragers deplete seed resources more quickly than individual foragers (Whitford 1976). When such colonies have collected sufficient seeds to fill their granaries, they cease foraging (Whitford 1976). Satiated colonies of group foragers remain inactive aboveground for long periods even when climatic conditions are conducive to surface activity (Whitford and Ettershank 1975). Colonies of individual foragers initiate foraging activity in the spring before seed production occurs and continue to forage until weather conditions force cessation of activity in the autumn.

Ant colony entrances may be obscured in desert grasslands ungrazed by domestic livestock because the grass foliar cover may exceed 80 percent. Desert fire ants, for example, tend to build nests in the bases of grass clumps where they are difficult to detect unless a line of foragers can be traced back to the nest entrance. In these situations, species may be undersampled if nests alone are recorded, as tests with bait showed. On the other hand, assessments of the composition and relative abundance of an assemblage of ants based on bait data must be treated with caution. The aggressive desert fire ant comes rapidly to baits and excludes other species. In our study of exclosures and grazed areas (table 6.4), the seed trays were in place

Table 6.4. Species of ants collected from seed trays in grazed and ungrazed paired plots.

	Ungrazed Plots				Grazed Plots			
Species	NW	N	S	W	NW	N	S	W
Aphaenogaster cockerelli	5	5	0	0	5	0	0	10
Pheidole militicida	10	20	0	0	10	5	25	0
Pheidole xerophila	0	0	0	0	0	0	5	0
Pogonomyrmex desertorum	5	20	0	15	20	0	0	55
Pogonomyrmex rugosus	10	0	0	0	5	0	5	0
Solenopsis xyloni	75	60	0	95	70	90	5	45

NOTES: Data are expressed as percentage of feeding stations at which a species was removing seeds (20 stations/plot). Multiple species were frequent at a single tray.

NW = northwest; N = north; S = south; and W = west.

at one location for several days before we began to record species at the trays. Desert fire ant colonies in the vicinity had collected so many seeds from the trays prior to our initial data recording that ants were no longer visiting the baits. Despite large differences in grass cover and plant species composition between grazed plots and ungrazed plots, we found no comparable differences in the ant faunas of those areas (table 6.4).

Ground-dwelling arthropods. Flightless arthropods and those with limited flying abilities make up the assemblage generally known as the ground-dwelling arthropod fauna. Most are nocturnal, especially the predatory arachnids. Grids of pitfall traps were used to sample this fauna at the Jornada del Muerto. Although a number of orders, families, and species of ground-dwelling arthropods inhabit desert grasslands, the diversity and abundance of these animals is greater in desertscrub vegetation.

A small darkling beetle (*Araeoschizus decipiens*) was the most abundant ground-dwelling beetle, but it made a relatively insignificant contribution to the ground-dwelling arthropod biomass (table 6.5). In general, tenebrionid beetle numbers decreased between 1988 and 1992, probably as a result of lower rainfall and higher temperatures in the last two years. The detritivore beetle genus *Eleodes*, which accounted for a large fraction of the biomass of ground-dwelling arthropods, has been the subject of numerous

Table 6.5. Abundances of ground-dwelling arthropods averaged for the growing season based on pitfall trap grids in two desert grasslands: black grama–dropseed basin grassland (I) and watershed bunchgrass grassland (II).

	1988		1989		1990	
Taxon	I	II	I	II	I	II
Arachnida						
Araneae	7.5	8.5	7.3	5.3	6.8	5.3
Scorpiones	6.5	5.0	7.1	6	5.6	5.3
Solifugae	5.3	5	1.3	1	0.8	1
Coleoptera						
Carabidae						
Pasimachus duplicatus	6	2	3	1.8	3.5	0
Tetragonoderus pallidus	1	2	0.5	0	0.2	0.2
Hymenoptera						
Mutilidae (3 species)	23.1	34.5	5.4	4.8	7.3	2.8
Orthoptera						
Gryllacrididae						
Ceuthophilus sp.	17.0	10.5	4.1	4.8	2.4	5.3
Gryllidae						
Gryllus sp.	2	1	6	2	1.4	5.5
Polyphagidae	5.3	3	2.4	1.3	4.6	0.5
Scarabaeidae						
Canthon puncticollis	0	0	0	0	0.5	1
Diplotaxis sp.	3.3	4	1.6	0	1.9	1
Trox sp.	0	0	0	0	0	1.5
Tenebrionidae						
Araeoschizus decipiens	165	165	33	9	15	3
Eleodes extricatus	14.8	23.0	6.8	1.8	3.6	0.3
Eleodes gracilis	4.8	4	1.1	4.3	1	2.5
Eleodes hispilabrus	6.5	2	2.1	0.5	0.1	0

NOTE: Numbers are average per 0.36-ha grid for each sampling season (May–October).

behavioral and physiological studies (Crawford 1981). In desert grasslands, *Eleodes* can frequently be seen feeding on the remains of seeds in the refuse piles of harvester ants. These stink beetles, or *pinacates,* often use head-standing defensive behavior when disturbed, and most individuals produce a noxious defensive secretion from their anal region.

Scarab beetles, which feed on dung, are not abundant in desert grasslands. Perhaps their scarcity is related to the virtual absence of large grazing mammals in the desert grasslands for the past 11,000 years (Parmenter and Van Devender, this volume; Van Devender, this volume). The low numbers of common dung beetles such as *Canthon* spp. may be related to climatic conditions or simply to domestic livestock management practices that shifted animals about in pastures adjacent to the ungrazed study areas. Other grassland dung beetles include *Copris* spp., *Onthophagus* spp., and *Phanaeus* spp., the latter a common bright metallic green Mexican dung beetle found in desert grasslands in Santa Cruz County, Arizona (pers. comm., T. R. Van Devender, Arizona–Sonora Desert Museum, 1993).

Members of the order Orthoptera make up the other major group of insect detritivores in the desert grasslands. These include camel crickets (Gryllacrididae; fig. 6.5), common crickets (Gryllidae), and desert cockroaches (*Arenivaga* spp.: Polyphagidae). Orthopteran populations appear to vary greatly from one habitat to the next and from year to year (table 6.5). Most burrow in the sand and are active at night. While there are no data on their food habits, most orthopterans probably feed on small plant fragments.

The remaining ground-dwelling arthropods are predators such as ground beetles (Carabidae), centipedes, scorpions, and spiders. Most predators decline in numbers during dry years (table 6.5), but the scorpions maintained a relatively constant abundance and biomass over the three years of our study; perhaps because they can reduce their metabolic rate when deprived of food (Crawford 1981) and thus are not likely to starve.

Grass canopy arthropods. Not much is known about the insects that live in or on the canopies of desert grasses, or their effects on the host plants. In a three-year study of a black grama–mesa dropseed (*Sporobolus flexuosus*) desert grassland, Ellstrom (1973) found that the peak densities of insects (78–99/m^2) were 40 times lower than at a site in tall-grass prairie and 7.5 times lower than at a site in short-grass prairie in eastern Colorado. The biomass of arthropods in the desert grassland was also very low, with peaks ranging from 80 g/ha in a dry year to 3,330 g/ha in a wet year. The rela-

Figure 6.5. Camel cricket (Gryllacrididae). Photo courtesy of David Lightfoot.

tively low density and biomass of canopy-dwelling arthropods in desert grassland is a function of the variability in timing of growth and flowering of desert grasses in response to the relatively unpredictable wet season. The abbreviated growing season prevents many species of arthropods from attaining large population sizes.

Ellstrom (1973) found no consistent differences in canopy cover, arthropod densities, or biomass attributable to grazing by livestock. The lack of difference in vegetation cover and species composition probably accounted for the lack of differences in abundance and biomass of the canopy arthropods.

Populations of herbivorous arthropods increase when new vegetative biomass and flowering stalks are produced late in the growing season. The dominant herbivores in Ellstrom's study area included lace bugs (Tingidae), leafhoppers (Cicadellidae), seed bugs (Lygaeidae), and thrips (Thysanoptera) (Ellstrom 1973), but only lace bugs (*Corythucha morrilli*, *C. venusta*, and *Gargaphia opacula*) and the thrips (*Chirothrips simplex* and *Haplothrips haplophilus*) fed on black grama and other grasses. Ellstrom (1973) concluded that the low population levels of chewing herbivores could not be attributed to the sampling method because visible damage from lace bugs and thrips was not seen on grasses in the study area.

In an earlier study, Watts (1963) made irregular sweep-net collections of insects for a year in black grama grasslands. He reported nine orders, 55 families, 109 genera, and 120 species of arthropods. Four species of thrips accounted for considerably more than 50 percent of the total insect population; 97 percent of the thrips were *Chirothrips falsus*, which feeds on developing grass seed, causing immature or empty florets.

Using an insecticide, Watts (1963) markedly reduced the abundance of thrips and other herbivores. Black grama exhibited higher seed set in the treated plots, suggesting that the insects had a deleterious effect on the sexual reproduction of this grass. Since black grama mostly reproduces asexually by stolons, however, thrips probably have a minimal effect on its long-term survival.

Grasshoppers. Two types of grasshoppers are common in desert grasslands: terricoles, or soil mimics (the color pattern matches soil or rocks), and graminicoles, or grass mimics (the color pattern matches grass leaves) (Uvarov 1977; table 6.6, figs. 6.6 and 6.7). Cryptic grasshoppers—those able to match their substrate—suffer significantly less predation than other

Table 6.6. Characteristics of the grasshopper life form common in the desert grassland.

	Characteristics		
Life Form	Morphology	Appendages	Coloration
Terricoles	dorsoventrally compressed; disruptive	variable; tarsal arolei and prosternal sternal process absent	browns, grays, mottled pattern
Graminicoles	laterally compressed; elongate; face oblique	legs short; tarsal arolei present; prosternal process lacking	green, yellow or combination pattern of longitudinal striping

SOURCE: Uvarov 1977.

Figure 6.6. This grasshopper mimics grass leaves. Photo courtesy of David Lightfoot.

grasshoppers (Cox and Cox 1974; Gillett and Gonta 1978; Isley 1938). Grasshoppers in desert grassland are most diverse and abundant on patches of bare ground. Of the 28 species of grasshoppers recorded from the Jornada del Muerto Basin, only three grass mimics were relatively abundant, compared with seven abundant species of soil mimics (Light-

Figure 6.7. The ground-dwelling bandwinged grasshopper (*Trimerotropis pallidipennis*) matches the soil. Photo courtesy of David Lightfoot.

foot 1985; table 6.7). The most abundant grasshopper was the bandwinged grasshopper (*Trimerotropis pallidipennis*), a widespread soil mimic found in both desert grassland and desertscrub vegetation.

Grass-dwelling grasshoppers generally feed on a variety of grasses but frequently exhibit preferences for certain species (Joern 1979a; Mulkern 1980; Mulkern et al. 1964; Ueckert and Hansen 1971). Ground-dwelling grasshoppers are generalists that feed on a variety of low-growing plants, including annual and perennial forbs, grasses (Joern 1979a; Scoggan and Brusven 1973; Sheldon and Rogers 1978), lichens, algae, and moss (Sheldon and Rogers 1978). Because these foods vary both in the time they are available and in distribution, terricoles tend to be generalist feeders.

The diversity and density of grasshoppers in arid grasslands (Uvarov 1977) primarily reflect the high diversity and the temporal and spatial distributions of their plant foods, but the long growing season and dense vegetation cover where grasshoppers can escape from predators are certainly important factors as well. Lightfoot (1985) presented evidence that predation is the major selective agent that has structured desert grassland grasshopper assemblages. He demonstrated that substrate selection for hiding was of greater or equal importance to the selection of food plants.

Table 6.7. Common grasshopper species from the Jornada del Muerto Basin that occur in desert grasslands.

Species	Life Form Guild	Relative Abundance
Cibolacris parviceps	terricole	++
Derotmema haydenii	terricole	+
Opeia obscura	graminicole	+
Orphulella pelidna	graminicole	+
Parapomala pallida	graminicole	++
Phrynotettix robustus	terricole	+
Psoloessa delicatula	terricole	+
Hippopedon gracilipes	terricole	+
Trimerotropis pallidipennis	terricole	+++
Trimerotropis strenua	terricole	++

SOURCE: Lightfoot 1985.

NOTE: Relative abundances are as follows: + = more than 3 individuals in survey, ++ = more than 25 individuals in survey, +++ = more than 100 individuals in survey.

Assuming that desert grassland grasshoppers select microhabitats to escape predators, then competition for escape space may be important in determining grasshopper assemblages (Joern 1979b; Otte and Joern 1977).

Robber flies. For the past several years we have studied the robber flies (Asilidae: Diptera; fig. 6.8) in desert grassland on the Jornada del Muerto. Robber flies were sampled by aerial net in five grass communities and by systematic collections along transect belts approximately 6 m wide. The data were compiled to provide information on species composition and flight seasons (table 6.8). Forty-five species belonging to 21 genera were recorded; the number of species occurring at a single site ranged from 8 on the alkali sacaton-tobosa playa to 25 in bunchgrass on the lower bajada. Many of the species found in the small playa were strays from adjacent desertscrub habitats, especially early in the year. Approximately 45 percent of the known robber fly species in Doña Ana County occur in the desert grasslands.

One robber fly genus, *Efferia*, accounted for one-third of all species of desert grassland robber flies in the Jornada and appeared to be dominant at most of our collecting sites (Wilcox 1966). *Efferia* are generally adapted to

Figure 6.8. Robber flies (Asilidae) on a grass stalk. Photo courtesy of David Lightfoot.

arid and semiarid habitats and frequently constitute 25 percent or more of the robber fly fauna at lower elevations in the southwestern United States. One species, *E. benedicti*, is widespread in western North America and occurs in all the Jornada grasslands except for one playa.

Seven genera and four species found in the Jornada are also found in short-grass prairie in Colorado (Rogers and Lavigne 1972), and three genera

Table 6.8. Robber flies present in five grassland communities on the Jornada del Muerto Range, Doña Ana County, New Mexico.

	Grassland Community				
Species	Black Grama Grassland	Bunchgrass Grassland	Tobosa Playa	Isaacks Playa	College Playa
Ablautus flavipes	+	+			
A. rufotibialis		++		+	
Cerotainiops abdominalis		+			
C. lucyae				+	
Dicropaltum mesae			+		+
Efferia apache	+				
E. benedicti	++	++		+	+
E. bicolor		+			
E. cressoni	+	+		+	
E. helenae	+	+			
E. kelloggi		+	+	+	
E. luna		++		++	+
E. mortensoni		++			
E. ordwayae				+	
E. pallidula	++				
E. pilosa		++			
E. subarida		++			+
E. tuberculata				+	
E. tucsoni		++		++	
E. varipes	+				
Heteropogon cazieri			+	+	
H. johnsoni			++	+	
Leptogaster hesperis			+		
L. patula		+			
Machimus nr. *erythocenemius*					+
Mallophora fautrix					+
Megaphorus lascrucensis				++	
M. prudens		+			
M. pulchrus	+	++		+	++
Metapogon punctipennis		+			

Species	Grassland Community				
	Black Grama Grassland	Bunchgrass Grassland	Tobosa Playa	Isaacks Playa	College Playa
Omninablautus arenosus		+		+	
Polacantha composita				+	
Proctacanathella leucopogon	+	+	+	+	
Proctacanthus nearno	+	+		+	
P. nigrofemoratus	++		+		
Promachus giganteus		++		++	+
P. nigrialbus	++	++			++
Psilocurus sp.			+	+	
Regasilus blantoni				+	
Saropogon coquillettii	+	+			
Scleropogon duncani	+[a]	++		+	+
S. indistinctus	++				
S. picticornis				+	
Stichopogon fragilis	+				
Wilcoxia nr. *martinorum*		++			
Total species	16	25	8	22	10

NOTE: ++ = common, 10 or more seen in a survey session; + = present, recorded at least once from site.

[a] Small *Scleropogon* specimens from this site may include *S. coyote.*

and one species (*E. benedicti*) also occur in a mesic grassland in British Columbia (Cannings 1989). The faunas are more similar than these comparisons suggest, however, because a species found in one grassland is frequently replaced by a closely related species elsewhere.

Robber fly species have characteristic and predictable flight periods. Although temperature and moisture may alter daily flight times, the adults of a given species are usually found at the same time and place each year (fig. 6.9). A few anomalous species are known; for example, *Ospriocerus abdominalis*, a predator on meloid beetles, flies sporadically between May and October. *Heteropogon johnsoni* normally appears in late summer but briefly emerged following an unusually wet spring in June 1992. An undescribed species, *Psilocurus* sp., was found in the tobosa swales in June and again in

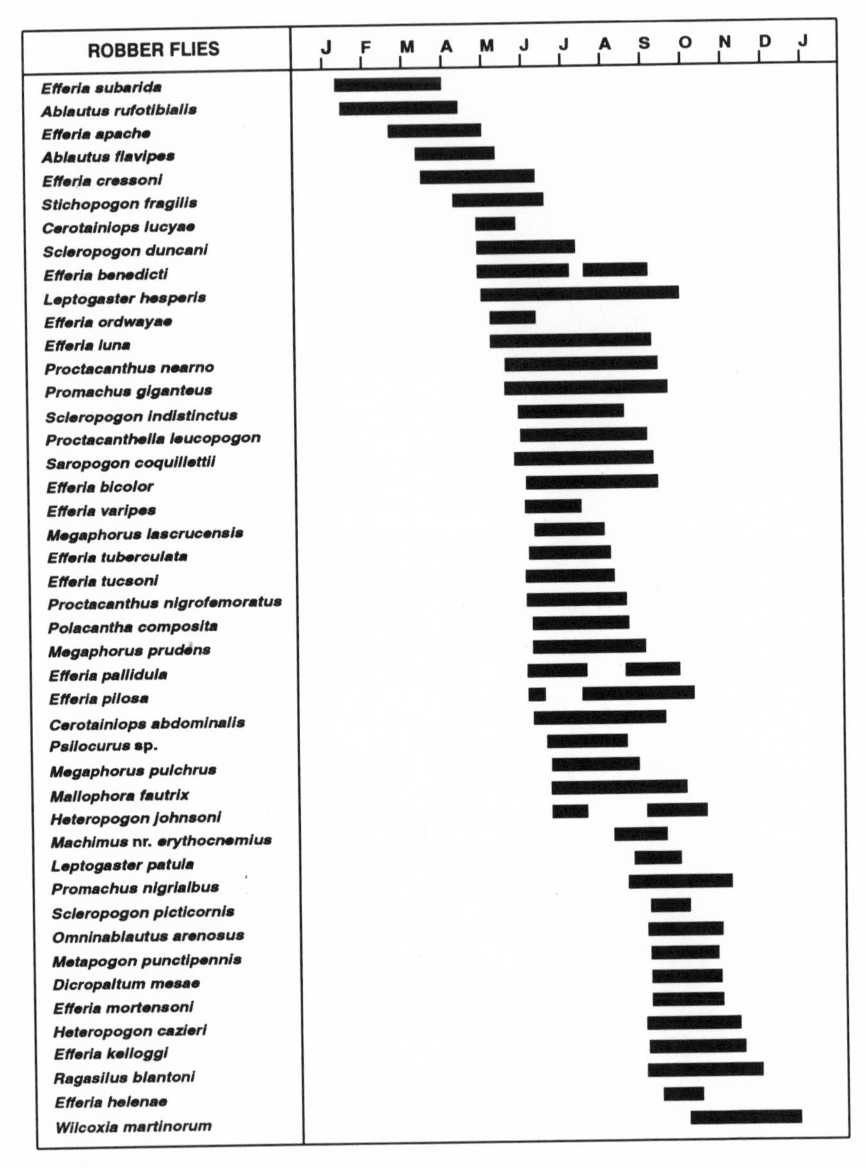

Figure 6.9. Seasonal flight periods of grassland robber flies (Asilidae) in the Jornada del Muerto, Doña Ana County, New Mexico.

September, probably in response to rainfall.

There are three seasonal assemblages of robber flies in the desert grassland. The six species active from late winter through late spring are mostly ground perchers unable to tolerate hot summer soil surfaces; nearly all of these perch on shrubs or forbs and thermoregulate by moving between the soil surface and vegetation. Thirteen species are active from autumn through early winter. Ground-perching species—including three species of *Efferia*, *Omninablautus*, *Regasilus*, and *Wilcoxia*—are prominent in this assemblage. The greatest number of species is active in the summer, especially during the period of summer rains. These species have the largest numbers of individuals per unit area.

Shrubby grasslands have a richer robber fly fauna than grassland sites with little shrub cover. Sites with heavy clay soils usually have few flies. Clay soils are not conducive to robber fly larval success or to the growth of forbs and shrubs that attract potential prey and provide perching sites for the adults. A number of robber flies are restricted to specific areas, and some were absent from adjacent sites of seemingly identical habitat. For example, an obligate shrub-perching species, *Efferia bicolor*, was not found in the watershed bunchgrass habitat, which has little shrub cover, but was present further upslope where shrub cover was greater. The species apparently requires denser shrub cover than is present in most of the desert grasslands on the Jornada. Another species, *E. ordwayae*, seems to be restricted to tobosa grass swales with partial shrub cover. An especially localized genus comprised of three species, *Heteropogon cazieri, H. johnsoni,* and *H. patruelis,* was sporadically distributed through tobosa swales. The localized nature of robber fly populations in tarbush vegetation leads to distinct species assemblages in adjacent grassland sites, where these species are occasional visitors.

The reasons for the restricted distributions of robber fly populations are not understood. Perhaps a particular microhabitat provides more resources (e.g., more abundant prey, preferred prey species, and better perching sites) than an apparently similar adjacent area; or perhaps the population represents the result of a newly established (founder) or a declining relictual population. In addition, the suites of species present in an area may change along with plant composition and cover changes caused by climatic changes or livestock grazing. Interspecific interactions such as competition for prey, predation by other species of robber flies, and cannibalism may also determine the abundance or absence of a species at a site.

The diverse robber fly fauna of the desert grasslands exists because the species partition the available resources temporally and spatially as well as behaviorally and morphologically. Intrafamily competition has been proposed as the driving force for niche separation in robber flies; it is known, for example, that differences in niche characteristics allow species to coexist in short-grass prairie (Rogers and Lavigne 1972). Fisher and Hespenheide (1992) considered cannibalism to be a major cause of species replacements over time, although they cautioned that the overall role of adult competition in niche differentiation has not been adequately demonstrated. Most of the intraspecific competition and mortality may occur among the subterranean larvae. The larval biology is unknown for most species.

The structure of the robber fly assemblages reported here reflects the contemporary habitat. Dramatic changes have occurred in desert grasslands since the 1880s, though. Several of the dominant species, including *Efferia benedicti, E. pallidula, Promachus nigrialbus,* and *Scleropogon indistinctus,* were probably widespread and dominant members of the robber fly faunas of the desert grasslands before Europeans arrived but are not common in shrubland communities. Many desertscrub species that were formerly probably restricted to riparian areas or to shrubby bajadas have been able to colonize the grasslands because of the presence of shrubs there. The robber fly fauna of desert grasslands must be considered to be as dynamic as the plant species composition (Van Devender, this volume). Whatever the changes that occur in plant species composition, however, robber flies will remain among the most important arthropod predators in desert grasslands.

Invertebrates of ephemeral aquatic environments. Ephemeral lake basins (playas) are common features in desert grasslands. Small playas and drainage channels are frequently modified into livestock watering tanks that capture and store runoff water. While these areas can flood at any time, flooding is most frequently the result of intense summer convectional storms that produce overland flow that fills arroyo channels. Playas can be virtually devoid of vegetation because of intense use by livestock.

A diverse invertebrate fauna populates playas and water tanks immediately after they are flooded. Loring et al. (1988) classified the aquatic invertebrates into two groups based on life history characteristics: fast onset with immediate development, including mosquitoes (*Aedes* spp.), clam shrimp (*Eulimnadia texana*: Conchostraca), and tadpole shrimp (*Triops longicaudatus*: Notostraca); and fast onset with prolonged development, including water fleas (*Moina wierzejskii*: Cladocera) and fairy shrimp (*Streptocephalus texanus, Thamnocephalus platyurus*: Anostraca). Clam shrimp and mosquitoes can complete their entire life cycles in two weeks or less. The desiccation-resistant eggs of the clam shrimp remain in the soil when the playas dry and hatch immediately when they flood. Mosquitoes disperse to newly flooded areas from more permanent aquatic habitats such as metal water tanks. Fast onset and immediate development give clam shrimp access to the organic particulates in the ephemeral lakes before the larger competitors reduce the availability of these food resources (Loring et al. 1988). Within 20 to 40 days after flooding, the larger competitors—fairy shrimp, rotifers (Rotifera), and water fleas—dominate the playa waters. The populations of clam shrimp and mosquitoes are regulated primarily by moisture and temperature conditions, whereas the fairy shrimp and rotifers are regulated by bird and tadpole predation and by drying of the ponds.

The diversity and composition of playa lake invertebrate communities are influenced by the length of time that the playa retains water. Playas that remain flooded for longer than one month attract a number of aquatic insects. Adult dragonflies and damselflies (Odonata) may be seen flying around the edges of a flooded playa, and back swimmers (Notonectidae), predaceous diving beetles (Dytiscidae), and bloodworm fly larvae (Chironomidae) become abundant in the water (Whitford 1972).

The species diversity and community composition of aquatic invertebrates in playa lakes also depend on the amount of dead plant material in the playa, which determines the quality and quantity of suspended particulates and plankton populations. Large quantities of decomposing detritus immobilize the nutrients required by phytoplankton. Large rotifer and tadpole shrimp populations occur in playas with an abundance of dead grasses, whereas pools that are largely devoid of detritus have large populations of fairy shrimp and fewer rotifers and tadpole shrimp. The quantity of detritus has no apparent effect on the water fleas (Medland 1988).

The invertebrate fauna of the few permanent ponds in desert grasslands consists of back swimmers and the larvae of dragonflies, damselflies, mosquitoes, and flies. The adults of these insects disperse to ephemeral ponds during the wet season.

Functional Roles

The importance of invertebrate animals to ecosystem properties and processes is frequently underemphasized. Despite their abundance, invertebrate herbivores consume less than 10 percent of the live biomass in most terrestrial ecosystems (Chew 1974). The smallness of their fraction of the energy flow in ecosystems would seem to relegate invertebrates to an inconsequential status in ecosystem structure and dynamics. When the indirect effects of invertebrates on nutrient cycling processes and their direct and indirect effects on soil heterogeneity are considered, however, their importance becomes more evident. Indeed, ecosystem properties like resilience (the ability to recover following disturbance) are directly affected by the activities of key invertebrate species. Several groups of invertebrates found in desert grasslands produce nutrient-enriched, water-enhanced patches near their nests. Water infiltration, water storage, and rates of nutrient turnover are increased in such patches.

The limited data on the functional roles of invertebrates in their ecosystems are best interpreted by examining the species and species groups we know the most about. Data regarding these groups can be used to address a variety of questions about the functional roles of invertebrates in desert grasslands, including questions about the habitat specificity of the processes.

Soil invertebrates. Soil invertebrates play an essential role in decomposition and nutrient-cycling processes by feeding on bacteria and fungi. In order for microbes to grow on decomposing organic matter, most of which has a carbon:nitrogen ratio of 70–100:1, they must concentrate nutrients (especially nitrogen) in their biomass and create C:N ratios of 10–35:1 (Anderson et al. 1984). Microbes compete directly with plant roots for nitrogen and other nutrients. The accumulation of nutrients in microbial biomass, called immobilization, results in a very slow release of essential nutrients in mineral form—the form necessary for uptake by plant roots. Invertebrates that ingest microbes excrete mineral nutrients that are immediately available to plants. Additionally, invertebrates that have well-developed gut microfloras may process dead plant material rapidly and excrete minerals in a form usable by plants. Thus, by changing the relationship between immobilization and mineralization of nutrients, invertebrates have an important effect on nutrient cycling.

The population densities of nanorchestid and tydeid soil mites (fig. 6.1) vary seasonally, but not with the magnitude seen in other taxa (Cepeda-Pizarro and Whitford 1989; Steinberger and Whitford 1985). These mites are active even when the soils are dry, and they continuously graze on fungal hyphae and excrete mineral nitrogen and other elements. When mites were chemically excluded from decaying roots, significantly more nitrogen was immobilized by the fungi growing on the roots (Parker et al. 1984).

Termites. Subterranean termites (Isoptera) are probably the most abundant and functionally most important arthropods in desert grasslands, and they influence the abundance and structure of the soil microfauna as well. Densities of cryptostigmatid mites were 2.5 times greater on plots where termites had been chemically excluded (Silva et al. 1989).

The biomass of subterranean termites in desert grasslands may exceed the biomass of domestic livestock (Haverty and Nutting 1975; Johnson and Whitford 1975; table 6.9), although these abundant insects are difficult to observe unless their foraging galleries are constructed around food materials or large dry dung pats on the surface, which can be broken open to reveal the insects inside. The most effective method of estimating termite abundance and distribution is a bait technique using toilet paper rolls distributed on a grid (La Fage et al. 1973; fig. 6.10). The termites enter the rolls and eat the paper, and an instantaneous estimate of feeding termites can be obtained by shaking the rolls and counting the termites that fall out. Estimates based on this method yielded abundances in the thousands per square meter.

Subterranean termites are very important consumers of dead plant material and animal dung. Direct measurements of dead grass and dung removed by termites were made by comparing plots where termites had been chemically excluded with untreated plots. The amount of mass lost from dung (1–4 percent), dead grass (15–25 percent), and grass roots (2–12 percent) was much lower in the absence of termites (table 6.10).

Subterranean termites are also responsible for decomposing dry dung and incorporating it into the soil. The dry dung of domestic livestock

Table 6.9. Biomass of subterranean termites on two desert grassland sites (estimated by toilet paper bait rolls) versus that of domestic cattle.

	Chihuahuan Desert		Sonoran Desert	
	Termites	Livestock	Termites	Livestock
Average live biomass (g/m^2)	3.6	0.8	3.1	0.8
Termite/livestock biomass	4.4		3.8	

SOURCES: The termite data are from Haverty and Nutting 1975; and Johnson and Whitford 1975.

Figure 6.10. Bait roll showing surface galleries and paper eaten by subterranean termites.

Table 6.10. Organic material removed by termites versus mass lost in the absence of termites.

	Percentage Removed	
Material	Termites Present	Termites Absent
Annual plants (leaves & stems)	40–85	20–40
Annual plants (dead roots)	50–70	10–30
Perennial grass	40–80	15–25
Perennial grass roots	50–70	2–12
Creosotebush leaves	0–90	60–85
Dead wood	<1–5	—
Cattle dung	60–100	<1–4
Rabbit dung	15–50	—

SOURCES: Data from MacKay et al. 1987; Silva et al. 1989; Whitford et al. 1982, 1988; and unpublished data.

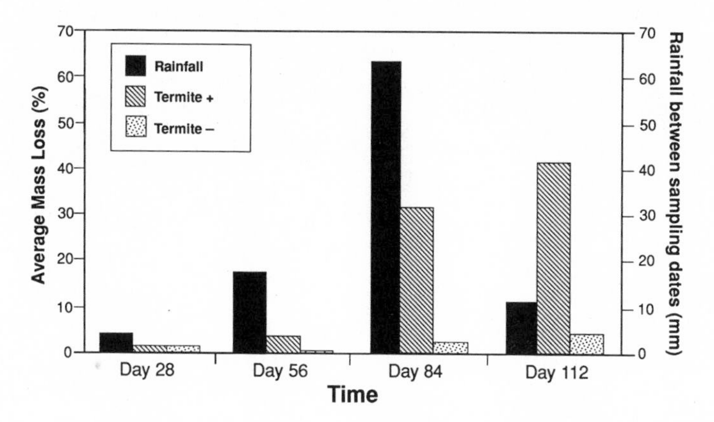

Figure 6.11. Comparison of cow dung decomposition on plots with termites present (Termite +) and excluded (Termite −).

decomposes very slowly (fig. 6.11). If subterranean termites were not present in desert grasslands grazed by domestic livestock, a considerable fraction of the soil surface would eventually be covered with dried dung (Parmenter and Van Devender, this volume). Dried dung suppresses the growth of plants and, if a sufficiently large area is covered, reduces the productivity and carrying capacity of the ecosystem. Using Whitford et al.'s (1982) data on stocking rates and assumed defecation rates of cattle, we calculated that over a 50-year period approximately 20 percent of the soil surface would be covered by dried dung if it were not broken down by termites. A 20 percent reduction of the surface area would result in a significant loss of productivity.

When one takes into consideration all the plant litter, standing dead plant material, and feces consumed by termites, it becomes evident that 50 percent or more of all photosynthetically fixed carbon is consumed by termites (Whitford 1991). Thus, most of the cycling of carbon and other nutrients in desert grasslands is a function of subterranean termite activity. In fact, the organic matter content of soils on a desert watershed studied by Whitford (1991) was negatively correlated with the abundance and activity of subterranean termites.

Large quantities of dead plant material are consumed and broken down in the termite gut. The minerals released from that breakdown are either used by the insect or its gut microflora or are deposited as feces, often at depths below the root zone of desert grasses. Nutrients are returned to the soil via the feces, gallery sheeting, and predators. Nutting et al. (1987) found that soil brought to the surface by subterranean termites (*Gnathamitermes perplexus* and *Heterotermes aureus*) in an Arizona desert grassland had higher levels of organic carbon and macronutrients than adjacent undisturbed soil profiles.

The gallery systems of subterranean termites affect soil pore spaces, water infiltration, and aeration of the soil. Water infiltration rates on termite-free plots were significantly lower (average 51.3 mm/hr) than they were on plots where termites were present (average 88.4 mm/hr) (Elkins et al. 1986). The soil bulk density (mass per unit volume) increases and water storage decreases when termites are removed (Gutierrez and Whitford 1987). The importance of termites in soil modification is greater in areas where soils are shallow and/or have relatively high clay content. The presence of galleries in such soils has a beneficial effect by increasing water infiltration and soil aeration. Aeration and infiltration are high in sandy soils without gallery structures.

Subterranean termites contribute to soil turnover even in sandy soils by constructing foraging galleries around dead grass stems and other food items. These galleries are made with materials transported to the surface from deep in the soil profile. The galleries eventually erode and are added to the surface soil. The quantity of soil added to the surface by these activities has been estimated at 43.9 kg per hectare per day in a basin bunchgrass grassland and 10.4 kg per hectare per day in a piedmont black grama grassland. These are conservative estimates based on termite activities late in the summer wet season (MacKay and Whitford 1988). Over time, this turnover can have a significant effect on soil properties and even on soil genesis.

Ants. The temporal and spatial characteristics of ant nests determine to a great extent the effect of ants on the ecosystem. The nests represent temporally variable small-scale disturbances that can maintain ecosystems in a dynamic nonequilibrium (ever-changing) state by providing colonization sites for competitively inferior plant species (Grubb 1977; Pickett and White 1985). Ant nests represent disturbances of a few square centimeters to as large as 5 m^2, on time scales of three to six weeks to as long as 70–80 years.

Nests of the black harvester ant are long lasting and relatively large. Nests larger than a meter in diameter tend to affect the soil nutrient concentrations at the edge of the nest disc, the area around the nest entrance that is kept clear of vegetation by the ants (Carlson and Whitford 1991; Culver and Beattie 1983; Di Marco 1987; Whitford 1988). The concentrated nutrients there and the relative importance of those nutrients to nearby

Table 6.11. Effects of black harvester ant (*Pogonomyrmex rugosus*) nests on selected soil chemical properties in three locations on a Chihuahuan Desert grassland watershed.

	Burrograss-Tobosa Basin		Desertscrub Low Slope		Bunchgrass Mid-slope	
Chemical	AN	OAN	AN	OAN	AN	OAN
Calcium	21.8	24.8*	13.8	15.1	5.9	7.4
Magnesium	2.5	3.0*	1.6	1.6	2.0	2.0
NO_3-N	8.2	2.6*	1.8	0.8	2.5	0.8
NH_4-N	1.0	1.3	0.6	1.0	1.7	0.7*
Total N	629.0	657.4*	497.0	465.2	432.7	388.8*
Inorganic P	12.4	7.4	4.7	4.6	20.4	8.0*

SOURCE: Data from Di Marco 1987.

NOTES: AN = nest perimeter, OAN = 5 m from the nest perimeter.

Concentrations: Ca++ and Mg++ are in meq/100 g of soil; and NO3-N, NH4-N, total N, and inorganic P are in g/kg soil.

*Indicates significant differences between concentration on ant nests and at 5 m distant.

plants vary spatially, even within the same watershed (table 6.11). The concentration of soil nutrients at the edge of the nest disc and the enhanced infiltration resulting from the nest tunnels and chambers promote the growth of a corona of dense vegetation around the nest disc that is both different in composition and more productive than vegetation in areas away from ant nests. These hot spots may be foci for the activities of other animals. We observed antelope (*Antilocapra americana*) preferentially feeding on Texas filaree (*Erodium texanum*) at the edges of black harvester ant nests in a burrograss area on the Jornada.

In their review, Lobry de Bruyn and Conacher (1990) stated that the few studies conducted to date do not justify the conclusion that ants have a negligible effect on soil turnover, because most researchers have limited their investigations to ants that occupy permanent nests and to time scales far shorter than those required for soil formation. In desert grasslands, the amount of soil moved to the surface by ants in the construction and maintenance of their nests varies with the soil type, species composition of the ant community, and differences in colony densities in the various habitats (table 6.3). Species such as the black harvester ant, the desert long-legged ant (*Aphaenogaster cockerelli*), and the hairless honeypot ant, which construct large, permanent nests, carry only small amounts of soil to the surface at the beginning of their activity period. Ants that move large quantities of soil include the crazy ant, the desert harvester ant, and the red piss ant (*Iridomyrmex pruinosum*). At the beginning of the growing season, the nest entrances of these species are easily identified by the cones of soil cleared from nest chambers (fig. 6.12). Soil movement by ants is an important factor in soil formation in several grassland habitats. An estimate of the amount of soil moved to the surface by ants in two black grama–dropseed habitats over 100 years was equal to a 1-cm layer of soil over the entire area (table 6.12). In the long term, grassland ants are cultivating the soils by transporting subsurface soil to the surface.

Nest-cone soils have a lower bulk density than intact soil and lack the surface crust that characterizes desert grassland soils, and are therefore readily eroded by wind and water. The redistribution of soils from ant nest cones by erosion has not been quantified. If there is significant wind erosion, the loss of clay-silt fractions may have a significant role in the development of the sandy soils characteristic of many desert grassland sites.

The seed-harvesting and consumption activities of desert ants have important implications for the area's vegetation, although there are few unequivocal data that demonstrate the impact of seed harvesting on the ecosystem. The exceptions are the studies by Brown et al. (1979), Davidson et al. (1984), and Inouye et al. (1981). Long-term studies showed that ants increase the plant species diversity by differentially harvesting the seeds of the dominant species. On plots where ants had been removed, ephemeral plants with small seeds dominated the community. Over the long term, however, the ephemeral plant assemblage exhibited density compensation, and this decline in the ephemerals shifted the assemblage back toward the original condition. Although seed harvesting by ants apparently affects the composition of the vegetation, especially ephemeral plants, harvesting does not result in the severe depletion of seeds that might be expected.

Yucca moths. Soaptree yucca, the tallest plant in many desert grasslands, is an important structural feature that provides nesting sites and perches for grassland birds (Kerley et al. 1992). The large masses of soaptree yucca flowers in early summer attract myriad insects. The pollination ecology of soaptree yucca was first described in 1872 and stands as a classic case of obligate mutualism. The plant and its insect pollinator are mutually depen-

Figure 6.12. Fresh soil cone at the entrance of a desert harvester ant (*Pogonomyrmex desertorum*) nest. Note the abundance of grass seed around the entrance.

dent; neither can survive without the other (Baker 1986). A female yucca moth (*Tegeticula yuccasella*) pollinates a yucca flower by pushing a small ball of sticky pollen down the stigma after she has laid an egg in the flower's ovary. The yucca flowers are pollinated and develop into fruits, and the moth larvae develop inside the fruits, consuming approximately one-sixth of the seeds. Each soaptree yucca fruit contains six rows of seeds separated

Table 6.12. Soil moved by ants in the cleaning and construction of nests on sandy–sandy loam soils versus soil moved by ants on clay–clay loam soils.

Soil type	Habitat	Soil Moved (kg/ha)
Sandy–sandy loam	Black grama–mesa dropseed	85.8
	Bunchgrasses, grazed	55.2
	Black grama, ungrazed	21.3
Clay–clay loam	Burrograss-tobosa basin	3.4
	Tobosa swale	0.2
	Alkali sacaton swale	0.1

by walls, however, which effectively keeps the yucca moth larva within one fruit segment and prevents it from eating all the seeds.

Yucca flowers attract many insect visitors, and the obligate nature of the yucca-moth relationship has been challenged by suggestions that other species of insects might pollinate the soaptree yucca flowers. James et al. (1993) bagged whole inflorescences to restrict either diurnal or nocturnal insect visitors. None of the flowers exposed to diurnal visitors set fruit, while those exposed to nocturnal visitors, including yucca moths, set fruit at the same rate as unbagged inflorescences. Yucca flowers that were hand pollinated by brushing pollen across the stigma opening did not set fruit; only the flowers that were pollinated by pushing small balls of pollen down the stigma opening (mimicking the behavior of the yucca moth) set fruit.

The relationship between soaptree yucca and the yucca moth is affected by cattle. The flowering stalks and flowers are nutritious and are quickly eaten by cattle, particularly if alternative food sources are not available. Cows will even push over the trunks of tall yuccas in order to eat the flowering stalks. Loss of these flowering stalks can result in local reductions in insect and bird diversity (Kerley et al. 1992).

Robber flies. Robber flies have been characterized as possibly the most important of all predators of insects (Londt 1991). Although the species richness of robber flies may not be as high in desert grassland as it is in some desertscrub vegetation or shrub-grassland ecotones, they are often abundant.

From the human viewpoint, robber fly predation on rangeland insects, especially grasshoppers, is considered beneficial. In Montana rangeland, Joern and Rudd (1982) found that predation by a *Proctacanthus milbertii* reduced grasshopper populations by 0.5–2 percent per day. A desert grassland robber fly, *Scleropogon* spp. (not to be confused with the burrograss plant), also feeds extensively on grasshoppers. On the other hand, robber fly predation on grasshopper parasites, notably flies, may also be significant. Lavigne et al. (1976), for example, recorded five bee fly genera (Bombyliidae) as the prey of *Efferia benedicti*. At least one other species of robber fly, *Heteropogon wilcoxi*, also preys extensively on bee flies (Lavigne and Holland 1969). Rees and Onsager (1985) partially removed robber flies from crested wheatgrass (*Agropyron cristatum*) plots in Montana and found a substantial rise in parasitism rates of grasshoppers by flies from four families.

The tendency for robber flies to prey on each other may reduce their impact on prey species. Lists of robber fly prey animals document considerable intrafamily predation (Rogers and Lavigne 1972). Our observations of *Efferia apache* in New Mexico indicate that low prey availability following the typically dry winters results in extensive cannibalism.

Summary

Desert grassland invertebrates are diverse and abundant and occupy all parts of the ecosystem. The soil fauna includes microscopic protozoans, nematodes, mites, and collembolans. Through their complex feeding relationships, these organisms regulate the growth of soil bacteria and fungi and control the availability of mineral nutrients for plants.

The soil is also home to large invertebrates that play key roles in the function of desert grassland ecosystems. Subterranean termites are abundant and important consumers of dead plant material and the dung of domestic livestock and wildlife. Termites affect soil properties such as bulk density, porosity, and texture by constructing underground foraging galleries and carrying soil to the surface to build foraging galleries around dead plants and other food items. Desert grassland ants contribute to the ecosystem as harvesters of seeds, predators, scavengers, and collectors of honeydew. The soil in the corona around ant nests is enriched with mineral nutrients. Nest galleries also affect soil porosity and water infiltration, resulting in patches of soil that support different or larger plants.

Yucca moths are obligate pollinators of soaptree yucca and depend on the yucca's fruits to complete their life cycle. Grasshoppers in desert grassland tend to either mimic grass color and feed on grasses, or mimic soil color and feed on annual plants, including grasses. The mimicry provides some protection from predation. Robber flies are important predators on grasshoppers and a variety of other insects in desert grassland ecosystems. Some species of robber flies are active throughout the year except for winter. Robber fly species that are grassland specialists have probably decreased in numbers as shrubs have increased in grass-dominated habitats.

Desert grassland invertebrates are not limited to terrestrial habitats. Ephemeral ponds are home to cladocerans, clam shrimp, fairy shrimp, rotifers, tadpole shrimp, damselflies, dragonflies, and mosquitoes. The rotifers and shrimp complete their life cycles within days and survive the long dry periods in the form of desiccation-resistant eggs. The aquatic insects migrate to the ephemeral ponds from permanent bodies of water such as livestock water tanks and permanent streams.

Despite the abundance and importance of invertebrates in desert grasslands, our knowledge of this fascinating fauna is scarce and fragmentary. In this chapter we have reviewed what is known about the biology of these animals, but our presentation also points out how much more remains to be learned.

Acknowledgments

Support for much of the research reported here and for the preparation of this chapter was provided by National Science Foundation grant no. 881160 and is a contribution of the Jornada Long-Term Ecological Research Program II. We thank Michelle Kerley for help in the field and David Lightfoot for helpful discussions and photographs. The preparation of this essay was cofunded by the United States Environmental Protection Agency, Environmental Monitoring Systems Laboratory–Las Vegas. The research described herein has not been subjected to the agency's peer and administrative review, and the conclusions and opinions given here are solely those of the authors and should not be construed to reflect the views of the agency. Mention of trade names or commercial products is for identification purposes only and does not constitute endorsement or recommendation for use.

Literature Cited

Anderson, J. M., A. D. M. Runnier, and D. W. H. Walton (eds.). 1984. Invertebrate-Microbial Interactions. Cambridge University Press, Cambridge, England.

Baker, H. G. 1986. Yuccas and yucca moths—a historical commentary. Annals of the Missouri Botanical Garden 73:556–564.

Brown, J. H., O. J. Reichman, and D. W. Davidson. 1979. Granivory in desert ecosystems. Annual Review of Ecology and Systematics 10:201–227.

Buffington, L. C., and C. H. Herbel. 1965. Vegetational changes on a semidesert grassland range. Ecological Monographs 35:139–164.

Cannings, R. A. 1989. The robber flies (Diptera: Asilidae) of a *Festuca* grassland in the Okanagan Valley, British Columbia. Journal of the Entomological Society of British Columbia 86:14–26.

Carlson, S. R., and W. G. Whitford. 1991. Ant mound influence on vegetation and soils in a semiarid mountain ecosystem. American Midland Naturalist 126:125–139.

Cepeda-Pizarro, J. G., and W. G. Whitford. 1989. Species abundance patterns of microarthropods in surface decomposing leaf litter and mineral soil on a desert watershed. Pedobiologia 33:254–268.

Chew, R. M. 1974. Consumers as regulators of ecosystems: an alternative to energetics. Ohio Journal of Science 72:359–370.

Cox, W. G., and D. G. Cox. 1974. Substrate color matching in the grasshopper *Circotettix rabula*. Great Basin Naturalist 34:66–70.

Crawford, C. S. 1981. Biology of Desert Invertebrates. Springer-Verlag, New York.

Culver, D. C., and A. J. Beattie. 1983. Effects of ant mounds on soil chemistry and vegetation patterns in a Colorado montane meadow. Ecology 64: 485–492.

Davidson, D. W., R. S. Inouye, and J. H. Brown. 1984. Granivory in a desert ecosystem: experimental evidence for indirect facilitation of ants by rodents. Ecology 65:1780–1786.

Di Marco, R. R. 1987. Effects of harvester ant nests on soil properties and vegetation. M.S. thesis, New Mexico State University, Las Cruces.

Elkins, N. Z., G. V. Sabol, T. J. Ward, and W. G. Whitford. 1986. The influence of subterranean termites on the hydrological characteristics of a Chihuahuan Desert ecosystem. Oecologia 68:521–528.

Ellstrom, M. A. 1973. Populations and trophic structure of a desert grassland invertebrate community. Ph.D. dissertation, New Mexico State University, Las Cruces.

Fisher, E. M., and H. A. Hespenheide. 1992. Taxonomy and biology of Central American robber flies with an illustrated key to genera. Pp. 611–632 *in* D. Quintero and A. Aiello (eds.), Insects of Panama and Mesoamerica: selected studies. Oxford University Press, Oxford.

Freckman, D. W., W. G. Whitford, and Y. Steinberger. 1987. Effect of irrigation on nematode population dynamics and activity in desert soils. Biology and Fertility of Soils 3:3–10.

Gillett, S. D., and E. Gonta. 1978. Locust as prey: factors affecting their vulnerability. Animal Behavior 26:282–289.

Gillis, J. E. 1982. Substrate colour-matching cues in the cryptic grasshopper *Circotettix rabula*. Animal Behavior 30:113–116.

Grubb, P. J. 1977. The maintenance of species richness in plant communities: the importance of the regeneration niche. Biological Reviews 52:107–145.

Gutierrez, J. R., and W. G. Whitford. 1987. Chihuahuan Desert annuals: importance of water and nitrogen. Ecology 68:2032–2045.

Haverty, M. I., and W. L. Nutting. 1975. Density, dispersion and composition of desert termite foraging populations and their relationship to superficial dead wood. Environmental Entomology 4:480–486.

Inouye, R. S., G. S. Byers, and J. H. Brown. 1980. Effects of predation and competition on survivorship, fecundity, and community structure of desert annuals. Ecology 61:1344–1351.

Isley, G. B. 1938. Survival value of acridian protective coloration. Ecology 19: 370–389.

James, C. D., M. T. Hoffman, D. C. Lightfoot, G. S. Forbes, and W. G. Whitford. 1993. Pollination ecology of *Yucca elata*: an experimental study of a mutualistic association. Oecologia 93:512–517.

Joern, A. 1979a. Feeding patterns in grasshoppers: factors influencing diet specialization. Oecologia 38:325–347.

———. 1979b. Resource utilization and community structure in assemblages of arid-grassland grasshoppers. Transactions of the American Entomological Society 105:253–300.

Joern, A., and N. T. Rudd. 1982. Impact of predation by the robber fly *Proctacanthus milberti* (Diptera: Asilidae) on grasshopper (Orthoptera: Acrididae) populations. Oecologia 55:42–56.

Johnson, K. A., and W. G. Whitford. 1975. Foraging ecology and relative importance of subterranean termites in Chihuahuan Desert ecosystems. Environmental Entomology 4:66–70.

Kerley, G. I. H., F. Tiver, and W. G. Whitford. 1992. Herbivory of clonal populations: cattle browsing affects reproduction and population structure of *Yucca elata*. Oecologia 93:112–117.

La Fage, J. P., W. L. Nutting, and M. I. Haverty. 1973. Desert subterranean termites: a method for studying foraging behavior. Environmental Entomology 2:954–956.

Lavigne, R. J., and F. R. Holland. 1969. Comparative behavior of eleven species of Wyoming robber flies (Diptera: Asilidae). University of Wyoming Agricultural Experiment Station Science Monograph 18. Laramie, Wyo.

Lavigne, R. J., L. Rogers, and F. Lavigne. 1976. Ethology of *Efferia benedicti* (Diptera: Asilidae) in Wyoming. Proceedings of the Entomological Society of Washington 78:145–153.

Lightfoot, D. C. 1985. Substrate utilization and guild structure in desert grasshopper assemblages. M.S. thesis, New Mexico State University, Las Cruces.

Lobry de Bruyn, L. A., and A. J. Conacher. 1990. The role of ants and termites in soil modification: a review. Australian Journal of Soil Research 28:55–93.

Londt, J. H. G. 1991. Afrotropical Asilidae (Diptera): observations on the biology and immature stages of *Damilis femoralis* Ricardo 1925 (Trigonomiminae). Annals of the Natal Museum 32:149–162.

Loring, S. J., W. P. MacKay, and W. G. Whitford. 1988. Ecology of small desert playas. Pp. 89–113 *in* J. L. Thames and C. D. Ziebell (eds.), Small water impoundments in semi-arid regions. University of New Mexico Press, Albuquerque.

MacKay, W. P., S. Silva, S. J. Loring, and W. G. Whitford. 1987. The role of subterranean termites in the decomposition of above ground creosotebush litter. Sociobiology 13:235–239.

MacKay, W. P., and W. G. Whitford. 1988. Spatial variability of termite gallery production in Chihuahuan Desert plant communities. Sociobiology 14: 281–289.

Medland, V. L. 1988. Influence of terrestrial vegetation on the production and community structure of a desert playa. M.S. thesis, New Mexico State University, Las Cruces.

Mulkern, G. P. 1980. Population fluctuations and competitive relationships of grasshopper species. Transactions of the American Entomological Society 106:1–42.

Mulkern, G. P., D. R. Toczek, and M. A. Brusven. 1964. Biology and ecology of North Dakota grasshoppers. II. Food habits and preference of grasshoppers associated with the sandhills prairie. North Dakota Agricultural Experiment Station Research Report 11.

Nutting, W. L., M. I. Haverty, and J. P. La Fage. 1987. Physical and chemical alteration of soil by two subterranean termite species in Sonoran Desert grassland. Journal of Arid Environments 12:233–239.

Otte, D., and A. Joern. 1977. On feeding patterns in desert grasshoppers and the evolution of specialized diets. Proceedings of the Academy of Natural Sciences of Philadelphia 128:89–126.

Parker, L. W., P. F. Santos, J. Phillips, and W. G. Whitford. 1984. Carbon and nitrogen dynamics during the decomposition of litter and roots of a Chihuahuan Desert annual, *Lepidium lasiocarpum*. Ecological Monographs 34:339–360.

Pickett, S. T. A., and P. S. White. 1985. The Ecology of Natural Disturbance and Patch Dynamics. Academic Press, New York.

Rees, N. E., and J. O. Onsager. 1985. Parasitism and survival among rangeland grasshoppers in response to suppression of robber fly (Diptera: Asilidae) predators. Environmental Entomology 14:20–23.

Rogers, L. H., and R. J. Lavigne. 1972. Asilidae of the Pawnee National Grasslands in northeastern Colorado. University of Wyoming Agricultural Experiment Station Science Monograph 25.

Scoggan, A. C., and M. A. Brusven. 1973. Grasshopper-plant community associations in Idaho in relation to the natural and altered environment. Melanderia 12:22–33.

Sheldon, J. K., and L. E. Rogers. 1978. Grasshopper food habits within a shrub-steppe community. Oecologia 32:85–92.

Silva, S. I., W. P. MacKay, and W. G. Whitford. 1989. Temporal patterns of microarthropod population densities in fluff grass (*Erioneuron pulchellum*) litter: relationship to subterranean termites. Pedobiologia 33:333–338.

Stanton, N. L. 1988. The underground in grasslands. Annual Review of Ecology and Systematics 19:573–589.

Steinberger, Y., and W. G. Whitford. 1985. Microarthropods of a desert tobosa grass (*Hilaria mutica*) swale. American Midland Naturalist 114:225–234.

Ueckert, D. N., and R. M. Hansen. 1971. Dietary overlap of grasshoppers on sandhill rangeland in northeastern Colorado. Oecologia 8:276–295.

Uvarov, B. 1977. Grasshoppers and Locusts: A Handbook of General Acridology, vol. 2. Centre of Overseas Pest Research, London.

Wallwork, J. A. 1982. Desert Soil Fauna. Praeger, New York.

Watts, J. G. 1963. Insects associated with black grama grass, *Bouteloua eriopoda*. Annals of the Entomological Society of America 56:374–379.

Whitford, W. G. 1972. Jornada validation site. U.S./IBP Desert Biome Research Memorandum 72-4. Utah State University, Logan.

———. 1974. Jornada validation site. U.S./IBP Desert Biome Research Memorandum 74-4. Utah State University, Logan.

———. 1976. Foraging behavior of Chihuahuan Desert harvester ants. American Midland Naturalist 95:455–458.

———. 1978a. Foraging in seed-harvester ants, *Pogonomyrmex* spp. Ecology 59:185–189.

———. 1978b. Structure and seasonal activity of Chihuahuan Desert ant communities. Insectes Sociaux 25:79–87.

———. 1988. Effects of harvester ant, *Pogonomyrmex rugosus*, nests on soils and a spring annual, *Erodium texanum*. Southwestern Naturalist 33:482–485.

———. 1989. Abiotic controls on the functional structure of soil food webs. Biology and Fertility of Soils 8:1–6.

———. 1991. Subterranean termites and long-term productivity of desert rangelands. Sociobiology 19:235–243.

Whitford, W. G., D. J. Depree, P. Hamilton, and G. Ettershank. 1981. Foraging ecology of seed-harvesting ants, *Pheidole* spp., in a Chihuahuan Desert ecosystem. American Midland Naturalist 105:159–167.

Whitford, W. G., and G. Ettershank. 1975. Factors affecting foraging activity in Chihuahuan Desert harvester ants. Environmental Entomology 4: 689–696.

Whitford, W. G., Y. Steinberger, and G. Ettershank. 1982. Contribution of subterranean termites to the "economy" of Chihuahuan Desert ecosystems. Oecologia 55:298–302.

Whitford, W. G., K. Stinnett, and J. Anderson. 1988. Decomposition of roots in a Chihuahuan Desert ecosystem. Oecologia 75:8–11.

Wilcox, J. 1966. *Efferia* Coquillett in America north of Mexico (Diptera: Ascillidae). Proceedings of the California Academy of Sciences, ser. 4, 34: 85–234.

York, J. C., and W. A. Dick-Peddie. 1969. Vegetation changes in southern New Mexico during the past hundred years. Pp. 157–166 *in* W. G. McGinnies and B. J. Goldman (eds.), Arid lands in perspective. University of Arizona Press, Tucson.

7

Diversity, Spatial Variability, and Functional Roles of Vertebrates in the Desert Grassland

Robert R. Parmenter and Thomas R. Van Devender

Vertebrates are common and conspicuous organisms in the Southwest, and they perform many important ecological functions in the desert grassland ecosystem. In this chapter we first address the patterns of species distributions across North America and then compare desert grassland species assemblages with those of other southwestern ecosystems. This is followed by a discussion of how vertebrates participate in the trophic interactions that form the food webs in desert grasslands. In addition, we examine the role of vertebrates in soil turnover, nutrient cycling, and subsequent effects on vegetation that influence the long-term dynamics and perpetuation of desert grasslands.

Spatial Patterns of Vertebrate Species

With regard to vertebrates the American Southwest is one of the most biologically diverse regions in the United States. The Basin and Range topography that characterizes the area provides tremendous variability in elevation, slope, and aspect, forming a wide variety of habitats, which in turn permit the existence of a wide range of species. Desertscrub, desert grassland, or plains grassland cover the valley floors, while interior chaparral or pinyon-juniper and oak woodlands occupy the lower mountain slopes. Ponderosa pine (*Pinus ponderosa*) and mixed conifer forests occur at higher elevations, giving way to subalpine meadows and alpine tundra on the highest peaks. Species richness (number of species per standardized unit area) contour maps on 165–250-km^2 grids have been compiled for mammals (Simpson 1964), birds (Cook 1969), and reptiles (Kiester 1971). The species richness of mammals in the United States peaks in the Southwest and is equaled only in central California. The greatest concentrations of bird species occur along the United States border with Mexico. The number of reptile species found in the Southwest is exceeded only by the number in south-central Texas. Amphibians and fish, which depend more than the others on water resources, reach their species richness maxima in the southeastern United States.

Desert Grassland Vertebrates

Desert grasslands form an important habitat for a large number of southwestern vertebrates—not only the specialists that live exclusively in the grasslands, but also the many species that occupy adjacent riparian zones, desertscrub, and pinyon-juniper woodland. Grasslands intermix extensively with these other biotic communities in a complex mosaic, creating a patchy landscape (Burgess, this volume). Many vertebrates, especially birds and mammals, routinely use shrubs and trees along rivers or on mountainsides for shelter or nest sites but make frequent forays into nearby desert grassland in search of food.

The nature of the desert grassland ecosystem imposes many restrictions on the life forms of its residents. Precipitation is scarce and unpredictable, and temperatures and wind velocities fluctuate greatly on daily, seasonal, annual, and decadal time scales. Protective cover from weather and predators is not abundant. Animals that survive in this environment must have morphological, physiological, and behavioral traits that allow them to tolerate drought, heat, and cold; they avoid predators through fleetness of foot or wing, camouflage, or subterranean escape.

The species richness of the desert grassland vertebrates is best evaluated by comparing lists of the species found in various habitat types from a single region in the Southwest. As an example we have selected a reasonably well known area in Socorro County in south-central New Mexico. The region has been intensively sampled by biologists with the Sevilleta Long-Term Ecological Research Program (LTER) and the Museum of Southwestern Biology at the University of New Mexico. A wide variety of vegetation is present there, including desert grassland, Chihuahuan and Great Basin desertscrub, pinyon-juniper woodland, montane forest and meadows, and riparian forest and wetlands along the Rio Grande.

Mammals. The desert grassland of central New Mexico supports more species of mammals (56) than any other major ecosystem in the region (table 7.1), partially because of the great diversity of rodents found there, particularly ground squirrels (Sciuridae), kangaroo rats (*Dipodomys* spp.), and mice. Rodents (20–32 species) are the dominant mammals in many desert grasslands (table 7.2), where they feed extensively on grasses and forbs as well as on the many insects and other invertebrates present. Grassland specialists include bannertail and Ord kangaroo rats (*Dipodomys spectabilis* and *D. ordi*), the black-tailed prairie dog (*Cynomys ludovicianus*; fig. 7.1A), and the spotted ground squirrel (*Spermophilus spilosoma*). Some mammals, such as porcupine (*Erethizon dorsatum*) and black bear (*Ursus americanus*), are present occasionally but are not normally thought of as grassland species. These animals tend to venture into grassland areas interspersed with mesquite and other shrubs. Elk (*Cervus elaphus*) populations today are generally restricted to montane areas, partly because of the pressures of human hunters and habitat alteration. During Spanish colonial times, however (and today in protected reserves), elk were one of the major seasonal (winter) herbivores in the northern desert grasslands. Similarly, mule deer (*Odocoileus hemionus*) and pronghorn (*Antilocapra americana*; fig. 7.2A) herds were once common in the grassland valleys of the Southwest but were widely reduced or extirpated during the 1800s and early 1900s. Fortunately, rigorous game management policies have restored many populations.

Birds. Bird species are far more numerous in desert grassland than any other group of vertebrates. Some species are occasionally observed in the desert grasslands adjacent to riparian habitats, and many more are clearly migrants and transients. In central New Mexico, the riparian zones along the Rio Grande harbor the greatest variety of bird species, as illustrated by the species list from Bosque del Apache National Wildlife Refuge (table 7.3). The refuge is a 24,000-ha artificial wetland established in 1939 by the United States Fish and Wildlife Service near Truth-or-Consequences in the Rio Grande valley. Each winter, visitors at the refuge are thrilled by the sight of tens of thousands of sandhill cranes (*Grus canadensis*) and snow

Table 7.1. The number of mammal species found in various habitats in Socorro County, central New Mexico.

Family	Habitat Type					
	Desert-scrub	Desert Grassland	Pinyon-Juniper	Montane Forest	Montane Meadow	Riparian Zone
Shrews (Soricidae)	2	1	2	1	2	1
Bats (Molossidae, Vespertilionidae)	9	10	7	12	12	8
Rabbits (Leporidae)	2	2	1	1	1	1
Squirrels (Sciuridae)	4	6	5	5	—	—
Gophers (Geomyidae)	4	4	1	1	1	1
Kangaroo rats, pocket mice (Heteromyidae)	5	5	3	—	—	—
Beaver (Castoridae)	—	—	—	—	—	1
Mice, rats (Arvicolidae, Muridae, Cricetidae, Zapodidae)	9	13	10	8	8	6
Porcupine (Erethizontidae)	1	1	1	1	1	—
Coyote, foxes (Canidae)	3	3	3	2	2	3
Bear (Ursidae)	1	1	1	1	1	1
Raccoon, ringtail (Procyonidae)	1	1	1	1	1	1
Weasels, badger, skunks (Mustelidae)	5	4	5	4	3	4
Cats (Felidae)	2	2	2	2	2	2
Deer, elk (Cervidae)	2	2	2	3	3	2
Sheep (Bovidae)	—	—	1	1	1	—
Pronghorn (Antilocapridae)	1	1	1	—	—	—
Total species	51	56	46	43	38	31

Table 7.2. Rodent species collected in three desert grassland locations in the southwestern United States.

Species	Common Name	Location		
		SEV	PORT	SRER
Cricetidae	New World rats and mice			
Baiomys taylori	Northern pygmy mouse		X	
Neotoma albigula	White-throated packrat	X	X	X
Neotoma micropus	Southern plains packrat	X		
Onychomys arenicola	Chihuahuan grasshopper mouse	X		
Onychomys leucogaster	Northern grasshopper mouse	X	X	
Onychomys torridus	Southern grasshopper mouse	X	X	X
Peromyscus boylei	Brush mouse	X		
Peromyscus eremicus	Cactus mouse	X	X	X
Peromyscus leucopus	White-footed mouse	X		X
Peromyscus maniculatus	Deer mouse	X	X	X
Peromyscus truei	Pinyon mouse	X		
Reithrodontomys fulvescens	Fulvous harvest mouse		X	X
Reithrodontomys megalotis	Western harvest mouse	X	X	X
Reithrodontomys montanus	Plains harvest mouse	X	X	X
Sigmodon arizonae	Arizona cotton rat	X		X
Sigmodon fulviventer	Tawny-bellied cotton rat	X	X	
Sigmodon hispidus	Hispid cotton rat	X	X	
Sigmodon ochrognathus	Yellow-nosed cotton rat		X	
Erethizontidae	Porcupine			
Erethizon dorsatum	Porcupine	X	X	
Geomyidae	Pocket gophers			
Geomys arenarius	Desert pocket gopher	X		
Geomys bursarius	Plains pocket gopher	X		
Pappogeomys castanops	Mexican pocket gopher	X		
Thomomys bottae	Valley pocket gopher	X	X	X
Heteromyidae	Pocket mice, kangaroo rats			
Chaetodipus baileyi	Bailey pocket mouse			X
Chaetodipus hispidus	Hispid pocket mouse		X	
Chaetodipus intermedius	Rock pocket mouse	X	X	
Chaetodipus penicillatus	Desert pocket mouse		X	X

Species	Common Name	Location		
		SEV	PORT	SRER
Heteromyidae (*continued*)				
Dipodomys merriami	Merriam kangaroo rat	X	X	X
Dipodomys ordi	Ord kangaroo rat	X	X	
Dipodomys spectabilis	Bannertail kangaroo rat	X	X	X
Perognathus amplus	Arizona pocket mouse			X
Perognathus flavescens	Plains pocket mouse	X		
Perognathus flavus	Silky pocket mouse	X	X	X
Muridae	Old World rats and mice			
Mus musculus	House mouse (introduced)	X		X
Sciuridae	Squirrels			
Ammospermophilus harrisi	Yuma antelope squirrel		X	X
Ammospermophilus interpres	Texas antelope squirrel	X		
Ammospermophilus leucurus	White-tailed antelope squirrel	X		
Cynomys gunnisoni	Gunnison prairie dog	X		
Spermophilus spilosoma	Spotted ground squirrel	X	X	
Spermophilus tereticaudus	Round-tailed ground squirrel			X
Spermophilus tridecemlineatus	Thirteen-lined ground squirrel	X		
Spermophilus variegatus	Rock squirrel	X		X
Total species		33	23	20

NOTE: SEV = Sevilleta National Wildlife Refuge, Socorro County, New Mexico; PORT = grassland/desertscrub transitional area near Portal in southeastern Arizona; SRER = Santa Rita Experimental Range near Tucson, Arizona.

geese (*Chen hyperborea*). Birds normally associated with wetlands can be found within desert grassland if local water sources such as springs and water developments are available. Wetland birds have also been reported from riparian zones in southeastern Arizona (Strong and Bock 1990).

Birds that live predominantly in desert grasslands include horned lark (*Eremophila alpestris*), lark bunting (*Calamospiza melanocorys*), meadowlarks (*Sturnella magna* and *S. neglecta*), scaled quail (*Callipepla squamata*), and a variety of sparrows such as Baird's (*Ammodramus bairdii*), black-throated (*Amphispiza bilineata*), and Cassin's (*Aimophila cassinii*). The permanent residents must cope with the general lack of vertical structure, and, unlike

Figure 7.1. Many grassland mammals and reptiles, like the common burrowing vertebrates shown here, seek underground refuge to avoid predators or unfavorable environmental conditions. A. Black-tailed prairie dog (*Cynomys ludovicianus*). B. Badger (*Taxidea taxus*). C. Desert grassland box turtle (*Terrapene ornata luteola*). D. Prairie rattlesnake (*Crotalus viridis viridis*).

desertscrub and woodland species, many nest on the ground, in particularly dense clumps of grasses, or in low-growing shrubs (fig. 7.3). As their name implies, burrowing owls (*Athene cunicularia*) nest below the ground. The coloration of desert grassland birds, especially the females, generally matches the browns, grays, and yellows of their surroundings because predator avoidance on the surface of the ground mostly depends on camouflage.

Bird community dynamics in desert grasslands are influenced greatly by the productivity of the vegetation, particularly seeds, which in turn is directly related to climate. Much of the annual rainfall comes during the summer monsoons, and most of the grasses respond with rapid growth during midsummer, setting seed in late summer or early fall (McClaran, this volume). Peaks in resident bird densities generally coincide with maximum seed production in late summer (Maurer 1985). Insect population dynamics also play a role in the timing of bird reproduction because many

Figure 7.2. A. Large ungulates such as the pronghorn (*Antilocapra americana*) are an important part of the desert grassland fauna. B. Newborn pronghorn fawns require the cover of tall grass for several days until they are able to run at full speed. C. Pronghorns create small soil disturbances by kicking out soil before urinating or defecating. D. When an animal dies, the nutrients from its carcass fertilize the adjacent soils, creating an island of lush grasses and forbs.

species utilize insect prey as an important protein source for their young. Many species of northern birds overwinter in the desert grasslands, where temperatures are relatively mild and grass seeds left over from the summer crop are widely available.

Reptiles. The diversity of reptiles is quite similar among the major southwestern ecosystems. The Socorro County desert grassland supports 44 species of reptiles, compared with 45 and 46 species in nearby desertscrub and woodland habitats (table 7.4). Fewer reptiles are found in the higher-elevation montane forests (26 species) and meadows (18 species).

Lizards and snakes are the most abundant reptiles in desert grassland, although the desert grassland box turtle (*Terrapene ornata luteola*) is common in certain areas (fig. 7.1C). Lizards typical of desert grassland sites

Table 7.3. The number of bird species in different locations in Socorro County, New Mexico.

Family	Location		
	SEV	BDA	MAG
Loons (Gaviidae)	—	2	—
Grebes (Podicipedidae)	2	5	—
Pelicans (Pelecanidae)	1	1	—
Cormorants (Phalacrocoracidae)	2	2	—
Bitterns, herons (Ardeidae)	7	11	—
Storks (Ciconiidae)	—	1	—
Ibises (Threskiornithidae)	1	2	—
Swans, geese, ducks (Anatidae)	18	32	—
American vultures (Cathartidae)	1	2	1
Kites, eagles, hawks (Accipitridae)	9	16	6
Caracara, falcons (Falconidae)	4	4	3
Pheasant, turkey, quail, grouse (Phasianidae)	4	5	5
Rails, gallinules, coots (Rallidae)	4	5	—
Cranes (Gruidae)	2	2	—
Plovers (Charadriidae)	2	5	—
Stilts, avocets (Recurvirostridae)	2	2	—
Sandpipers, phalaropes (Scolopacidae)	9	25	1
Gulls, terns, jaegers (Laridae)	3	12	—
Pigeons, doves (Columbidae)	6	2	—
Cuckoos, roadrunner, anis (Cuculidae)	1	3	1
Barn owls (Tytonidae)	—	1	—
Typical owls (Strigidae)	3	9	6
Goatsuckers (Caprimulgidae)	3	4	3
Swifts (Apodidae)	1	2	1
Hummingbirds (Trochilidae)	2	4	6
Kingfishers (Alcedinidae)	1	2	1
Woodpeckers (Picidae)	5	10	7
Tyrant flycatchers (Tyrannidae)	7	17	10
Larks (Alaudidae)	1	1	1
Swallows (Hirundinidae)	6	7	2
Jays, magpies, crows (Corvidae)	5	8	6

Family	Location		
	SEV	BDA	MAG
Titmice, chickadees (Paridae)	2	2	3
Verdins (Remizidae)	1	1	—
Bushtits (Aegithalidae)	1	1	1
Nuthatches (Sittidae)	—	3	4
Wrens (Troglodytidae)	4	9	2
Kinglets, gnatcatchers (Muscicapidae)	2	4	3
Thrushes (Muscicapidae)	4	9	6
Mockingbirds, thrashers (Mimidae)	5	7	1
Pipits (Motacillidae)	—	1	—
Waxwings (Bombycillidae)	—	1	1
Silky flycatchers (Ptilogonatidae)	1	1	—
Shrikes (Laniidae)	2	2	1
Starlings (Sturnidae)	1	1	1
Vireos (Vireonidae)	—	4	2
Wood warblers (Emberizidae)	5	24	12
Tanagers (Emberizidae)	2	4	2
Cardinals, grosbeaks (Emberizidae)	3	7	5
Sparrows (Emberizidae)	14	27	7
Blackbirds, orioles (Emberizidae)	8	14	4
Finches (Fringillidae)	3	9	5
Old World sparrows (Passeridae)	1	1	—
Total species	168	339	122

NOTE: SEV = Sevilleta National Wildlife Refuge (mixed grassland-shrubland-woodland); BDA = Bosque del Apache National Wildlife Refuge (riparian forest and wetlands); MAG = Magdalena Mountains (montane forests and meadows).

include lesser earless (*Holbrookia maculata*), side-blotched (*Uta stansburiana*), southern prairie (*Sceloporus undulatus consobrinus*), and one or more species of whiptail (*Cnemidophorus* spp.) lizards. In dry, gravelly arroyos one can usually find greater earless (*Cophosaurus texanus*) and zebra-tailed (*Callisaurus draconoides*) lizards. In southeastern Arizona and southwestern New Mexico, the Gila monster (*Heloderma suspectum*) is occasionally found in rock outcrops in desert grassland.

Figure 7.3. A. A black-throated sparrow (*Amphispiza bilineata*) nest in snakeweed (*Gutierrezia sarothrae*). B. This sparrow nest contains two nestlings.

Table 7.4. The number of reptile and amphibian species found in various habitats in Socorro County, New Mexico.

	Habitat Type					
	Desert-scrub	Desert Grassland	Pinyon-Juniper	Montane Forest	Montane Meadow	Riparian Zone
REPTILES						
Turtles (Chelydridae, Emydidae, Trionychidae)	1	1	1	—	—	4
Geckos (Gekkonidae)	1	1	1	—	—	—
Alligator lizards (Anguidae)	1	1	—	1	—	1
Lizards (Iguanidae)	13	11	12	6	2	3
Skinks (Scincidae)	2	2	2	2	1	2
Whiptails (Teiidae)	5	6	5	5	2	—
Nonvenomous snakes (Colubridae, Leptotyphlopidae)	16	18	19	9	5	15
Rattlesnakes (Viperidae)	5	5	5	3	3	5
Total reptiles	45	44	46	26	18	32
AMPHIBIANS						
Salamanders (Ambystomatidae)	1	1	1	1	1	1
Toads (Bufonidae)	4	5	4	1	1	5
Spadefoot toads (Pelobatidae)	3	3	1	1	1	3
Treefrogs (Hylidae)	2	2	2	2	1	2
True frogs (Ranidae)	4	4	3	4	3	4
Total amphibians	14	15	11	9	7	15

The horned lizards (*Phrynosoma* spp.) are an interesting group of ant eaters with flattened bodies and (in most species) spiny scales on the body and bony horns on the head. Round-tailed (*P. modestum*) and Texas (*P. cornutum*) horned lizards are common in desert grassland from western Texas to southeastern Arizona. The regal horned lizard (*P. solare*), a common species in tropical and subtropical communities from southern Sonora to Arizona, enters desert grassland in southeastern Arizona. The short-horned lizard (*P. douglassi*), found predominantly in montane woodlands and forests in the Southwest and the Sierra Madre Occidental, locally enters desert grassland at the lower edge of woodlands in Arizona, Chihuahua, New Mexico, and Texas.

Common nonvenomous snakes in desert grassland include the bullsnake or gopher snake (*Pituophis melanoleucus affinis*, and *P. m. sayi*), coachwhips (*Masticophis flagellum cingulum*, *M. f. lineatulus*, *M. f. piceus*, and *M. f. testaceus*), desert grassland kingsnake (*Lampropeltis getulus splendida*), Great Plains ratsnake (*Elaphe guttata emoryi*), trans-Pecos ratsnake (*E. subocularis*), western hognose snake (*Heterodon nasicus kennerlyi*), and western hook-nosed snake (*Gyalopion canum*). The brown vine snake (*Oxybelis aeneus*), green ratsnake (*E. triaspis*), and Mexican hook-nosed snake (*G. quadrangulare*) are tropical species occasionally found in desert grassland in the Atascosa and Santa Rita Mountains of Pima and Santa Cruz Counties, south-central Arizona.

Four venomous snakes are widespread in the desert grassland: Mohave (*Crotalus scutulatus*), prairie (*C. viridis viridis*; fig. 7.1D), and western diamondback (*C. atrox*) rattlesnakes, and the desert grassland massasauga (*Sistrurus catenatus edwardsi*). In addition, the diminutive Arizona coral snake (*Micruroides euryxanthus*), a member of the cobra family (Elapidae), reaches desert grassland in southeastern Arizona and southwestern New Mexico.

Reptiles, like other vertebrates, partition habitats according to their individual needs for food and shelter, although competition between various species may also influence their relative abundance at any given site. For example, in Hidalgo County, southwestern New Mexico, desert spiny (*Sceloporus magister*), side-blotched, tree (*Urosaurus ornatus*), and western whiptail (*Cnemidophorus tigris*) lizards prefer habitats with greater densities of shrubs; other species, including the greater earless, leopard (*Gambelia wislizeni*), round-tailed horned, and zebra-tailed lizards, prefer more open areas with few shrubs (Baltosser and Best 1990).

Mitchell (1979) examined habitat, food, and behavioral factors to determine how four very similar species of whiptails could coexist in southeastern Arizona and found that each species used a slightly different part of the habitat. Little striped whiptail lizards (*Cnemidophorus inornatus*) inhabited grass-dominated areas, while western whiptails preferred areas dominated by mesquite. Two all-female (parthenogenetic) species were found in the transition zones between habitats: desert grassland whiptails (*C. uniparens*) inhabiting the ecotone between grass- and mesquite-dominated areas, and Sonoran whiptails (*C. sonorae*) occupying the ecotone between mesquite habitats and Arizona Upland Sonoran desertscrub.

Asexual reproduction is relatively uncommon in vertebrates and is not found at all in mammals and birds. Some 10 species of whiptail lizards are the only southwestern reptiles that venture outside the usual sexual patterns of reproduction (Cole 1984). All the individuals of these species are female and reproduce by an autofertilization method called parthenogenesis. Referring to all-female whiptail populations as species, even if they are morphologically distinct and have large geographic ranges, presents conceptual problems because all the individuals are essentially isolated from each other with no exchange of genetic material. These unisexual species are polyploids (they have more than one set of chromosomes) that were produced through hybridization of two sexual species. For example, the bisexual parents of the all-female New Mexican whiptail (*C. neomexicanus*) are thought to have been the diminutive striped and the larger western whiptails. Such hybrid species are the result of evolutionary shortcuts, not natural selection. Each individual in the species is genetically identical to the original hybrid offspring.

All-female whiptail species typically occupy transitional ecotones between the habitats where their parent species occur. Wright and Lowe (1968) called such groups "weeds." In any event, the desert grassland is an evolutionary center for all-female whiptails. Seven species are mostly or completely restricted to this habitat. Checkered (*C. tesselatus*), Chihuahuan (*C. exanguis*), and New Mexican whiptails are common in desert grassland in Texas and New Mexico, while desert grassland and Sonoran whiptails are more common in southeastern Arizona. The Gila spotted whiptail (*C. flagellicaudus*) is common in desert grassland–interior chaparral mosaics below the Mogollon Rim in central Arizona. The plateau whiptail (*C. velox*) lives in Great Basin grasslands on the Colorado Plateau above the Mogollon Rim.

Amphibians. Desert grassland is not usually thought of as typical habitat for amphibians. Certain toads, however, including Couch's (*Scaphiopus couchi*), plains (*S. bombifrons*), and western (*S. hammondi*) spadefoot toads, and green (*Bufo debilis*), Great Plains (*B. cognatus*), and southwestern Woodhouse's (*B. woodhousei australis*) toads, can be quite common (table 7.4). The true frogs, such as Chiricahua (*Rana chiricahuensis*), lowland (*R. yavapaiensis*), and plains (*R. blairi*) leopard frogs, and the introduced bullfrog (*R. catesbeiana*), are generally limited to permanent water, often in artificially developed water sources.

Occasionally one finds tiger salamanders (*Ambystoma tigrinum*) in desert grassland water developments as well, although their presence has oftentimes been facilitated by humans. Tiger salamanders living in isolated ponds may exhibit a condition known as neoteny, in which the animal becomes mature at an earlier stage in life than usual. The salamanders reproduce as aquatic larvae and may never transform into terrestrial adults. In some cases, neotenic larvae develop large heads and feed on other larvae rather than algae.

Some form of permanent or ephemeral water must be present to facilitate amphibian reproduction in the desert grassland. Livestock water developments and ponds are reliable water sources that are readily colonized by amphibians. In addition, summer thunderstorms routinely fill small playas and pools with water. When this happens, spadefoot toads, stimulated by the low-frequency sound of the raindrops striking the earth (Dimmitt and Ruibal 1980), come forth from their underground burrows and converge on these ponds to mate. The larvae must develop rapidly, as the tadpoles are literally in a race against time; the playas and pools will soon dry up in the desert heat. As with tiger salamanders, some tadpoles develop unusually large heads and mouths and become cannibalistic, feeding on their fellow tadpoles rather than on algae. As the water temperature rises, the tadpoles' physiological processes accelerate and growth becomes more rapid (Zweifel 1968). If the water remains long enough, the tadpoles will successfully metamorphose into young toadlets, leave the water, and begin terrestrial life.

Trophic Interactions

Desert grassland plants, vertebrates, and invertebrates interact in a variety of ways, including herbivory, granivory, scavenging, and predation. Trophic interactions and other, abiotic variables drive the nutrient cycles and energy flow in the ecosystem. Populations rise and fall in concert with food resources, the abundance of prey and/or predators, and changes in the abiotic environment. The reproductive efforts of many species coincide with the maximum production of food; this in turn depends on the amount and timing of precipitation, which can be extremely variable (McClaran, this volume). Droughts and frosts have dramatic and sometimes devastating effects on plant and animal populations.

Vertebrate herbivores found in desert grasslands include elk (central New Mexico), javelina (*Tayassu tajacu*), black-tailed prairie dog, cotton rats (*Sigmodon* spp.), ground squirrels (*Ammospermophilus* spp., *Spermophilus* spp.), various mice and packrats (*Neotoma* spp.), rabbits (*Lepus* spp., *Sylvilagus* spp.), mule deer, pronghorn (fig. 7.2A), and domestic livestock (cattle, burros, goats, horses, and sheep). Other species such as the omnivorous desert grassland box turtle (fig. 7.1C) consume plants and small insects. Even belowground plant parts are not safe because Mexican (*Pappogeomys castanops*) and valley (*Thomomys bottae*) pocket gophers excavate long feeding tunnels in search of roots, bulbs, and corms. In summer droughts, javelina and jackrabbits (*Lepus alleni* and *L. californicus*) feed on the shallow, succulent roots of cacti and shrubs. In spite of their abundance, native herbivores generally harvest only a small percentage of the total plant production in any given year. In years when animal populations are high, however, vegetation consumption can increase markedly. On the Santa Rita Experimental Range in southern Arizona, desert cottontails (*Sylvilagus auduboni*) and jackrabbits can consume as much as 40 percent of the forage production (Reynolds and Martin 1968; Voorhies and Taylor 1933).

The most profound impacts on southwestern grasslands are exerted by black-tailed and Gunnison (*Cynomys gunnisoni*) prairie dogs. Vernon Bailey, chief naturalist of the United States Biological Survey from 1887 to 1924, had extensive experience with prairie dogs in New Mexico (Bailey 1931). Shortly after the turn of the century, prairie dog towns were estimated to have covered 2.5 million ha of choice grazing land in the state. Because 32 prairie dogs consume as much grass as one sheep, and 256 prairie dogs eat as much as one cow, prairie dogs were perceived by ranchers to be a serious threat to their herds. As a result, mostly during the 1930s, prairie dogs were exterminated in vast areas of Arizona and New Mexico as part of massive poisoning campaigns by the United States Predator and Rodent Control, a Biological Survey program that used strychnine-laced rolled oats for population control (B. Van Pelt, Arizona Department of Fish and Game, pers.

comm., 1994; fig. 7.4). The last surviving black-tailed prairie dog colonies in southwestern New Mexico near Silver City were poisoned in the 1960s (R. Wallace, New Mexico State Heritage Program, pers. comm. 1994).

The black-tailed prairie dog was once present in desert grasslands from trans-Pecos Texas to northwestern Chihuahua and southeastern Arizona. The Arizona prairie dog (*Cynomys ludovicianus arizonensis*) was found from southwestern New Mexico as far west as the Huachuca and Patagonia Mountains in Arizona (Hoffmeister 1986). In 1885, Edgar A. Mearns observed "immense colonies of Arizona prairie dogs in the region contiguous to the Southern Pacific Railroad in southeastern Arizona, extending as far west as the town of Benson on the San Pedro River" (Mearns 1907). Prairie dogs along the upper San Pedro River and in the Sulphur Springs Valley were poisoned from 1926 until 1931 (Brandt 1951; Hoffmeister 1986). The last two Arizona prairie dogs in southeastern Arizona were collected in 1938, 10 km southeast of Fort Huachuca, by Charles T. Voorhies, a zoologist at the University of Arizona (B. Van Pelt, Arizona Department of Game and Fish, pers. comm., 1994). In 1966 and 1974, the Arizona Game and Fish Department attempted unsuccessfully to reintroduce the black-tailed prairie dog onto Fort Huachuca and the Meadow Valley Flat in the San Rafael Valley. At the time of this writing the department is investigating the possibility of reintroducing them on Fort Huachuca.

Bailey (1931) wrote an early account of the effects of Arizona prairie dogs in southwestern New Mexico:

> In August, 1908, on a trip from Deming to Hachita and through the Playas and Animas valleys, the writer found the prairie dogs numerous in many localities, especially along the elevated and more open margins of the valleys, but extensive colonies were also seen in the bottoms of the great desert valleys. Animas Valley was an almost continuous prairie dog town for its whole length and breadth. In many places where rain had missed a part of the valley the prairie dogs had taken all the season's vegetation and had made barren deserts miles in extent. . . . Ten prairie dogs to an acre were estimated as a fair allowance for their numbers in the colonies, and apparently these colonies covered at least a third of Grant County, or approximately 1,000 square miles, which would give 6,400,000 prairie dogs to one county. In this arid region, one is usually struck by the extreme bareness of the prairie-dog towns, for the vegetation has been killed out for a considerable distance from the burrows. In places, the distance from the burrows to the grassland has become too great for safety to the animals, and the towns have been abandoned and new pastures sought by the colonies. It is often necessary in the arid regions for the prairie dogs to adapt themselves to unusual kinds of food, such as cactus, and the bark of mesquite

Figure 7.4. A total of 1,641 Gunnison prairie dogs (*Cynomys gunnisoni*) were poisoned using strychnine-laced rolled oats on 7 September 1917 by Agents D. A. Gilchrist and D. Stonier of the U.S. Biological Survey on the north boundary of Coconino National Forest in Arizona. Photograph property of Prescott National Forest in the Sharlot Hall Museum Library/Archives, Prescott, Arizona.

> and other available bushes growing within reach of their homes. . . . The colonies are generally in the best grazing localities, and in many places on the stock range, the grass has been greatly thinned or entirely destroyed over miles of country. Where the prairie dogs are forced to move to new pastures, the old prairie-dog towns grow up to worthless weeds or equally worthless foxtail grass, which furnishes forage for neither prairie dogs nor cattle.

Although the Arizona prairie dog had been exterminated from the Animas and Playas Valleys by the early 1950s (Findley et al. 1975), extensive colonies similar to those described by Bailey still occur in desert grassland near Janos in northwestern Chihuahua, Mexico. Colonies occur within 2 km of the New Mexico border near Antelope Wells. The Loma Los Ratones (Chihuahua) prairie dog town surveyed in 1985 by Yar Petryszyn and Joel S. Brown of the University of Arizona covered an area of 56 km^2. The population was estimated at 172,000 adults and young in May and 117,000 adults in November; that is, 2,089–3,014 animals per square kilometer. Considering that adults weigh about a kilogram, the biomass of prairie dogs (ca.

2,090 kg/km^2) in desert grassland exceeded that of livestock (800 kg/km^2) but not termites (3,600 kg/km^2) (Whitford et al., this volume). The soil surface was highly disturbed within the Los Ratones colony, and there were 71.6 entrances to underground burrow complexes per hectare. The vegetation had essentially been removed by prairie dog activities. On nearby bajadas, dense honey mesquite (*Prosopis glandulosa*) had invaded the desert grassland, possibly with the aid of bannertail kangaroo rat seed caches. These observations bring to mind the intriguing possibility that expansions of honey mesquite into desert grasslands since 1930 may have been in part related to the removal of natural control by prairie dogs.

An extinct bison (*Bison antiquus*) with massive widespread horns was widespread in Arizona and Sonora, Mexico, during the last ice age, the Wisconsin glacial period (Agenbroad and Haynes 1974). The modern bison (*B. bison*) appears to have evolved from its late Pleistocene ancestor by a simple reduction in body size. Populations of bison have fluctuated dramatically during the 11,000 thousand years of the present interglacial (the Holocene) and probably were at their peak in the Great Plains when Europeans first arrived in the sixteenth century.

A prolonged drought in the mid-nineteenth century, in concert with greater hunting pressures from Native Americans and American settlers and exacerbated by increasing competition with domestic livestock (notably horses and cattle) for riparian winter grazing lands, brought the bison to the brink of extinction. The last confirmed hunting record of native bison in eastern New Mexico was in 1884 (Findley et al. 1975). Bison have been introduced into Arizona and New Mexico since the early 1900s (Findley et al. 1975; Hoffmeister 1986). Today, there are approximately 30,000 bison in North America, and several bison herds now exist in the desert grasslands of the southwestern United States.

Spanish and American explorers in the Southwest failed to find bison in desert grassland west of the Pecos River in eastern New Mexico (Bailey 1931; Findley et al. 1975); however, the archaeological record indicates their presence in southern Arizona. The River People of the Hohokam were farmers who constructed large-scale irrigation systems along the Gila and Salt Rivers above their junction and south along the Santa Cruz River into Tucson (Haury 1976). Bison horn cores, teeth, and bones were recovered in the excavation of Snaketown, a Hohokam settlement in the Gila River valley in Pinal County south of Phoenix that was occupied from before the birth of Christ until A.D. 1200 (Haury 1937). Bison bones associated with the years A.D. 1200 and 1380 were identified from two rooms in a 1968 excavation of the Hohokam Las Colinas site in Maricopa County near Phoenix (Johnson 1981). If the bones represent animals that lived in the area rather than material transported there from elsewhere, then bison and saguaro (*Carnegiea gigantea*) would have been found together between Phoenix and Tucson at that time.

Bison bones, some of them painted, found in a Babocomari Village excavation were dated to A.D. 1200–1450 (Di Peso 1951). The Babocomari River, a tributary of the San Pedro River, flows through desert grassland on the north end of the Huachuca Mountains east of Elgin in Cochise County, Arizona. In the 1970s, a bison skull was found eroding out of sediments in the same area (K. B. Moodie, University of Arizona, pers. comm. 1994).

Murray Springs, a site in desert grassland in the San Pedro River valley near Sierra Vista on the eastern base of the Huachucas in Cochise County, is a well-known Paleo-Indian site. The site is best known for the dramatic recovery in the 1970s of in situ Clovis spear points among mammoth bones in the 11,000-year-old lower sediments. Bones of extinct bison were found in the same deposit. This desert grassland site was likely dominated by grasses prior to 1890 but is presently covered with velvet mesquite (*Prosopis velutina*). Incidental to the Paleo-Indian excavation, the skeleton of a female bison with a near-term fetus inside was discovered (Agenbroad and Haynes 1974) and radiocarbon-dated to A.D. 1700. Not only were bison part of the desert grassland fauna, then, they were present at least into the seventeenth century! Johnson (1981) postulated that the bison expanded its range into Arizona from the desert grassland valleys of northwestern Chihuahua during cooler, moister climatic fluctuations. We can only wonder why bison disappeared from desert grasslands before European trappers, mountain men, and soldiers began recording their observations of wildlife in southeastern Arizona in the 1820s (Davis 1982).

Granivores, or seed eaters, have an important influence on the vegetation composition of desert grasslands. Granivorous rodents include bannertail, Merriam (*D. merriami*), and Ord kangaroo rats; desert (*Chaetodipus penicillatus*), hispid (*Perognathus hispidus*), and silky (*P. flavus*) pocket mice; certain species of opportunistic omnivorous rodents, especially deer mice (*Peromyscus* spp.), grasshopper mice (*Onychomys arenarius, O. leucogaster,* and *O. torridus*), and harvest mice (*Reithrodontomys megalotis* and *R. montanus*); and spotted ground squirrels. Seed consumption in the desert grassland can be high—up to 95 percent of the total annual seed crop of some species (Borchert and Jain 1978; Chew and Chew 1970; Nelson and Chew 1977; Soholt 1973). In addition to rodents, granivorous birds (larks, sparrows, wrens) consume large quantities of seeds (Pulliam and Brand 1975), particularly during the winter months when insects are rare. In fact, the pressure of bird granivory on the available seed population may exceed that of rodents and harvester ants (Parmenter et al. 1984).

A set of experiments conducted in southeastern Arizona clearly illustrates the importance of granivores in desert grassland (Brown and Heske 1990; Heske et al. 1993). When kangaroo rats were removed for 13 years from a site dominated by Chihuahuan Desert shrubs near Portal in southeastern Arizona, annual and perennial grasses became dominant (fig. 7.5). Apparently, seed selection by the rodents and the physical soil disturbances caused by their foraging activities inhibit grasses. Kangaroo rats prefer larger grass and forb seeds, and they collect vast quantities of them. In addition, the rodents' nightly digging and caching activities disturb the soil, which acts to favor annual and perennial forbs. When kangaroo rats were removed, more grass seeds survived to germinate, and they grew with less competition with short-lived forbs for water, nutrients, and space. Brown and Heske (1993) called the kangaroo rats "keystone species" because of the importance of their foraging activity in regulating the plant composition of these desert ecosystems. In this particular study, the dominant invading plants were six-weeks threeawn (*Aristida adscensionis*), an annual, and Lehmann lovegrass (*Eragrostis lehmanniana*), a naturalized perennial originally from southern Africa. Lehmann lovegrass is an aggressive, adventive species that has invaded a number of other desert grassland areas in southeastern Arizona (Anable et al. 1992). Similar studies are now being conducted in New Mexico to further ascertain the effect of kangaroo rats on native grasses and shrubs.

Important predators in the desert grassland include the coyote (*Canis latrans*), hawks, owls, and snakes, along with the insectivorous desert shrew (*Notiosorex crawfordi*), grasshopper mouse, lizards, and toads. Local distributions and numbers of predators can often be linked to prey abundance. For example, studies of coyote and jackrabbit populations in the Great Basin desert showed that jackrabbit abundance rises and falls on a decadal cycle, and that coyote populations track the jackrabbit cycle with a one- to two-year lag time (Stoddart 1987a,b). Predators are also known to actively regulate prey populations. Range caterpillar populations have been reduced by 50 percent by rodent predators (Bellows et al. 1982), as have various darkling (Tenebrionidae) and ground beetles (Carabidae) (Parmenter and MacMahon 1988a,b); and birds have the same effect on grasshoppers (Joern 1986).

Figure 7.5. Kangaroo rats (*Dipodomys* spp.) are keystone species in desert grasslands and desertscrub. A. The area left of the fence, which has gates to allow entry of kangaroo rats, has not been invaded by grasses. B. An area from which kangaroo rats have been excluded (left side of fence) is being invaded by tall grasses.

Carrion is another important food source in desert grasslands. Turkey vultures (*Cathartes aura*), well-known scavengers on decaying carcasses, locate dead animals via their keen sense of smell and excellent eyesight. Chihuahua and common ravens (*Corvus corax* and *C. cryptoleucus*) can be seen patrolling southwestern highways for roadkills during the early morning hours before the rising thermals allow their competitors, the turkey vultures, to forage. Golden eagles (*Aquila chrysaetos*) are occasionally seen scavenging carcasses on roads as well. Other vertebrates, such as badgers (*Taxidea taxus*; fig. 7.1B) and coyotes, feed on carrion opportunistically but tend to avoid badly decomposed flesh. Vertebrate carrion is generally a more important food source for invertebrates (Whitford et al., this volume) than for vertebrates.

Vertebrates and Soils

In addition to trophic interactions with plants and other animal species, vertebrates influence grassland soils by digging and burrowing, or simply by tracking across the soil surface (fig. 7.6). The mixing of subsurface materials with surface soils, plant litter, and animal feces fertilizes and aerates the soil, benefiting some plants and soil microorganisms. Pocket gopher activity is perhaps the best-known example of this phenomenon. In addition, animal burrows carry oxygen deep into the ground, aerating the soil around roots. Burrow systems also enhance infiltration of water into soils so that the critical plant root zones receive moisture during intense summer thunderstorms.

Soil disturbances influence the plant species composition of the desert grassland. Many plants need disturbed sites for germination and growth, and the mounds of several vertebrate species (badger, ground squirrels, kangaroo rats, pocket gophers, and prairie dogs) fill these requirements very well. Bannertail kangaroo rats create large mounds (5–10 m in diameter) on which a variety of annual and perennial forbs grow (fig. 7.7). The mounds are inherited by subsequent generations and exist for many decades in the grassland landscape. An analysis of kangaroo rat mounds observed in aerial photographs of New Mexico desert grasslands taken in 1935 and 1984 revealed that 69 percent of the 532 mounds were still visible and presumably still occupied by kangaroo rats after 50 years (B. Musick, University of New Mexico, pers. comm., 1994). The continual occupation of these mounds results in constant soil disturbance that promotes the

Figure 7.6. Burrowing animals have many direct and indirect effects on the grassland's soils and vegetation. A. Rodent burrows aerate soils, alter underground moisture regimes, and mix subsurface and surface soil layers. B. Ord kangaroo rats (*Dipodomys ordi*) and silky pocket mice (*Perognathus flavus*) cache seeds in shallow soil pits, not only creating soil disturbances but also facilitating seed germination. C. Valley pocket gophers (*Thomomys bottae*) tunnel extensively below the surface, building large surface mounds with the excavated earth. D. Animals such as the coyote (*Canis latrans*) leave tracks and scats that break up the soil surface and locally fertilize the soil.

growth of many species of annual plants (Moroka et al. 1982). There can be up to six bannertail kangaroo rat mounds per hectare in suitable habitat, making the mounds an important and widespread feature of the desert grasslands.

The profound soil disturbances in desert grassland valleys wrought by prairie dogs before 1930 were discussed above.

In addition to influencing the plant community, vertebrate burrows provide excellent shelter for a wide variety of other animals. Bannertail kangaroo rats create labyrinths of tunnels beneath their mounds in which cli-

A

B

C

matic conditions are ameliorated and food resources (microarthropods, feces, fungi, seeds, and plant leaves and stems) are concentrated. The burrows are focal points for foraging and mating, thereby enhancing the overall faunal diversity of the desert grassland. Burrow systems are preferentially used by such grassland vertebrates as the desert grassland massasauga, Great Plains skink (*Eumeces obsoletus*), and side-blotched lizard, along with many arthropods including click beetles (Elateridae), darkling beetles, desert cockroaches (*Arenivaga* spp.: Polyphagidae), and sun spiders (Solifugae) (Hawkins and Nicoletto 1992). Other southwestern reptiles commonly found in kangaroo rat burrows include bullsnakes, coachwhips, Mohave, prairie and western diamondback rattlesnakes, western hognose snakes, and various whiptail lizards.

Soil nutrients such as calcium, nitrogen, phosphorus, and potassium are important in determining the overall productivity of the vegetation, and animals contribute to the availability of nutrients not only through soil tillage but also through fecal deposition and death. Some of the nutrients originally consumed by the animals pass through their bodies and are returned to the soil via feces and urine, creating fertility islands (figs. 7.2 and 7.7). Animal feces deposited on the soil surface are decomposed by bacteria, fungi, insects (especially termites; see Whitford et al., this volume), and physical processes including freeze-thaw cycles, raindrop impact, and wind movements. While the overall importance of vertebrate feces in grasslands may be small, local effects on individual plants can be significant.

Vertebrate carrion is an extremely nutrient-rich resource for foraging arthropods, particularly flies and beetles. After feeding and breeding on the carcass, these wide-ranging insects disperse the carcass's nutrients across the landscape. As the carcass decomposes, certain nutrients—especially nitrogen, potassium, sodium, and sulphur—become available to soil microorganisms and plants. Plant growth in the highly fertilized vicinity of vertebrate carcasses can be extremely lush (fig. 7.2D). The minerals associated with bone, such as calcium and phosphorus, reach the system more slowly and probably have little impact on the nutrient status of surrounding soils

Figure 7.7. (*Opposite page*) Mounds made by bannertail kangaroo rats affect the desert grassland flora. A. A bannertail kangaroo rat (*Dipodomys spectabilis*). B. Bannertail kangaroo rat nest mound in a desert grassland site. Note the differences in vegetation on the mound compared with the grassland areas in the background. C. Close-up of mound showing large numbers of snakeweed (*Gutierrezia sarothrae*) and tumbleweed (*Salsola australis*) plants growing on it.

and vegetation. However, some gnawing vertebrates, notably rabbits and rodents, use bones as a source of calcium.

Fire and Vertebrates

Fires are natural disturbances that act to stabilize desert grasslands, although their frequency and intensity may be altered by human activity (McPherson, this volume). Unlike less mobile invertebrates, few vertebrates are killed during rapidly moving desert grassland fires. They escape the flames and smoke by entering underground burrows, running, or flying. Periodic fires, although superficially destructive, actually help maintain grasslands by removing encroaching shrubs and cacti. Many woody plants and succulents resprout following fire, however, and a high frequency of fire may be needed to maintain a grass-dominated desert grassland (Burgess, this volume; McPherson, this volume).

Fires also cycle many important nutrients back into the soil (McPherson, this volume). A sudden flush of fresh, new plant growth invariably follows a grassland fire, and this in turn attracts vertebrates. Larger, more mobile herbivores, such as pronghorn and mule deer, readily locate burned areas and browse on the new shoots of forbs and shrubs that grow there. On the other hand, animals such as cotton rats and javelina that use areas of tall grass (e.g., alkali sacaton, *Sporobolus airoides*) for cover may be displaced until the grasses regain their full height (Bock and Bock 1978, 1979). Grasses and forbs are quick to respond to the increase in available nutrients after a fire (McPherson, this volume), and herbivorous mammals (deer, prairie dogs, pronghorn, and rabbits) and granivorous birds and rodents benefit from the enhanced production in burned grasslands.

Humans and Grassland Invertebrates

Humans have interacted with vertebrate populations in the southwestern United States for a very long time. The early indigenous peoples were hunter-gatherers who may have exerted considerable influence on the abundance of their prey species. Some paleoecologists have suggested that overhunting by Paleo-Indians was responsible for the extinctions of many large mammals at the end of the Ice Age, about 11,000 years ago (Martin 1973). Ironically, a number of widespread herbivores—including camel (*Camelops hesternus*), flat-headed peccary (*Platygonus compressus*), horses (*Equus* spp.), mammoth (*Mammuthus columbi*), mountain deer (*Navahoceros fricki*), pronghorns (*Capromeryx* spp., *Stockoceros onusroagris*), Shasta ground sloth (*Nothrotheriops shastense*), shrub ox (*Euceratherium collinum*), and tapir (*Tapirus* sp.)—disappeared just as Ice Age forests and woodlands retreated and Great Plains grasslands and southwestern desert grasslands developed (Martin 1975; Van Devender, this volume). The use of fire by early inhabitants to modify and manage desert grassland habitats would have enhanced the processes that occur naturally in lightning-caused fires (McPherson, this volume).

With the development of permanent settlements and agriculture in the last few thousand years came changes in vertebrate diversity. While certain species were continually hunted for food and clothing, others apparently benefited from the presence of humans. Emslie's (1981) archaeological study of bird bones indicated that the diversity of bird species in desert grassland near New Mexican pueblos was actually greater hundreds of years ago than it is today. The bones were not only from common grassland and riparian birds but also from species that no longer occur in the region. The greater diversity was attributed to the Pueblo peoples' agricultural systems, which (1) provided "accessible, nonrestricting, and uniform habitat of highly edible plants" and large numbers of insects; (2) broke down the "natural habitat barriers in disturbed areas," especially along irrigation ditches, to allow successful colonization of new areas; and (3) established ecotones between agricultural fields and desert grasslands and desertscrub with more varied habitat plant architecture and greater plant diversity (Emslie 1981).

More recent human developments in desert grasslands have resulted in large-scale changes in wildlife populations and communities (fig. 7.8). The introduction of livestock and the growth of the livestock industry over the last century altered the character of many desert grasslands by contributing to their conversion to desertscrub (Bahre, this volume; Burgess, this volume; Roundy and Biedenbender, this volume), and this in turn led to changes in the vertebrate assemblages there. For example, in former desert grasslands near Portal, Arizona, an invasion of mesquite over the last 30 years was correlated with a reduction or elimination of glossy snake (*Arizona elegans*), desert grassland massasauga, and Mohave and prairie rattlesnake populations (Mendelson and Jennings 1992). In contrast, nine snake species more typical of desertscrub became more common there. Quantitative analyses of bones from Howell's Ridge Cave, a desert grassland area in

Figure 7.8. Domestic livestock can affect desert grasslands. A. Livestock grazing in desert grasslands. B. Fenceline contrasts of grazed (left) and ungrazed (right) rangelands illustrate the influence of grazing on vegetation cover. C. Fresh cattle feces. Nutrient cycles in grasslands may be altered by the slowly decomposing fecal materials. D. Cattle feces after 21 months.

southwestern New Mexico, recorded similar increases in desert reptiles during previous shrub expansions about 3,900, 2,500, and 990 years ago (Van Devender, this volume).

Recent efforts to reestablish grassland wildlife in restored grass-dominated habitats have met with limited success. For example, on the Santa Rita Experimental Range near Tucson, velvet mesquite was removed from several pastures by chaining in 1976. The cleared sites were compared with an unchained mesquite-dominated site and a grass-dominated site restored in 1955 by killing the mesquite with diesel fuel (Germano and Hungerford 1981; Germano et al. 1983). The grassland area restored in 1955 had the lowest abundance and fewest species of mammals, birds, and reptiles. This result suggests that many vertebrates prefer some degree of shrub cover and that a mosaic of shrub and grass habitats maximizes wildlife productivity.

The impact of introduced plants and animals on desert grasslands is perhaps best demonstrated by the practice of rehabilitating overgrazed rangelands with nonnative grasses. This management technique, initiated during the 1930s, often uses two species of perennial grasses native to southern Africa, Lehmann and Boer (*E. curvula* var. *conferta*) lovegrasses. They have been successfully established throughout the southwestern United States, but the replacement of native mixed-species grasslands with stands of nonnative lovegrasses has also resulted in changes in the desert grassland flora and fauna. For example, on the Appleton-Whittell Research Sanctuary in southeastern Arizona, the abundances of 26 species (5 birds, 3 rodents, 8 grasshoppers, and 10 plants) was significantly lower in areas dominated by lovegrasses than in native grasslands (Bock et al. 1986). The only vertebrate species that increased in the nonnative grass stands were the Arizona cotton rat (*Sigmodon arizonae*) and Botteri's sparrow (*Aimophila botterii*), both of which prefer tall-grass cover. In central Arizona, breeding bird densities in 40-ha plots dropped from 103 pairs in 9 species in chaparral to 24 pairs in 4 species in former chaparral that had been cleared and replanted with lovegrasses (Szaro 1981). However, the seeded grass areas supported breeding pairs of rock wren (*Salpinctes obsoletus*) and rufous-crowned sparrow (*Aimophila ruficeps*), which were not found in the chaparral. It remains to be seen whether the spread of nonnative grasses will constitute a major regional threat to desert grassland wildlife.

Summary

Some of the most species-rich assemblages of vertebrates in the United States are found in the Southwest, and a considerable proportion of these are in desert grasslands. In terms of species numbers, the desert grassland supports as many or more species as the other ecosystem types in the area with the exception of Arizona Upland desertscrub. Desert grassland vertebrates include many habitat specialists as well as more wide-ranging species. Herbivorous vertebrates influence the structure, composition, relative abundance, and productivity of plant species. Burrowing vertebrates also affect the species composition of the animal community by providing relatively safe underground shelter to a variety of other vertebrates and invertebrates. Vertebrates may also cause substantial soil disturbances, and their feces and urine (and carcasses, when they die) influence nutrient cycles. Some vertebrates appear to be well adapted to fire and human activities, while others decline in numbers or are eliminated by them. The introduction of nonnative plant species into the desert grassland results in shifts in the vertebrate community structure and composition as a result of qualitative and structural changes in the vegetation.

Acknowledgments

Support for the preparation of this chapter was provided by the University of New Mexico's Sevilleta LTER Program (contribution no. 36). James H. Brown, University of New Mexico, kindly provided the photographs for figure 7.5. Yar Petryszyn, University of Arizona, provided the list of rodents on the Santa Rita Experimental Range presented in table 7.2 and unpublished data on the black-tailed prairie dog population near Janos, Chihuahua. Bill Van Pelt, Arizona Game and Fish Department, and Roby Wallace, New Mexico Heritage Program, provided information on prairie dogs in Arizona, Chihuahua, and New Mexico. Harley G. Shaw, formerly of the Arizona Game and Fish Department, and Mona McKloskey, of the Sharlot Museum, provided the prairie dog photograph in figure 7.4. Larry D. Agenbroad, Northern Arizona University, and Paul C. Johnson, Pima Community College, provided information and literature on the archaeological record of bison in Arizona. Seymour Levy shared his knowledge of desert grassland and Sonoran Desert birds. George L. Bradley, University of Arizona, provided information on the ecology of the all-female whiptail lizards in central Arizona. Sandra L. Lehman processed the manuscript.

Literature Cited

Agenbroad, L. D., and C. V. Haynes. 1974. *Bison bison* remains at Murray Springs, Arizona. Kiva 40:309–313.

Anable, M. E., M. P. McClaran, and G. B. Ruyle. 1992. Spread of introduced Lehmann lovegrass (*Eragrostis lehmanniana* Nees.) in southern Arizona, USA. Biological Conservation 61:181–188.

Bailey, V. 1931. Mammals of New Mexico. North American Fauna 53. U.S. Department of Agriculture, Washington, D.C.

Baltosser, W. H., and T. L. Best. 1990. Seasonal occurrence and habitat utilization by lizards in southwestern New Mexico. Southwestern Naturalist 35: 377–384.

Bellows, T. S., Jr., J. C. Owens, and E. W. Huddleston. 1982. Predation of range caterpillar, *Hemileuca oliviae* (Lepidoptera: Saturniidae), at various stages of development by different species of rodents in New Mexico during 1980. Environmental Entomology 11:1211–1215.

Bock, C. E., and J. H. Bock. 1978. Response of birds, small mammals, and vegetation to burning of sacaton grassland in southeastern Arizona. Journal of Range Management 31:296–300.

———. 1979. Relationship of the collared peccary to sacaton grassland. Journal of Wildlife Management 43:813–816.

Bock, C. E., J. H. Bock, K. L. Jepson, and J. C. Ortega. 1986. Ecological effects of planting African lovegrasses in Arizona. National Geographic Research 2:456–463.

Borchert, M. I., and S. K. Jain. 1978. The effect of rodent seed predation on four species of California annual grasses. Oecologia 33:101–113.

Brandt, H. 1951. Arizona and Its Bird Life. Bird Research Foundation, Cleveland.

Brown, J. H., and E. J. Heske. 1990. Control of a desert-grassland transition by a keystone rodent guild. Science 250:1705–1707.

Chew, R. M., and A. E. Chew. 1970. Energy relationships of mammals of a desert shrub (*Larrea tridentata*) community. Ecological Monographs 40:1–21.

Cole, C. J. 1984. Unisexual lizards. Scientific American 250:94–100.

Cook, R. E. 1969. Variation in species density of North American birds. Systematic Zoology 18:63–84.

Davis, G. P., Jr. 1982. Man and wildlife in Arizona: the American exploration, 1824–1865. Ed. N. B. Carmony and D. E. Brown. Arizona Game and Fish Department, Scottsdale.

Dimmitt, M. A., and R. Ruibal. 1980. Environmental correlates of emergence in spadefoot toads (*Scaphiopus*). Journal of Herpetology 14:21–29.

Di Peso, C. C. 1951. The Babocomari Village site on the Babocomari River, southeastern Arizona. Amerind Foundation Publication no. 5. Dragoon, Ariz.

Emslie, S. D. 1981. Birds and prehistoric agriculture: the New Mexican pueblos. Human Ecology 9:305–328.

Findley, J. S., A. H. Harris, D. E. Wilson, and C. Jones. 1975. Mammals of New Mexico. University of New Mexico Press, Albuquerque.

Germano, D. J., and C. R. Hungerford. 1981. Reptile population changes with manipulation of Sonoran Desert shrub. Great Basin Naturalist 41:129–138.

Germano, D. J., R. Hungerford, and S. C. Martin. 1983. Responses of selected wildlife species to the removal of mesquite from desert grassland. Journal of Range Management 36:309–311.

Haury, E. W. 1937. Stone palettes and ornaments. Pp. 121–134 *in* H. S. Gladwin, E. W. Haury, E. B. Sayles, and N. Gladwin (eds.), Excavations at Snaketown. Material culture. Medallion Papers 25. Gila Pueblo, Globe, Ariz.

———. 1976. The Hohokam. Desert Farmers and Craftsmen. University of Arizona Press, Tucson.

Hawkins, L. K., and P. F. Nicoletto. 1992. Kangaroo rat burrows structure the spatial organization of ground-dwelling animals in a semiarid grassland. Journal of Arid Environments 23:199–208.

Heske, E. J., J. H. Brown, and Q. Guo. 1993. Effects of kangaroo rat exclusion on vegetation structure and plant species diversity in the Chihuahuan Desert. Oecologia 95:520–524.

Hoffmeister, D. F. 1986. Mammals of Arizona. University of Arizona Press, Tucson.

Joern, A. 1986. Experimental study of avian predation on coexisting grasshopper populations (Orthoptera: Acrididae) in a sandhills grassland. Oikos 46: 243–249.

Johnson, P. C. 1981. Mammalian remains from Las Colinas. Pp. 269–289 *in* L. C. Hammack and A. P. Sullivan (eds.), The 1968 excavations at Mound 8, Las Colinas Ruins Group, Phoenix, Arizona. Arizona State Museum Archeological Series 154. Arizona State Museum, Tucson.

Kiester, A. R. 1971. Species density of North American amphibians and reptiles. Systematic Zoology 20:127–137.

Martin, P. S. 1973. The discovery of America. Science 179:969–974.

———. 1975. Vanishings, and future of the prairie. Geoscience and Man 10:39–49.

Maurer, B. A. 1985. Avian community dynamics in desert grasslands: observational scale and hierarchical structure. Ecological Monographs 55:295–312.

Mearns, E. A. 1907. Mammals of the Mexican boundary of the United States. Smithsonian Institution Bulletin 56. Smithsonian Institution, Washington, D.C.

Mendelson, J. R., and W. B. Jennings. 1992. Shifts in the relative abundance of snakes in a desert grassland. Journal of Herpetology 26:38–45.

Mitchell, J. C. 1979. Ecology of southeastern Arizona whiptail lizards (*Cnemidophorus*: Teiidae): population densities, resource partitioning, and niche overlap. Canadian Journal of Zoology 57:1487–1499.

Moroka, N., R. F. Beck, and R. D. Pieper. 1982. Impact of burrowing activity of the bannertail kangaroo rat on southern New Mexico desert rangelands. Journal of Range Management 35:707–710.

Nelson, J. F., and R. M. Chew. 1977. Factors affecting seed reserves in the soil of a Mohave Desert ecosystem, Rock Valley, Nye County, Nevada. American Midland Naturalist 97:300–320.

Parmenter, R. R., and J. A. MacMahon. 1988a. Factors influencing species composition and population sizes in a ground beetle community (Carabidae): predation by rodents. Oikos 52:350–356.

———. 1988b. Factors limiting populations of arid-land darkling beetles (Tenebrionidae): predation by rodents. Environmental Entomology 17:280–286.

Parmenter, R. R., J. A. MacMahon, and S. B. Vander Wall. 1984. The measurement of granivory by desert rodents, birds, and ants: a comparison of an energetics approach and a seed dish technique. Journal of Arid Environments 7:75–92.

Pulliam, H. R., and M. R. Brand. 1975. The production and utilization of seeds in plains grassland of southeastern Arizona. Ecology 56:1158–1166.

Reynolds, H. G., and S. C. Martin. 1968. Managing grass-shrub cattle ranges in the Southwest. U.S. Department of Agriculture, Agricultural Handbook 162.

Simpson, G. G. 1964. Species density of North American Recent mammals. Systematic Zoology 13:57–73.

Soholt, L. F. 1973. Consumption of primary production by a population of kangaroo rats (*Dipodomys merriami*) in the Mojave Desert. Ecological Monographs 43:357–376.

Stoddart, L. C. 1987a. Relative abundance of coyotes, jackrabbits and rodents in Curlew Valley, Utah. Final Report, U.S. Department of Agriculture Project DF-931.07, Denver Wildlife Research Center.

———. 1987b. Relative abundance of coyotes, lagomorphs, and rodents on the Idaho National Engineering Laboratory. Final Report, U.S. Department of Agriculture Project DF-931.07, Denver Wildlife Research Center.

Strong, T. R., and C. E. Bock. 1990. Bird species distribution patterns in riparian habitats of southeastern Arizona. Condor 92:866–885.

Szaro, R. C. 1981. Bird population responses to converting chaparral to grassland and riparian habitats. Southwestern Naturalist 26:251–256.

Voorhies, C. T., and W. P. Taylor. 1933. The life histories and ecology of jack rabbits, *Lepus alleni* and *Lepus californicus* ssp., in relation to grazing in Arizona. University of Arizona Agriculture Experiment Station Technical Bulletin 49.

Wright, J. W., and C. H. Lowe. 1968. Weeds, polyploids, parthenogenesis, and the geographical and ecological distribution of all-female *Cnemidophorus*. Copeia 1968:128–138.

Zweifel, R. G. 1968. Reproductive biology of anurans of the arid Southwest, with emphasis on adaptation of embryos to temperature. Bulletin of the American Museum of Natural History 140:1–64.

8

Human Impacts on the Grasslands of Southeastern Arizona

Conrad J. Bahre

The semidesert and plains grasslands, which make up about 45 percent of the vegetation cover of southeastern Arizona (D. E. Brown and Lowe 1980; fig. 8.1), have been the lifeblood of southern Arizona's cattle industry. In fact, most of the recorded changes in the grasslands of southeastern Arizona occurred after the beginning of large-scale cattle ranching and fire exclusion in 1870 (Bahre 1991). Although the grasslands have been occupied by humans for at least 11,000 years (Haynes 1992), we know very little about human impacts on these ecosystems before 1730, the year Hispanic settlers first arrived there (Officer 1987).

This chapter chronicles the succession of cultures that have occupied southeastern Arizona and examines how different perceptions and uses of the land have affected the grasslands. According to the historical record, the desert and plains grasslands of southeastern Arizona were comparatively brush-free before Anglo-Americans settled there (Bahre 1991). In the 1850s and 1860s, even after purported large-scale Mexican cattle ranching in the late 1820s and early 1830s, the grasslands were open and largely free of brush. Grass was plentiful, the streams and rivers dissecting the grasslands were in parts unchanneled and lined with galeria forests and marshes (*ciénagas*, often spelled *cienega*), wildfires were common, and ante-

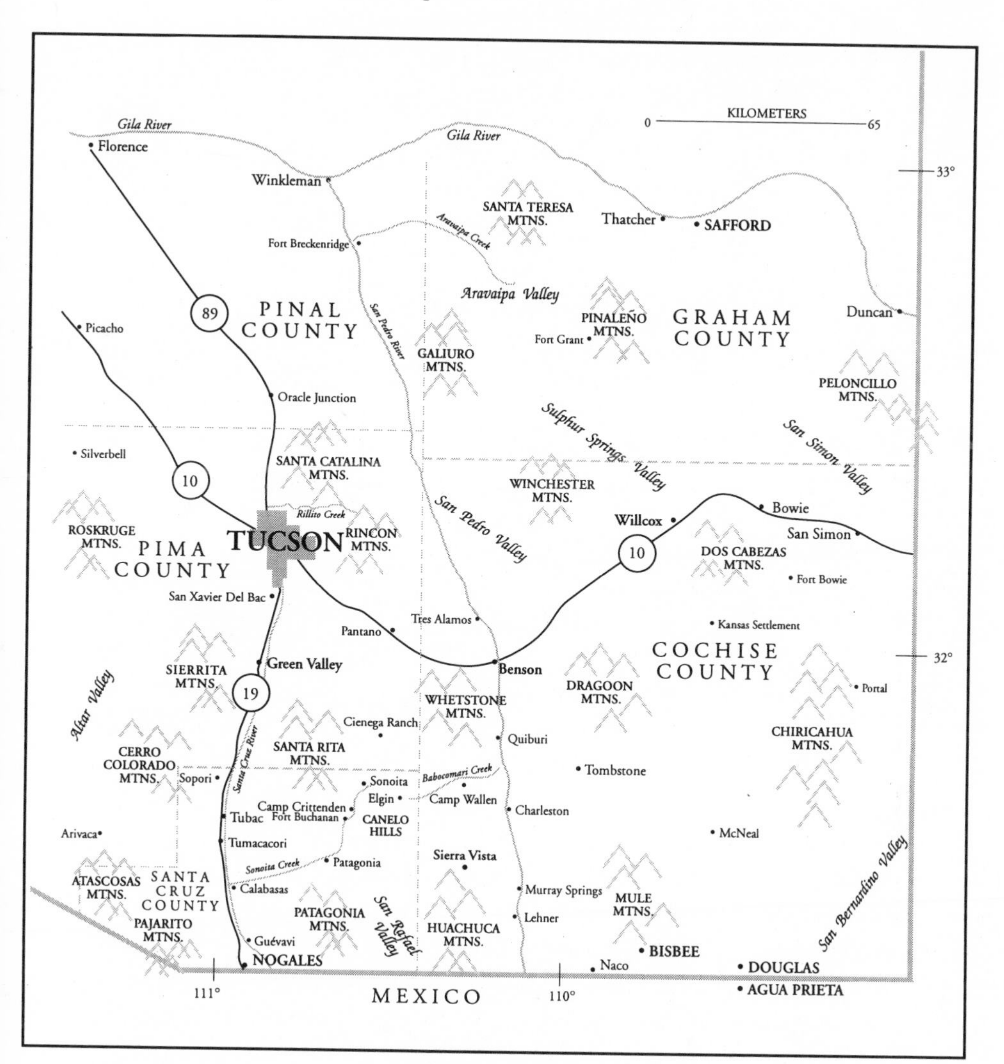

Figure 8.1. Place-name map of southeastern Arizona.

lope (*Antilocapra americana*), prairie dog (*Cynomys ludovicianus*), and Mexican wolf (*Canis lupus baileyi*) were abundant. Today, the landscape is different. The native grasses have declined, and in many areas nonnative species have replaced them; wildfires are rare; erosion is commonplace; and several grassland predators and herbivores have been eliminated. Probably the two most dramatic changes in the grasslands are the extensive increases in woody shrubs and trees and the landscape fragmentation resulting from localized urban and rural settlements.

Amerindian Occupation, 9300 B.C.–A.D. 1536

Both the archaeological record and early contact ethnographies indicate great differences in the technological capabilities and land-use practices of the prehistoric Amerindians of southeastern Arizona. The extent of their impact on the grasslands is largely unknown. Not only is there no evidence that the grasslands even existed thousands of years ago, there is little evidence that grasslands, especially the plains grasslands, were of much use to any Amerinds except possibly the Paleo-Indian Clovis big game hunters and Archaic period grass-seed collectors. The Clovis big game hunters hunted Pleistocene megafauna in southeastern Arizona 11,000 years ago and are best known for their kill sites in the upper San Pedro Valley at Naco (Haury et al. 1953), Lehner (Haury et al. 1959; Haynes and Haury 1982), Escapule (Hemmings and Haynes 1969), and Murray Springs (Hemmings 1970). Bifacially flaked lanceolate (Clovis) projectile points and other artifacts of chipped stone related to the tool kits of hunters and skinners have been found at these sites along with the skeletons of large Pleistocene animals such as bison (*Bison antiquus*), camel (*Camelops hesternus*), dire wolf (*Canis dirus*), horse (*Equus* cf. *conversidens*, *E.* cf. *niobrarensis*, *E.* cf. *occidentalis*), mammoth (*Mammuthus columbi*), and tapir (*Tapirus* sp.). The late Pleistocene and early Holocene climate of southeastern Arizona was supposedly wetter and cooler than it is today (Haynes 1991; Van Devender, this volume).

The eclipse of the Clovis big game hunters occurred about 10,000 years ago along with the extinction of the megafauna (Haynes 1991). It is not known, however, whether the extinction was a result of environmental change, overhunting, or both. A period of marked desiccation occurred in the region about 11,000 years ago (Haynes 1991). If the Clovis big game hunters were responsible for the extinction of the megafauna, as Paul Schultz Martin (1975) believes, then early humans may have affected the grasslands that have developed over the last 10,000 years by removing a large segment of the grazing fauna. It is also conceivable that the use of fire drives by the big game hunters contributed to the demise of the megafauna and possibly maintained and extended the grasslands (Sauer 1944).

The Paleo-Indian big game hunters were followed by bearers of the Cochise Culture, whose lithic complex suggests that they depended primarily on wild vegetal foods (Sayles and Antevs 1941). Members of the Cochise Culture lived in small nomadic groups that foraged for plants and hunted small game. The Cochise Culture, which appears to have changed very little for thousands of years, was predominant roughly between 8500 B.C. and A.D. 1, during the so-called Archaic period (Huckell 1984). Our understanding of subsistence patterns and culture history during the Archaic is still in its infancy, however, and this may account for the fact that we perceive the Cochise Culture to have changed little over many millennia. After 2500 B.C. the Cochise peoples became more sedentary, living in pit houses and experimenting with agriculture (Whalen 1971). According to Sharon Urban and Lisa Huckell (University of Arizona, pers. comms., September 1992), the Cochise peoples may have depended to a large extent on grass seeds for food, which might explain the abundance of Cochise sites in plains grasslands. Charles Di Peso (1956) argued against a long occupation for the Cochise Culture and believed that a different culture, an offshoot of the Cochise Culture that he named the O'otam, has inhabited southeastern Arizona since 5000 B.C. He postulated that the O'otam were the ancestors of the historic Pima Indians encountered by the first Spanish *entradas* in southeastern Arizona.

About A.D. 1, a new culture—the Hohokam—appeared in southeastern Arizona west of the San Pedro Valley and along the Santa Cruz, middle Gila, and Salt Rivers (Gumerman 1991). Hohokam origins are attributed to either an internal development from Cochise Culture or to an immigrant group from Mexico. They lived in small, clustered villages, made ceramics, and depended on irrigated agriculture (Haury 1976). Their crops were agave (*Agave* spp.), barley grass (*Hordeum* spp.), beans (*Phaseolus vulgaris*), cotton (*Gossypium* spp.), maize (*Zea mays*), squash (*Curcubita* spp.), and tobacco (*Nicotiana* spp.). The Hohokam had well-developed irrigation systems, and many of their villages—like those in the Salt River valley near present-day Phoenix—were sustained by large and complex canal systems. Much of the Hohokam diet, however, centered on wild plant collecting and hunting. Cactus fruits, mesquite (*Prosopis* spp.) seeds, and agave stems were dietary staples along with rabbit (Leporidae), deer (*Odocoileus* spp.), and bighorn sheep (*Ovis canadensis*). For the most part, the northern and eastern boundaries of the Hohokam coincided with the distribution of saguaro (*Carnegiea gigantea*) and mesquite (Fish and Nabhan 1991).

The modern-day Pima and Tohono O'odham (Papago) are thought to be the descendants of the Hohokam, although Di Peso (1956) believed that the Hohokam were invaders from Mexico who dominated the O'otam, transforming their culture with intensive canal irrigation, elaborate shell and stone carving, new pottery modes, copper bells, and the Mesoamerican ball game. Massive crop failures, O'otam rebellion, or other unknown causes led to the major Hohokam sites being abandoned between 1300 and 1450 (Di Peso 1956; Gumerman 1991). According to Di Peso (1956), present-day O'otam (Tohono O'odham) myths tell how the Hohokam were driven from the region as a result of crop losses and the taxing of the O'otam to the point of rebellion.

The Mogollon, contemporaries of the Hohokam, inhabited southeastern Arizona east of the San Pedro Valley and south of the Gila River. Like the Hohokam, they may have developed from the Cochise Culture (P. S. Martin 1979). The Mogollon were primarily mountain people who lived in pit houses, built kivas, had multiroom pueblos and well-developed agriculture, and made pottery. Between 1000 and 1400 the major Mogollon sites were abandoned, possibly for the same reasons that caused the Hohokam sites to be deserted (P. S. Martin 1979).

The period from 1350 to 1536 in southeastern Arizona is not well understood archaeologically, but it is assumed that the Tohono O'odham, Gileños, Pima, and Sobaipuri, who inhabited the region at Spanish contact, were the modern relatives of either the Hohokam, the O'otam, or a mixture of O'otam and western Pueblo. Most of what we know about the prehistoric inhabitants of southeastern Arizona after 1350 is based largely on the archaeological work of Charles Di Peso (1951, 1953, 1956, 1958), but neither Di Peso nor later archaeologists have been able to shed much light on the relationship between prehistoric Amerinds and southeastern Arizona's grasslands. There is evidence that the Amerinds utilized some wild grass seeds for food, especially big sacaton (*Sporobolus wrightii*) and browntop panicgrass (*Brachiaria fasciculata*) seeds (Doebley 1984).

Sauer (1935) estimated that approximately 30,000 Amerindians lived in southeastern Arizona during the period of initial Spanish contact (1536–1730). By the late 1690s, however, Old World diseases and attacks by Apaches had greatly reduced the population, and postcontact estimates of the population probably do not reflect the true prehistoric numbers.

The extent to which the prehistoric inhabitants of southeastern Arizona influenced the grassland ecology is unknown. There are few archaeological sites far from perennial water supplies in grassland environments, especially in the plains grasslands. The major use of grasslands was, probably, to collect vegetal foods from the arborescent and succulent taxa in the desert grasslands and grass seeds in the plains grassland. In general, the grasslands were poorly suited for aboriginal agriculture because the Amerinds lacked plows before Spanish contact.

The accidental or intentional commuting of fire to the grasslands by

Amerindians—through fire drives, mescal roasting, smoke signals, abandoned campfires, fire during warfare, and so forth—may have contributed to the largely brush-free state of the grasslands when the first Europeans arrived.

The Spanish and Mexican Occupations, 1536–1854

Although Spanish explorers may have passed through southeastern Arizona as early as 1536 and Jesuit priests had established *visitas* and missions in the Santa Cruz Valley by 1701, Hispanic settlers did not arrive until the late 1730s, and then only in small numbers (Officer 1987). For the most part, Spanish and Mexican influences in southeastern Arizona were significant only in the upper Santa Cruz Valley (fig. 8.1).

The Spaniards, 1536–1821

Probably the first non-Indian to pass through southeastern Arizona was a North African named Estevanico (Bancroft 1889). It is uncertain whether he accomplished this feat with the Cabeza de Vaca expedition in 1536 or later with Fray Marcos de Niza, who, in search of the fabled Seven Cities of Cíbola (Zuni in present-day New Mexico), supposedly traveled along the San Pedro River in May 1539 (Bandelier 1890; Sauer 1932).

Marcos de Niza was followed by Francisco Vásquez de Coronado, who led an expedition along the entire length of the San Pedro River on his way to Cíbola in 1540 (Bolton 1949). None of these explorers, however, seemed much interested in the desert grassland, and their descriptions of the landscape of southeastern Arizona are too vague to ascertain the true nature of the grasslands in the sixteenth century. Between the Coronado expedition and 1691 there was no further Spanish penetration into southeastern Arizona except for an occasional slave raid. In 1691, Father Eusebio Francisco Kino visited the Amerindian village of Tumacácori (Bolton 1919) and soon afterward established several visitas and missions in the region then known as Pimería Alta. He also introduced Old World livestock and crops to the Sobaipuri and other Pima inhabitants. It is clear that between 1604 and the outbreak of the Pueblo Revolt in 1680, the Sobaipuri, who then lived in the upper Santa Cruz and San Pedro River valleys at Tres Alamos, Bac, Santa Catarina, Quiburi, and other villages, had trade relations with the Spanish colonists of the Rio Grande valley (Bolton 1952). Therefore, it is conceivable that the Sobaipuri and other Pima peoples may have obtained

European livestock—chickens (*Gallus gallus*), ducks (*Anser boshas*), and rabbits—and some crop plants before Father Kino arrived. In part, this theory is supported by the archaeological work of Di Peso (1951), who indicated that the Sobaipuri were growing peach (*Prunus persica*) and English walnut (*Juglans regia*) trees along Babocomari Creek long before Kino arrived.

At about the same time as Father Kino's entrada, Apaches began to advance into southeastern Arizona (Aschmann 1970; Di Peso 1956). They forced the Sobaipuri to abandon Quiburi and the rest of the San Pedro River valley by the end of the seventeenth century, leaving southeastern Arizona east of the Santa Cruz Valley in Apache hands. In 1705, with reassurances from the Spaniards, the Sobaipuri resettled Quiburi and remained there until the Apaches drove them out again in 1762 (Nentvig 1980).

By 1740, Spanish settlers had cattle herds in the region south of Guévavi and around Arivaca and Sopóri (Kessell 1970). When the Jesuits were expelled from the New World by the Spanish Crown in 1767, almost the entire Spanish population of Arizona (about 500 persons) lived in Tubac. The first Spanish settlers in Tucson arrived in 1775 when the presidio of San Augustín de Tucson was built about 16 km north of the mission of San Xavier. One year later, in an effort to contain the Apaches in the San Pedro Valley, the Spaniards established the presidio of Santa Cruz de Terrenate at Quiburi (it was abandoned around 1789; Officer 1987).

A fairly long period of decreased hostilities between the Spaniards and the Apaches began in 1786, when Viceroy Bernardo de Gálvez bribed the Apaches with a fixed stipend of liquor and meat, and lasted until Mexican independence in 1821 (Moorhead 1968; Sheridan 1986). Although peace with the Apaches encouraged a number of Hispanic ranchers and miners to move to Arizona, their numbers were never very large and their activities appear to have had little impact on the grasslands. There was some small-scale farming, ranching, and mining in the Santa Cruz Valley between Tucson and Guévavi and in the surrounding mountains. Livestock, especially cattle and horses, may have substantially degraded the rangelands adjacent to the presidios and villages, but there is no evidence that livestock numbers were very large. In 1804, for example, only 3,500 cattle, 2,600 sheep, and 1,200 horses were reported at Tucson, then the largest Spanish settlement in Arizona with a population of 1,015; at Tubac, the other major settlement, only 1,000 cattle and 5,000 sheep were counted (McCarty 1976).

Between 1691 and 1821 a number of settlements were established in southeastern Arizona (mostly in the Santa Cruz Valley): missions at San Xavier del Bac, Tumacácori, and Guévavi; visitas at Tucson, Arivaca, and Calabasas; presidios at Tubac (established in 1751), Tucson, and Santa Cruz de Terrenate; and Amerindian villages at Santa Catarina, Tres Alamos, Quiburi, and Sopóri (Brinckerhoff 1967; Officer 1987; Wagoner 1975; fig. 8.1).

The introduction of Old World technology, crops, and livestock by the Spaniards eventually led to major changes in the grasslands of southeastern Arizona. Because southeastern Arizona was on the margin of New Spain's northern frontier, however, Spanish settlement was sparse and concentrated in the upper Santa Cruz Valley. There is no doubt that livestock, especially cattle, horses, and sheep, were abundant near the presidios and missions, but their effect on regional grasslands was probably minimal.

The Mexicans, 1821–1854

After Mexico gained its independence in 1821, the Apaches resumed raiding in southeastern Arizona, with devastating impact (Weber 1982). Despite the Apache harassment, however, several large stock-raising grants were established between 1821 and 1827 near the present international boundary (Mattison 1946; Officer 1987; fig. 8.2). These grants, abandoned by the 1830s or early 1840s because of Apache attacks (Mattison 1946; Officer 1987), now represent some of the largest private landholdings in southeastern Arizona. In their heyday, thousands of horses, cattle, and sheep were supposedly run on these grants. According to Bartlett (1854), there were 140,000 cattle on the Babocomari and San Bernardino grants alone.

The small Hispanic population of Arizona during the Mexican period, isolated and subject to frequent Apache attacks, was not much inspired toward agriculture or mining development. During the Mexican occupation, the Hispanic population was as low or lower than it was during the Spanish occupation. In 1831, only 465 persons, not counting friendly Apaches, were reported living at Tucson, and only 303 were at Tubac (Officer 1987). In 1848, there were no Hispanics at Tumacácori; Tubac had only 249 people, of whom 200 were friendly Apaches; and Tucson had a population of 760 (Officer 1987).

The major towns in southeastern Arizona during the Mexican occupation were Tucson and Tubac, followed by the small settlements of San Xavier, Tumacácori, and Calabasas (Wagoner 1975). Even Tucson and Tubac suffered major population shifts at that time. A cholera epidemic killed hundreds of Tucsonans in 1855, and Tubac was completely abandoned when Charles D. Poston, an Anglo-American miner, reestablished the settlement in 1856 (Officer 1987).

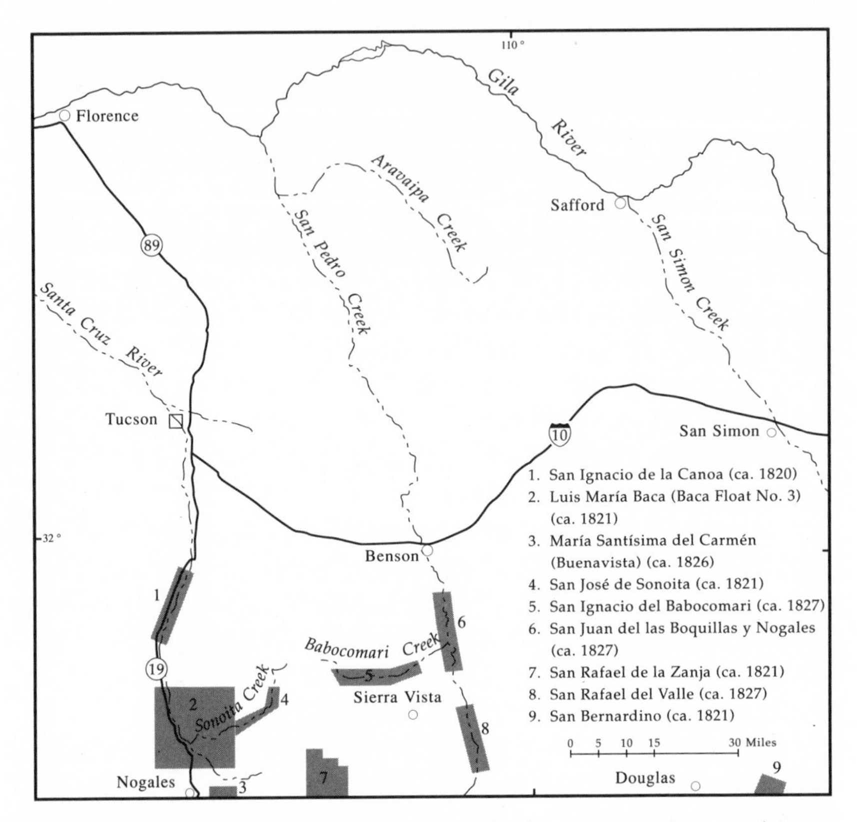

Figure 8.2. Authenticated Spanish and Mexican land grants in southeastern Arizona. From Wagoner 1975.

Beginning in 1846, when war broke out between the United States and Mexico, and continuing through 1854, when Mexican jurisdiction over the region ended with the Gadsden Purchase, several American military and scientific expeditions and thousands of immigrants heading for the California goldfields crossed southeastern Arizona, first traveling southwest from the San Bernardino Valley into Mexico and then north down either the San Pedro or Santa Cruz River valley to the Gila River. Many of these travelers and expedition members wrote fine accounts of the grasslands of southeastern Arizona (Bartlett 1854; P. St. G. Cooke 1938; Emory 1857; Goetzmann 1959).

A number of accounts written between 1846, when Lieutenant Colonel Philip St. George Cooke led the Mormon Battalion through southeastern Arizona, and 1853 mentioned wild cattle on the abandoned Mexican land grants, especially in the San Bernardino and San Pedro Valleys (Hastings 1959; Hastings and Turner 1965). Cooke's battalion was attacked by wild cattle near the junction of Babocomari Creek and the San Pedro River (P. St. G. Cooke 1938). When John Russell Bartlett (1854), commissioner of the United States–Mexican Boundary Survey, entered southern Arizona with the survey team in 1851, he described the San Bernardino Valley as desolate and covered with cattle trails. Near present-day Agua Prieta, his party used cattle dung for fuel because of the lack of firewood in the area. On the same trip Bartlett encountered a party of 30 or 40 Mexicans camped at the confluence of Babocomari Creek and the San Pedro River to hunt wild cattle for tallow, hides, hooves, and meat (Bartlett 1854). From 1850 to 1853, large numbers of cattle were driven across southeastern Arizona by immigrants to California. According to the records, few if any of these cattle were left behind (Brady n.d.; Cameron 1896), but they must have affected the grass along the major trails in their passage. Christiansen (1988) estimated that there may have been as many as 100,000 wild cattle in southeastern Arizona during the 1840s and early 1850s, but this figure is based on circumstantial evidence; records of the number of cattle present during the 1820s are based almost wholly on oral histories. Christiansen further pointed out that the wild cattle had disappeared by 1854.

Except for the settlements of the upper Santa Cruz Valley and the Pima villages along the middle Gila River, the Apaches controlled all of southeastern Arizona east of the Santa Cruz drainage during the Mexican occupation. In 1853, only 32 years after Mexico won its independence, the Mexican government sold to the United States all of present-day Arizona between the Gila River and the present international boundary. The Gadsden Purchase was signed in Mexico City on 23 December 1853 and proclaimed on 30 June 1854.

The principal cultural impact bearing on the grasslands of southeastern Arizona during the Mexican occupation was the introduction of large-scale cattle and sheep ranching, especially on the Mexican land grants. The sizes of these herds are in all probability grossly exaggerated in the historical accounts, although large numbers of cattle may have grazed the Santa Cruz, San Pedro, and San Bernardino Valleys near the present-day international boundary. Considering the general lack of livestock water developments then (there were no windmills or stock tanks) and the intermittent nature of most of the streams, it is difficult to believe that large numbers of cattle could have been supported by the grass and browse in the rangelands adjacent to major sources of perennial water. Furthermore, large-scale cattle ranching during the 1820s would have been curtailed by Apache depredations, which presumably brought ranching operations to a standstill almost as soon as the land grants were established (Haskett 1935). Even if there had been a large number of cattle in the region in the 1820s and 1830s, there is no written evidence of overgrazing, although cattle grazing probably weakened the grass cover in areas near water or adjacent to the major settlements. R. U. Cooke and Reeves (1976) pointed out that erosion and stream entrenching, possibly resulting from agricultural disturbances and cattle raising, occurred along the Santa Cruz River near Tumacácori in the 1850s.

Historical records and some fire scar data indicate that wildfires were fairly common in the grasslands of southeastern Arizona during the Spanish and Mexican occupations. Dobyns (1981), Pyne (1982), and Bahre (1985, 1991) used information from early Spanish and Mexican records, ethnographies, and nineteenth-century newspapers as evidence that Amerindians, especially the Apaches, set wildfires for a variety of reasons both before and after Anglo-American settlement. But the incidence of lightning fire is so high there that the addition of burning by Amerindians may have contributed little to the natural incidence of fires (Bahre 1991).

A large number of Spanish and Mexican archival records still remain to be gleaned for information on the grasslands of southeastern Arizona in the eighteenth and early nineteenth centuries. Spanish and Mexican settlement of the region was sparse and concentrated for the most part in the upper Santa Cruz Valley. Probably the most significant human impact on the grasslands during this period was large-scale ranching. The numbers of animals and their effects, if any, are unknown. No doubt some grasslands were severely affected, but, given the descriptions of the lush grasslands in southeastern Arizona in the 1850s and 1860s, ranching during the Mexican occupation seems to have caused little long-lasting and extensive disturbance.

Anglo-American Occupation, 1854–Present

After the Gadsden Purchase, the United States government moved quickly to contain the Apaches, establishing Fort Buchanan on Sonoita Creek in 1857, Fort Breckenridge near the mouth of Aravaipa Creek in 1859, and Fort Bowie in Apache Pass in 1862 (Brandes 1960; Serven 1965; Wagoner 1975). Aside from a few Anglo-American ranchers, miners, and farmers, who settled mostly in the upper Santa Cruz Valley shortly after 1854, major

Anglo settlement of southeastern Arizona did not begin until the 1870s. According to the United States Census (U.S. Department of Commerce 1864), the non-Amerindian population of all of southeastern Arizona in 1860 was only 1,635, most of whom were in Tucson and Tubac. Soon after the Civil War, a few Anglo settlers moved into the Babocomari and Sonoita drainages to farm, and Camps Wallen and Crittenden were established in 1866 and 1867, respectively, to protect settlers in those drainages (Brandes 1960).

Anglo-American settlement expanded in the 1870s and early 1880s, largely because of four interdependent, though distinct, events: (1) the subjugation of the Apaches in 1872 (although some renegade Apaches, including Geronimo and his band, continued to raid throughout the region until 1886); (2) the completion of the Southern Pacific Railroad from Tucson to El Paso in 1881 (Greever 1957; Myrick 1975); (3) the development of mining in Tombstone for silver and in Bisbee for copper in 1878 (*Bisbee Review*, 8 Aug. 1923; Gird 1907); and (4) the boom in cattle ranching. The subjugation of the Apaches and the completion of the railroad in particular facilitated the expansion of the ranching industry and heavy grazing of the grasslands. In the 1860s and 1870s, before the railroad was finished, ranchers had begun moving large herds of sheep and cattle into the region, especially after a series of severe droughts in California (Haskett 1935; Morrisey 1950; Wagoner 1952).

By 1885, so much investment capital, both foreign and domestic, had poured into southeastern Arizona's ranching industry that cattle numbers exceeded all expectations. By 1891, cattle numbers in the region had reached nearly 400,000, before the severe drought of 1891–1893 caused a massive die-off (Bahre 1991). After the drought, major changes in the grasslands became apparent, many of which have persisted to the present. The ranching industry had stabilized by 1900, but overstocking and overgrazing continued (fig. 8.3).

The early twentieth century saw a major nationwide push to conserve range and forest resources, and in 1902 national forest reserves were established in southeastern Arizona in the Santa Rita, Chiricahua, Santa Catalina, and Pinaleño Mountains (Allen 1989; Baker et al. 1988; Lauver 1938). Shortly thereafter, other reserves were established in the Huachuca, Dragoon, and Tumacácori Mountains. In 1907 these reserves were renamed national forests, and in 1917 all of southeastern Arizona's national forest lands were consolidated as the Coronado National Forest. The United States Department of Agriculture Forest Service curtailed the free use of

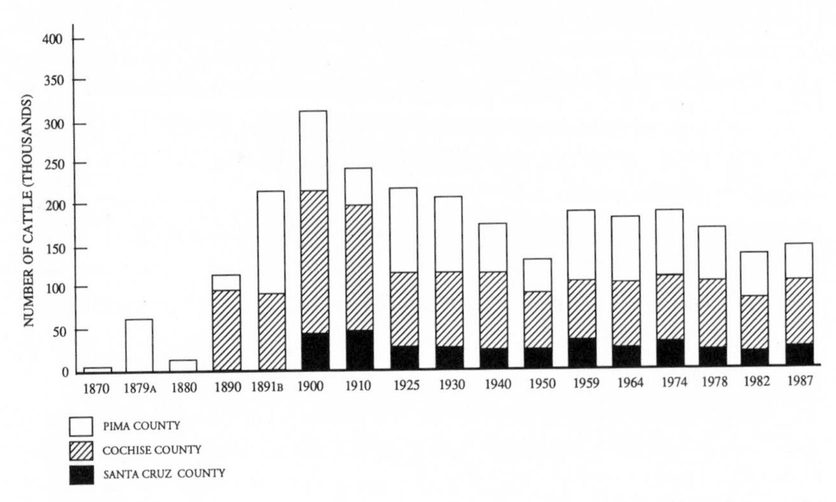

Figure 8.3. The number of cattle in Pima, Cochise, and Santa Cruz Counties, southeastern Arizona, 1870–1987. Includes dairy, feedlot, irrigated pasture, and range cattle. Pima County was established in 1863 and then included the modern counties of Cochise and Santa Cruz. Cochise County was established from Pima County in 1881, and Santa Cruz County was established in 1899. Information for all years except 1879 and 1891 is from U.S. Department of Commerce, Bureau of the Census agricultural censuses.

[a]*Arizona Weekly Star*, 1 Jan. 1879.

[b]Annual Report of the Governor (1893), 53d Cong. 2d Sess., H. ex. Doc. I, Serial 3211. Wagoner (1961) noted: "By 1891 cattle production had reached a peak in the Arizona Territory. In that year there were nearly 720,000 head on the tax roles. It is the opinion of men who understood the methods of listing property in those days that there were at least twice that number on the grasslands."

forage resources in the national forests and advocated grazing control and fire suppression.

The Taylor Grazing Act was enacted in 1934 to prevent overgrazing, and consequent soil and water deterioration, as well as to stabilize the livestock industry dependent on public rangelands (Voigt 1976). Grazing districts were established and grazing permits were issued. The Forest Service, the Department of the Interior's Grazing Service (now the Bureau of Land Management), and other federal agencies charged with range and watershed protection introduced changes in the grasslands largely intended to improve them for cattle.

Between 1910 and 1940, agriculture in southeastern Arizona depended largely on surface water irrigation from the Santa Cruz, San Pedro, and Gila Rivers, although some areas were irrigated with water from artesian wells and by groundwater pumping in areas with high water tables. In the 1910s and 1920s, some homesteaders tried unsuccessfully to prove up their claims by dry farming in the plains grasslands around Kansas Settlement and in the Sonoita-Elgin area. With the development of efficient, low-cost irrigation pumps in the 1940s, thousands of hectares of grassland were cleared for irrigated agriculture. Irrigated agriculture has declined since the 1970s, however, and thousands of hectares of former cropland have been abandoned.

The rapidly growing population of Arizona, especially since the 1940s, has led to expanded urban and rural development on privately owned lands and is threatening federal and state trust lands, which are continually being sold to the public for development. Since 1950 the number of rural subdivisions being built in southeastern Arizona has exploded, especially in Cochise County, which has both the largest area of grassland and the largest amount of private land (41 percent)—almost all of it being used for agriculture, ranching, or subdivision development (Hecht and Reeves 1981). Cattle numbers in the county have declined, in part because of tract development in former rangelands, and much private grazing and agricultural land has been purchased by people seeking rural retirement or investment opportunities. In 1981 Santa Cruz and Cochise Counties were first and second, respectively, in the sale of remote subdivision lots in Arizona. The population capacity of the subdivisions in Cochise County alone in 1974 was 254,000 (Hamernick and Brown 1975); the county's total population in 1990 was 97,624. Tighter restrictions on developers in 1975 forced many major rural subdivisions into bankruptcy (Hecht and Reeves 1981), leaving behind thousands of cleared hectares with eroding, unused bulldozed roads.

Although several agricultural and mining towns in southern Arizona have suffered major population declines since 1930, due either to the Great Depression or to more recent major economic downturns in regional copper mining and irrigated agriculture, regional population growth has generally been steady and continuous. According to the 1990 Census, the total population of southeastern Arizona was about 700,000, about 124 times the population of 5,648 reported in 1880 (U.S. Department of Commerce 1883, 1992). One city in the grassland, Sierra Vista, grew from 3,121 inhabitants in 1960 to 32,983 in 1990. Cities of more than 10,000 in southeastern

Arizona in 1990 included Douglas (12,822), Green Valley (13,231), Nogales (19,489), Sierra Vista (32,983), and Tucson (405,390) (U.S. Department of Commerce 1992). The landscape fragmentation that has resulted from extensive abandoned farmland combined with increasing urban and rural settlements may isolate areas of desert grassland and hinder the dispersal of species and the spread of fires.

What follows is a more detailed examination of some of the major land-use patterns that have affected the grasslands of southeastern Arizona since the Gadsden Purchase. Livestock grazing, wildfire exclusion, nonnative plant introductions, farming, groundwater pumping, and wild hay harvesting initiated and have perpetuated most of the vegetation changes witnessed in the grasslands since the 1870s.

Livestock Grazing

Few areas of grassland in southeastern Arizona have escaped the impact of cattle ranching. In spite of this, we know surprisingly little about the impact of cattle and other livestock on the evolution of the grasslands. Generally, only the changes that result in lower carrying capacities for cattle have been identified; even then, the relation of cattle to some of the changes is uncertain. Are cattle, for example, responsible for the increase of woody plants in the grasslands since the 1890s? Most of the major recorded changes in southeastern Arizona's grasslands followed the arrival of large-scale ranching and the overgrazing prevalent near the turn of the century.

In 1901, D. A. Griffiths (1901), chief botanist in charge of grass and forage plant investigations for the Arizona Experiment Station in Tucson, observed that the rangelands of southern Arizona were more degraded than any others he had seen in the western United States. This observation was made at a time when livestock numbers had been greatly reduced (possibly by as much as 75 percent) as a result of the 1891–1893 drought (Cameron 1896). But the rangelands were badly degraded even before the drought. In 1891, for example, J. W. Toumey, the botanist who preceded Griffiths at the Arizona Experiment Station, wrote: "There are valleys [in southeastern Arizona] over which one can ride for several miles without finding mature grasses sufficient for herbarium specimens without searching under bushes or in other similar places" (1891b). Large-scale cattle ranching began in the 1870s, and it is remarkable that southeastern Arizona's rangelands, described as among the best in the state in the 1850s and 1860s, had become so overgrazed in less than 20 years.

In 1872, H. C. Hooker, the most prominent Anglo rancher in southeastern Arizona, had 11,000 cattle in Sulphur Springs Valley. At the time it was the largest cattle herd east of the Santa Cruz Valley in southeastern Arizona (Haskett 1935; Morrisey 1950). There were also cattle in the Santa Cruz Valley south of Tucson, in parts of the San Pedro Valley, along Babocomari Creek between its junction with the San Pedro River and present-day Sonoita, in the Altar Valley, around Arivaca, along the Gila River from Duncan to Thatcher, and along Sonoita Creek west of present-day Patagonia to Calabasas (Cameron 1896; Haskett 1935; Morrisey 1950; Wagoner 1951, 1952, 1961). In 1881 the Southern Pacific Railroad advertised for settlers, and soon ranchers from Texas, New Mexico, and the Mexican states of Durango, Chihuahua, and Sonora began moving their herds into southeastern Arizona. The transformation of the grasslands occurred rapidly: by 1884 every running stream and permanent spring had been claimed and the adjacent ranges stocked with cattle (Cameron 1896; fig. 8.3). In his annual report for 1883, Governor F. A. Tritle of Arizona boasted that there were 14 million ha of grasslands in the Arizona Territory—enough pasture, he claimed, for 7.68 million cattle (U.S. House of Representatives 1883).

Between 1885 and 1890, investments continued to pour into the cattle industry in southern Arizona, and by 1890 cattle were nearly everywhere. The *Arizona Daily Star* (5 Jan. 1890) reported: "Conservative cattlemen estimate that at the lowest calculations there are now more than 150,000 head of cattle in Pima County . . . , others claim as many as 250,000." The *Daily Star* for 19 June 1888 listed the Pima County cattle barons and their livestock as follows:

> Pima County cattle barons . . . W. L. Vail and Associates have not less than 20,000 head. Last year they branded over 4,000 calves. Maish and Driscoll have in the neighborhood of 23,000 head and they also branded over 4,000 calves last year. The Land and Hays Company of the Barbacumbrie [Babocomari] grant, have about 15,000 head, a number of which are in Cochise County. They branded 3,000 calves. The Cameron Bros. on the San Rafael Ranch, have almost 12,000. They handled nearly 3,000 calves last season. Sabino Otero has about 7,000 head and handled over 2,200 calves. Gen. Pusch has about 6,000 head. Richardson and Gormley have between 7,000 and 8,000 head. They branded 2,100 calves. They recently purchased 100 head from Ashburn Bros. Arnatha has about 2,500 head on his Santa Cruz ranch, and there are about a dozen others in the Santa Cruz Valley who have from 500 to 1,500 head. Then there are a number on the San Pedro who have between 500 to 1,000 head, and also a number of stockmen through the Tanque Verde, Santa Ritas, Sierias

[Sierritas] and ranges west of Tucson, who have bunches of cattle ranging from 500 to 1,000 head.

Before 1892, sheep were also important in southeastern Arizona; according to census reports, they actually outnumbered cattle. The *Arizona Weekly Star* (1 Jan. 1879) reported 78,500 sheep in Pima County versus 68,600 cattle. In 1880, the Cienega Ranch just west of the Whetstone Mountains was reported to have 23,000 sheep (U.S. Department of Commerce 1883), and the *Arizona Daily Star* for 20 March 1892 reported 20,000 sheep in the Chiricahua Mountains. Between the late 1870s and the early 1890s there were large numbers of sheep in the Sonoita Valley, Santa Rita Mountains, lower Santa Cruz Valley, and along the west side of the Chiricahua Mountains (*Arizona Daily Star*, 22 Oct. 1884, 29 March, and 28 May 1892; *Arizona Weekly Citizen*, 6 Dec. 1873, 6 June 1874, and 5 Dec. 1874, 13 March 1875, 4 Aug. 1877; Potter 1902).

In his 1891 annual report (U.S. House of Representatives 1893), Governor J. N. Irwin of Arizona listed 121,377 cattle in Pima County, which then included present-day Santa Cruz County, and 95,850 cattle in Cochise County. These numbers, conservative because few ranchers reported the total number of their cattle to the tax assessors, were present at the beginning of the 1891–1893 drought (Haskett 1935). In 1891 and 1892 almost no summer rains fell, and in the first months of 1893 the combined effects of drought and overgrazing led to the deaths of an estimated 50–75 percent of the livestock. Summer rains in July 1893 saved the cattle industry from complete ruin, but overstocking and overgrazing continued. Hendrickson and Minckley (1984) estimated that shortly before and during the 1891–1893 drought, 377,474 cattle grazed southeastern Arizona, more than twice the number of cattle presently grazing the rangelands of Pima, Cochise, Santa Cruz, and Graham Counties combined (Bahre 1991).

After 1893, major changes in the grasslands were recorded: many areas were completely denuded of grass cover and topsoil was eroded; ciénagas were destroyed and sections of the San Pedro, San Simon, Santa Cruz, and Babocomari Rivers ceased perennial flow; headward cutting (channel deepening progressing upstream) and channel entrenching began along most of the major streams; the hills were covered with cattle trails; oaks (*Quercus* spp.) and other trees had browse lines; and woody shrubs and nonnative weeds began to increase in the rangelands (U.S. House of Representatives 1893). Between 1893 and enactment of the Taylor Grazing Act in 1934, most of the region—with the possible exception of the forest reserves—was overstocked and overgrazed, although never again did livestock numbers approach their 1891 level (Wagoner 1952).

Until 1900, most of the area was open range and cattle moved at will, almost always within 5–8 km of water. The only major fences followed railroad rights-of-way. Most national forest lands were not fenced until the 1930s, and public lands were among the most abused. Some ranchers resisted the introduction of fences; even today, old-timers say that fencing the open range resulted in lower carrying capacities because cattle were concentrated in limited areas, causing even worse overgrazing. The Stock Raising Act of 1916, under whose provisions any citizen could apply for a cattle homestead of 267 ha, initiated the closing of the open public range in southeastern Arizona. Before that, most ranchers had depended on small private landholdings, usually with permanent water sources, as their base of operations and grazed their cattle on the open range (Rodgers 1965; Wagoner 1952). The increased number of ranching homesteads led to a decline in the amount of open range and forced large ranchers to accumulate more patented land to ensure adequate forage for their cattle. After 1920, the number of ranches larger than 420 ha increased greatly (Rodgers 1965). Overgrazing of Forest Service lands was largely curtailed after 1906, although overgrazing continued on other federal lands until the Taylor Grazing Act was enacted. For the most part, cattle numbers have declined since 1900 (fig. 8.3).

The Forest Service and other government agencies initiated a series of federally sponsored livestock programs that led to a new set of changes in the grasslands, sometimes at the expense of the plants least palatable to cattle. The Forest Service and the Bureau of Land Management initiated contour plowing, fencing, rotational and deferred grazing, prescribed burning, fire suppression, control of woody plants by herbicides, introduction of nonnative forage plants, construction of check and spreader dams, predator and rodent control, and weed control, and began chaining and bulldozing border pinyon (*Pinus discolor*), juniper (*Juniperus* spp.), mesquite, and oaks (Bahre 1977). These practices and a host of other activities designed to protect watersheds and improve the livestock industry culminated in the landscape seen today in the grasslands of southeastern Arizona.

Probably the most dramatic change in the grasslands since the 1890s has been the rapid increase of scrubby trees and shrubs (Bahre and Shelton 1993; Harris 1966; Humphrey 1956, 1958; Humphrey and Mehrhoff 1958; Leopold 1924; Parker and Martin 1952; Thornber 1910; Wright 1980). These woody increasers, whose spread has been influenced, if not caused, by over-

Figure 8.4. Santa Rita Experimental Range in (A) 1902 (courtesy of the U.S. Geological Survey) and (B) 1989 (photograph by D. Lippert), looking east toward the Santa Rita Mountains from the top of Huerfano Butte. The Santa Rita Forest Reserve was established in 1902, and the Experimental Range was set aside in 1903 (Bahre 1991). The area

grazing and fire exclusion, include mesquite, acacia (*Acacia* spp.), creosotebush (*Larrea divaricata*), juniper, oak, burroweed (*Isocoma tenuisecta*), thread-leaf groundsel (*Senecio longilobus*), and snakeweed (*Gutierrezia sarothrae*). Areas of former grassland that were once cut for wild hay and are now dominated by scrubby trees and shrubs in southeastern Arizona include the lower slopes of the Santa Rita Mountains (Humphrey 1958; fig. 8.4), the Aravaipa Valley, the area from the foot of the Sierra San José to Greenbrush Draw on the east side of the San Pedro Valley near the international boundary (Hastings 1959), the Croton Springs region southwest of Willcox (Hastings 1959), the east-facing alluvial fans of the Huachuca and Whetstone Mountains (Robert 1869; Rodgers 1965), the plains east and west of the Dragoon and Chiricahua Mountains (Robert 1869), large sections of the San Simon Valley (Hinton 1878), and the Altar Valley east of the Baboquivari Mountains.

Darrow (1944) estimated that by 1944, mesquite and acacia had invaded 0.7 million ha of grassland in Cochise County alone; and Parker and Mar-

was a major source of wild hay in the 1880s and 1890s (Bahre 1987). The 1902 image shows an overgrazed rangeland that was still fairly brush-free. By 1989, mesquite and acacia had increased dramatically. The mining area in the middle background is Helvetia.

tin (1952) reported that by 1952, fully half of the 3.8 million ha supporting mesquite in southern Arizona had been colonized by that species since 1850. Although mesquite has continued to increase on some rangelands, the most dramatic increases seem to have occurred between 1900 and 1934, the year the Taylor Grazing Act began to curtail overgrazing on public lands.

According to Rodgers (1965), ranchers in the upper San Pedro Valley recorded a two- to threefold increase in brush and mesquite on their rangelands between 1900 and 1934. Nevertheless, mesquite has not increased everywhere within its range (see Burgess, this volume; McAuliffe, this volume), and the range of mesquite in southeastern Arizona, as in the rest of the Southwest, seems to have remained relatively stable over the past 150 years (Bahre 1991; Bahre and Shelton 1993; Johnston 1963; Malin 1953).

The causes identified for the mesquite increase include (1) a reduction in the frequency and intensity of wildfires due largely to overgrazing following Anglo-American settlement and fire suppression; (2) a decline in natural perennial grasses, which, when healthy and dense, can reduce mesquite seedling establishment; (3) increased dissemination by livestock and/or Merriam kangaroo rats of scarified mesquite seed; (4) hoof damage to ground cover and soil compaction by livestock resulting in reduced moisture in the upper layers of soil, which hinders grass establishment and growth; (5) a lessening of the effectiveness of invertebrate seed predators; (6) land clearing and cultivation; (7) decreased activity of jackrabbits and wood rats; and (8) climatic change (Bahre and Shelton 1993; Bogusch 1952; A. L. Brown 1950; J. R. Brown and Archer 1989; Cox et al. 1993; Glendening and Paulsen 1955; Haskell 1945; Humphrey 1958; Humphrey and Mehrhoff 1958; Wright 1980).

The Bureau of Land Management's 1978 environmental impact report on the nearly 1.3 million ha that make up the upper Gila–San Simon area (the entire San Simon watershed and the lower San Pedro Valley from the Second Standard Parallel South north to the Gila River) exemplifies both real and perceived changes in the grasslands resulting from cattle and livestock management programs in southeastern Arizona. The San Simon Valley, often regarded as one of the most seriously disturbed environments in the United States (Peterson 1950), has been a focus of land management and conservation efforts since 1950. In stark testament to the long period of overstocking and overgrazing and channelization of San Simon Creek by nineteenth-century irrigation practices (Dobyns 1981), the valley's once fairly lush grassland was characterized in 1970 by extensive soil erosion and gullying, impoverished vegetation, and an annual sediment yield of as much as 9 ha-m/km^2 over 10 percent of the area (Jordan and Maynard 1970). Historical descriptions of the valley in the 1880s emphasize the ideal conditions for cattle. Bottomlands were covered with 65-cm-tall grasses, open areas were dominated by grama (*Bouteloua* spp.), and water was abundant on the valley floor (*Arizona Daily Star*, 29 Nov. 1882; Barnes 1936; Hinton 1878). More than 50,000 cattle grazed the valley in the 1880s (Thornber 1910); at present, fewer than 5,000 graze there.

In 1978, almost 42 percent of the upper Gila–San Simon area suffered moderate to high erosion, and 91 percent of the rangeland was judged to be in poor or fair condition (U.S. Department of Interior 1978). Since Anglo settlement, this seriously degraded land has seldom been used for anything other than livestock grazing.

Probably no single land use has had a greater effect on southeastern Arizona's grasslands than livestock grazing and range management programs. The rangelands have been managed for cattle for so long that we are uncertain about their pregrazing condition. Grazing since the 1870s has led to soil erosion, destruction of those plants most palatable to livestock, changes in grassland fire ecology, the spread of nonnative plants, and a steady increase in the density of woody shrubs and brush.

Wildfire

Before the 1890s, wildfires were fairly common in the grasslands of southeastern Arizona wherever there was sufficient fuel (Bahre 1991). The major ignition source was probably lightning (the Mogollon Rim country has among the highest incidence of lightning fires in the continental United States), but wildfires have been set by humans in the grasslands for millennia (Bahre 1991; Komarek 1968; Pyne 1984; Schroeder and Buck 1970). With the arrival of Anglo-American settlers, the frequency of wildfires declined dramatically as a result of both overgrazing and deliberate fire exclusion. Even in the early 1880s, however, wildfires were still common in the grasslands (Bahre 1991). Clara Spalding Brown (1893), one of the first female residents of Tombstone, described a prairie fire along the road from Charleston to the Huachuca Mountains in the early 1880s:

> Then we came upon a prairie fire which we had observed raging in the distance. It was moving swiftly along our right, but as the grass on our left had been previously destroyed, it did not menace us with danger. It was a weird sight, not without beauty, as the remorseless line advanced, darting up vivid forks of flame, and presently became entangled with a whirlwind, a dense black cloud like a waterspout moving along in the track of the fire, and ever and anon revealing in its dusky heart tongues of flashing light, a startling contrast to their inky background. The burning grass, though brown and sear [*sic*], was full of nutrition, and was a great boon to cattle in a country where only the river-bottoms and certain moist valleys like the San Simeon [Simon] yield verdure.

Since the establishment of the first forest reserves in Arizona, Forest Service administrators have followed a policy of almost total fire exclusion and suppression. For example, Leopold (1924) pointed out that Forest Service administrators in Arizona at the turn of the century encouraged overgrazing to reduce fire hazards and promote the growth of trees.

Local newspaper accounts of wildfires in southeastern Arizona between 1859 (the year the first newspaper in southeastern Arizona was printed) and 1890 demonstrate that (1) wildfires were much larger in areal extent in the grasslands during that period than they are now; (2) the occurrence of large grassland fires seems to have declined after the 1880s; (3) the cessation of major grassland fires preceded the brush invasion of the 1890s; (4) Amerindians, especially Apaches, set wildfires; (5) wildfire suppression was favored by early Anglo settlers; and (6) wildfires in the grasslands were fairly common before 1890 (Bahre 1985, 1991).

Even today some range managers advocate prescribed burning of the grasslands to destroy brush (S. C. Martin 1975; Wright and Bailey 1982), but the meager grass cover in many areas does not provide sufficient fine fuel to sustain fire. In any case, many mesquite trees are now too large to be killed by grass fires (Humphrey 1958: McPherson, this volume).

Anglo-Americans changed the frequency and ecological role of wildfire in the grasslands of southeastern Arizona by purposeful fire exclusion and overgrazing. The native plants and animals of the desert and plains grasslands evolved under conditions of frequent fires (C. E. Bock and Bock 1978; J. H. Bock et al. 1976; Cable 1967; Humphrey 1958; Wright 1980). Landscape fragmentation will further reduce the likelihood that fires will be as extensive as they were in the past, and the lack of fire may influence the dynamics of these grasslands.

Nonnative Plants

Since 1860, nonnative plants have been introduced into southeastern Arizona's grasslands for a variety of reasons, but mostly to reseed degraded rangelands and to control erosion. According to Cox et al. (1984), since the 1880s, ranchers and range managers in southeastern Arizona have reseeded nearly 300 species of nonnative and native plants, most of these in the grasslands. Several nonnative plants have become naturalized and have displaced native ones, but for the most part nonnative plants prefer and become established in disturbed areas.

The number of nonnatives reported in the flora of Arizona has nearly doubled in the past 50 years. In 1942, approximately 190 of Arizona's 3,200 plant species were nonnatives (Kearney and Peebles 1942); today the number is 330 (Burgess et al. 1991).

Two of the earliest nonnative plants intentionally planted in the rangelands of southeastern Arizona were Bermuda grass (*Cynodon dactylon*) and filaree (*Erodium cicutarium*). Bermuda grass was planted to control erosion along Cienega Creek in 1902 (Potter 1902), and filaree, supposedly introduced from California in the 1860s, was first planted as forage in the Sulphur Springs and Aravaipa Valleys in the late 1860s and throughout other parts of Arizona by the 1880s (*Arizona Daily Star*, 10 Feb. 1880, 13 June 1886, 14 Dec. 1887; Thornber 1906).

Brought to Arizona from southern Africa in 1932, Lehmann lovegrass (*Eragrostis lehmanniana*), a warm-season, perennial bunchgrass, has been used extensively by the U.S. Department of Agriculture Soil Conservation Service and the Arizona State Department of Transportation to reseed degraded rangelands and control erosion along transportation rights-of-way. In 1940 it began to appear in areas that had not been seeded (Cable 1971; Cox et al. 1987). Lovegrass invaded the grasslands of southeastern Arizona, and by 1986 it had come to dominate about 146,000 ha of grassland and former grassland in the San Pedro, Sulphur Springs, and Santa Cruz Valleys (Anable et al. 1992; Cox and Ruyle 1986; Crider 1945). Lehmann lovegrass appears to have a competitive advantage over native grasses because cattle prefer to graze the latter, reducing their vigor and allowing lovegrass to take over during the summer growing season (Humphrey 1959; Martin 1983). On the other hand, densities of Lehmann lovegrass in 70-year-old livestock exclosures were similar to those in adjacent grazed areas (McClaran and Anable 1992). According to C. E. Bock et al. (1986), the impact of Lehmann lovegrass spread has been dramatic and largely negative. Those authors concluded that sites dominated by a mixture of native perennial grasses support a greater collective variety and abundance of indigenous animals than do monocultures of areas planted to lovegrass. Another nonnative that is expanding its range in southeastern Arizona at elevations below 700 m is buffelgrass (*Pennisetum ciliare*). Although nonnatives are significant members of the modern grassland flora of Arizona, the majority have been introduced only since the turn of the century.

Agriculture

Clearing land for irrigated agriculture has destroyed thousands of hectares of grassland, has created favorable conditions for weeds, and has caused erosion. Although the number of hectares irrigated from surface waters in southeastern Arizona has remained fairly constant since the 1920s, the number of hectares irrigated by groundwater, which skyrocketed after 1940, declined precipitously after 1973 when higher fuel costs, greater depths to groundwater, low agricultural commodity prices, and the higher value of urban land and water sent the agricultural industry dependent on groundwater into decline (Kennedy et al. 1986; Meitl et al. 1983). Meitl et al. (1983) estimated that about 125,000 ha of agricultural land were retired in southern Arizona between 1973 and 1983. It is unlikely that native grasses will reestablish on these cleared lands without direct seeding efforts.

Wild Hay Harvesting

The harvesting of wild, or mesa, hay in the plains and desert grasslands of the major valleys of southeastern Arizona was once a significant economic activity (Bahre 1987). It is one of many historic land uses that either points to or may have resulted in major vegetation changes in the grasslands. Between 1850 and 1920, local wild hay met much of the demand for horse, mule, and burro fodder in the cities, towns, mining camps, and military posts (*Arizona Weekly Star*, 23 May 1878; Bahre 1991; Griffiths 1901; Smith 1910; Willson 1966; Wooton 1916; Rockfellow 1955). In addition, ranchers put up large amounts of wild hay for livestock feed. So significant was wild hay harvesting at one time that hay roads and hay corrals were frequently noted in surveyors' field notes, especially in Cochise County (Surveyors' Field Notes, books 942 [1898], 943 [1898], and 958 [1885]). Today, most of the valleys in which wild hay was cut have been cleared for irrigated agriculture and urban development or taken over by brush and scrubby trees—so much so that it is difficult to believe that one hay contractor in 1902 had 20 mowing machines cutting hay near Fort Grant in Sulphur Springs Valley (*Arizona Daily Star*, 14 Oct. 1902). Mexicans and Amerindians used sickles, knives, and hoes to cut or dig up wild hay as well (Hastings 1959; Rockfellow 1955; Willson 1966). The preferred hay grasses were big sacaton, bristlegrass (*Setaria* spp.), feather fingergrass (*Chloris virgata*), grama grasses, little bluestem (*Schizachyrium scoparium*), red threeawn (*Aristida longiseta*), tobosa (*Hilaria mutica*), and vine mesquite (*Panicum obtusum*); the grama grasses were the most favored (Griffiths 1901; Toumey 1891a).

The largest amounts of wild hay were probably cut before 1899, the year of the first census of wild hay harvested in Arizona. According to the censuses, Arizona farmers harvested 9,524 tons of wild hay in 1899, 8,168 tons in 1909, and 7,802 tons in 1919. Between 1909 and 1919, about 60 percent of this hay was harvested in southeastern Arizona, and nearly 80 percent of that was in Cochise County. These census figures do not include hay harvested by ranchers, Amerindians, and entrepreneurs other than farmers. Yields per hectare ranged from 325 to 350 kg, although yields up to 750 kg/ha were reported for some riparian areas (Bahre 1987; *Pinal Drill*, 25 Sept. 1880; U.S. Department of Commerce 1902, 1913, 1922).

The demand for wild hay in southeastern Arizona declined rapidly after 1910 for a variety of reasons, including the development of motorized transport, the growth of irrigated agriculture, overgrazing of the grasslands, fencing of the public domain, the expansion of private landholdings, clearing of the grasslands, and the invasion of the grasslands by woody plants, which impeded mowing in many formerly harvested areas. For example, Government Draw near Tombstone, which at one time supplied tons of wild hay for Bisbee and Tombstone (*Arizona Daily Star*, 8 Sept. 1905), is now mostly covered by mesquite and acacia. Likewise, once-extensive grasslands along the lower slopes of the Santa Rita, Dragoon, Huachuca, and Mule Mountains used for haying are now covered by woody plants (Bahre 1991; Robert 1869).

That wild hay harvesting itself led to major vegetation changes is conjectural; nevertheless, Smith (1910) believed that hay harvesting contributed significantly to erosion along Rillito Creek in the 1880s. A contemporary newspaper account describes how hay cutters harmed the grama grasses near Silver Bell (*Arizona Daily Star*, 23 Feb. 1882):

> A party just in from Silver Bell District reports that the vast plain of gramma [*sic*] grass west of Tucson is being dug out by the roots, thus totally destroying the hope of the grass starting where it has been cut out. Many tons have been dug out by the roots and brought to this city for sale. The gramma grass of Arizona is the finest pasturage known, and is a source of great wealth in the growth of stock. This grass can be cut without killing the roots, and to this there cannot be urged any objection, and while our solons are passing game laws, it would be wise for them to look after our pasture lands. Gentlemen, give us a law to stop what might be termed barbarous practice of totally destroying our native pasture lands. Unless something is done, the gramma grass will soon be a thing of the past in Arizona.

Summary

The nearly 10,000 years of prehistoric Amerindian occupation of southeastern Arizona's grasslands surely brought about some changes in the native grasslands, but the nature and extent of those changes are unknown. For example, hunting may have been significant in the evolution of the grasslands, especially if it resulted in the extinction of the Pleistocene megafauna, and anthropogenic fires may have caused and/or expanded grasslands and maintained them brush-free. Additional changes may have resulted from agriculture and the selective gathering and spreading of wild plants.

The impact of the Hispanic occupation on the grasslands is similarly difficult to assess because the land-use records from the eighteenth and early nineteenth centuries are both meager and muddled. The impact was probably limited in extent because Hispanic settlement in the region was sparse and confined largely to the upper Santa Cruz Valley. Large-scale cattle ranching in the 1820s and 1830s in the San Bernardino and upper Santa Cruz and San Pedro Valleys appears to have been the only major cultural impact on the grasslands during the Hispanic occupation.

Most of the documented vegetation changes in the grasslands followed the initiation of major Anglo-American settlement in 1870 and included the following: (1) a dramatic increase in brush and scrubby trees, (2) a decline in native grasses, (3) rapid expansion of nonnative plants, and (4) clearing of large areas of grassland for urban and rural development (especially for agriculture and remote subdivisions). Livestock management, soil and water conservation programs, wildfire exclusion, and population increases have all played major roles in these changes. Because much of the grassland in southeastern Arizona, especially in Cochise County, is privately owned, more and more of it is being cleared for remote subdivisions, small hobby ranches, and farming enterprises such as vineyards. The resulting landscape fragmentation may influence the extent of natural processes like fire and water drainage patterns as well as the dispersal of grassland plants and animals. No area of grassland in southeastern Arizona is without human influence.

Literature Cited

Allen, L. S. 1989. Livestock and the Coronado National Forest. Rangelands 11: 14–20.

Anable, M. E., M. P. McClaran, and G. B. Ruyle. 1992. Spread of introduced Lehmann lovegrass (*Eragrostis lehmanniana* Nees.) in southern Arizona, USA. Biological Conservation 61:181–188.

Aschmann, H. H. 1970. Athapaskan expansion in the Southwest. Yearbook of the Association of Pacific Coast Geographers 32:79–97.

Bahre, C. J. 1977. Land-use history of the Research Ranch, Elgin, Arizona. Journal of the Arizona Academy of Science 12 (suppl. 2):1–32.

———. 1985. Wildfire in southeastern Arizona between 1859 and 1890. Desert Plants 7:190–194.

———. 1987. Wild hay harvesting in southern Arizona: a casualty of the march of progress. Journal of Arizona History 28:69–78.

———. 1991. Legacy of Change: Historic Human Impact on Vegetation in the Arizona Borderlands. University of Arizona Press, Tucson.

Bahre, C. J., and M. L. Shelton. 1993. Historic vegetation change, mesquite increases, and climate in southeastern Arizona. Journal of Biogeography 20:489–504.

Baker, R. D., R. S. Maxwell, V. H. Treat, and H. C. Dethloff. 1988. Timeless heritage: a history of the Forest Service in the Southwest. U.S. Department of Agriculture, Forest Service FS-409.

Bancroft, H. H. 1889. History of Arizona and New Mexico, 1530–1888. History Company Publishers, San Francisco.

Bandelier, A. F. 1890. Hemenway Southwestern Archaeological Expedition: contributions to the history of the southwestern portion of the United States. Papers of the Archaeological Institute of America, American series, vol. 5. John Wilson & Son, Cambridge.

Barnes, W. C. 1936. Herds in San Simon Valley. American Forests 42:456–457, 481.

Bartlett, J. R. 1854. Personal Narrative of Explorations and Incidents in Texas, New Mexico, California, Sonora, and Chihuahua Connected with the United States and Mexico Boundary Commission during the Years 1850, '51, and '53. 2 vols. D. Appleton & Company, New York.

Bock, C. E., and J. H. Bock. 1978. Response of birds, small mammals, and vegetation to burning sacaton grasslands in southeastern Arizona. Journal of Range Management 31:296–300.

Bock, C. E., J. H. Bock, K. L. Jepson, and J. C. Ortega. 1986. Ecological effects of planting African lovegrasses in Arizona. National Geographic Research 2:456–463.

Bock, J. H., C. E. Bock, and J. R. McKnight. 1976. A study of the effects of grassland fires at the Research Ranch in southeastern Arizona. Journal of the Arizona Academy of Science 11:49–57.

Bogusch, E. R. 1952. Brush invasion on the Rio Grande plains of Texas. Texas Journal of Science 4:85–90.

Bolton, H. E. (trans. and ed.). 1919. Kino's Historical Memoir of Pimeria Alta. 2 vols. A. H. Clark, Cleveland.

———. 1949. Coronado, Knight of Pueblos and Plains. Whittlesey House, New York.

———, (ed.). 1952. Spanish Exploration in the Southwest, 1542–1706. Barnes & Noble, New York.

Brady, P. R. n.d. MS 89, Brady Collection. Arizona Historical Society, Tucson.

Brandes, R. 1960. Frontier Military Posts of Arizona. Dale Stuart King, Globe, Ariz.

Brinckerhoff, S. B. 1967. The last years of Spanish Arizona, 1786–1821. Arizona and the West 9:5–20.

Brown, A. L. 1950. Shrub invasion of southern Arizona desert grassland. Journal of Range Management 3:172–177.

Brown, C. S. 1893. An adventure in the Huachucas. Overland Monthly 21:524–528.

Brown, D. E., and C. H. Lowe. 1980. Biotic communities of the Southwest. U.S. Department of Agriculture General Technical Report RM-78.

Brown, J. R., and S. Archer. 1989. Woody plant invasion of grasslands: establishment of honey mesquite (*Prosopis glandulosa* var. *glandulosa*) on sites differing in herbaceous biomass and grazing history. Oecologia 80:19–26.

Burgess, T. L., J. E. Bowers, and R. M. Turner. 1991. Exotic plants at the Desert Laboratory, Tucson, Arizona. Madrono 38:96–114.

Cable, D. R. 1967. Fire effects on semidesert grasses and shrubs. Journal of Range Management 20:170–176.

———. 1971. Lehmann lovegrass in the Santa Rita Experimental Range, 1937–1968. Journal of Range Management 24:17–21.

Cameron, C. 1896. Report of Colin Cameron, Esq. Pp. 222–231 *in* U.S. House of Representatives, Annual Report of the Governor of Arizona, 1896. 54th Cong., 2d sess. H. ex. Doc. 5. Serial 3490.

Christiansen, L. D. 1988. The extinction of wild cattle in southern Arizona. Journal of Arizona History 29:89–100.

Cooke, P. St. G. 1938. [Cooke's journal of the march of the Mormon Battalion, 1846–1847.] Pp. 63–240 *in* R. P. Bieber and A. B. Bender (eds.), Exploring Southwest trails, 1846–1854. Southwest Historical Series, vol. 7. Arthur H. Clark, Glendale, Calif.

Cooke, R. U., and R. W. Reeves. 1976. Arroyos and Environmental Change in the American South-West. Clarendon Press, Oxford.

Cox, J. R., A. De Alba-Avila, R. W. Rice, and J. N. Cox. 1993. Biological and physical factors influencing *Acacia constricta* and *Prosopis velutina* establishment in the Sonoran Desert. Journal of Range Management 46:43–48.

Cox, J. R., R. D. Madrigal, and G. W. Frasier. 1987. Survival of perennial grass transplants in the Sonoran Desert of the southwestern U.S.A. Arid Soil Research and Rehabilitation 1:77–87.

Cox, J. R., H. L. Morton, T. N. Johnsen, Jr., G. L. Jordan, S. C. Martin, and L. C. Fierro. 1984. Vegetation restoration in the Chihuahuan and Sonoran Deserts of North America. Rangelands 6:112–116.

Cox, J. R., and G. G. Ruyle. 1986. Influence of climatic and edaphic factors on the distribution of *Eragrostis lehmanniana* Nees. in Arizona, USA. Journal of the Grassland Society of South Africa 3:25–29.

Crider, F. J. 1945. Three introduced lovegrasses for soil conservation. U.S. Department of Agriculture Circular 730.

Darrow, R. A. 1944. Arizona range resources and their utilization. I. Cochise County. University of Arizona Agricultural Experiment Station Technical Bulletin 103:311–366.

Di Peso, C. C. 1951. The Babocomari Village Site on the Babocomari River, Southeastern Arizona. Amerind Foundation, Dragoon, Ariz.

———. 1953. The Sobaipuri Indians of the Upper San Pedro River Valley, Southeastern Arizona. Amerind Foundation, Dragoon, Ariz.

———. 1956. The Upper Pima of San Cayetano del Tumacácori. Amerind Foundation, Dragoon, Ariz.

———. 1958. The Reeves Ruin of Southeastern Arizona: A Study of a Prehistoric Western Pueblo Migration into the Middle San Pedro Valley. Amerind Foundation, Dragoon, Ariz.

Dobyns, H. F. 1981. From Fire to Flood: Historic Human Destruction of Sonoran Desert Riverine Oases. Ballena Press Anthropology Papers 20. Socorro, N. Mex.

Doebley, J. F. 1984. Seeds of wild grasses: a major food of southwestern Indians. Economic Botany 38:52–64.

Emory, W. H. 1857. Report on the United States and Mexican Boundary Survey. 2 vols. 34th Cong., 1st sess. S. ex. Doc. 108. A. O. P. Nicholson, Washington, D.C.

Fish, S. K., and G. P. Nabhan. 1991. Desert as context: the Hohokam environment. Pp. 39–60 *in* G. J. Gumerman (ed.), Exploring the Hohokam: prehistoric peoples of the American Southwest. University of New Mexico Press, Albuquerque.

Gird, R. 1907. True story of the discovery of Tombstone. Out West 27:39–50.

Glendening, G. E., and H. A. Paulsen, Jr. 1955. Reproduction and establishment of velvet mesquite as related to invasion of semidesert grasslands. U.S. Department of Agriculture, Forest Service Technical Bulletin 1127.

Goetzmann, W. H. 1959. Army Exploration in the American West, 1803–1863. Yale University Press, New Haven.

Greever, W. S. 1957. Railway development in the Southwest. New Mexico Historical Review 32:151–203.

Griffiths, D. A. 1901. Range improvement in Arizona. U.S. Department of Agriculture, Bureau of Plant Industry Bulletin 4.

Gumerman, G. J. 1991. Understanding the Hohokam. Pp. 1–28 *in* G. J. Gumerman (ed.), Exploring the Hohokam: prehistoric peoples of the American Southwest. University of New Mexico Press, Albuquerque.

Hamernick, D. M., and B. A. Brown. 1975. Arizona's remote subdivisions: an inventory. Office of Economic Planning and Development, Office of the Governor of Arizona, Phoenix.

Harris, D. R. 1966. Recent plant invasions in the arid and semiarid Southwest of the United States. Annals of the Association of American Geographers 56:408–422.

Haskell, H. S. 1945. Successional trends on a conservatively grazed desert grassland range. Journal of the American Society of Agronomists 37:978–990.

Haskett, B. 1935. Early history of the cattle industry in Arizona. Arizona Historical Review 6:3–42.

Hastings, J. R. 1959. Vegetation change and arroyo cutting in southeastern Arizona. Journal of the Arizona Academy of Science 1:60–67.

Hastings, J. R., and R. M. Turner. 1965. The Changing Mile: An Ecological Study of Vegetation Change with Time in the Lower Mile of an Arid and Semiarid Region. University of Arizona Press, Tucson.

Haury, E. W. 1976. The Hohokam: Desert Farmers and Craftsmen. Excavations at Snaketown, 1964–1965. University of Arizona Press, Tucson.

Haury, E. W., E. Antevs, and J. R. Lance. 1953. Artifacts with mammoth remains, Naco, Arizona. American Antiquity 19:1–24.

Haury, E. W., E. B. Sayles, and W. W. Wasley. 1959. The Lehner mammoth site, southeastern Arizona. American Antiquity 25:2–30.

Haynes, C. V. 1991. Geoarchaeological and paleohydrological evidence for a Clovis-Age drought in North America and its bearing on extinction. Quaternary Research 35:438–450.

———. 1992. Contributions of radiocarbon dating to the geochronology of the peopling of the New World. Pp. 355–374 *in* R. E. Taylor, A. Long, and R. S. Kra (eds.), Radiocarbon after four decades. Springer-Verlag, New York.

Haynes, C. V., and E. W. Haury. 1982. Archaeological investigations at the Lehner site, Arizona, 1974–1975. National Geographic Society Research Reports 14:325–334.

Hecht, M. E., and R. W. Reeves. 1981. Atlas of Arizona. Office of Arid Lands Studies, University of Arizona, Tucson.

Hemmings, E. T. 1970. Early man in the San Pedro Valley, Arizona. Ph.D. dissertation, University of Arizona, Tucson.

Hemmings, E. T., and C. V. Haynes. 1969. The Escapule mammoth and associated projectile points, San Pedro Valley, Arizona. Journal of the Arizona Academy of Science 5:184–188.

Hendrickson, D. A., and W. L. Minckley. 1984. Cienegas—vanishing climax communities of the American Southwest. Desert Plants 6:131–175.

Hinton, R. J. 1878. The Handbook to Arizona: Its Resources, History, Towns, Mines, Ruins, and Scenery. American News Company, New York.

Huckell, B. B. 1984. The archaic occupation of the Rosemont area, northern Santa Rita Mountains, southeastern Arizona. Arizona State Museum Archaeological Series no. 147, vol. 1. Arizona State Museum, University of Arizona, Tucson.

Humphrey, R. R. 1956. History of vegetational changes in Arizona. Arizona Cattlelog 11:32–35.

———. 1958. The desert grassland: a history of vegetational change and an analysis of causes. Botanical Review 24:193–252.

———. 1959. Lehmann's lovegrass, pros and cons. P. 28 *in* R. R. Humphrey (comp.), Your range—its management. University of Arizona Agricultural Experiment Station Special Report 2.

Humphrey, R. R., and L. A. Mehrhoff. 1958. Vegetation change on a southern Arizona grassland range. Ecology 39:720–726.

Johnston, M. C. 1963. Past and present grasslands of southern Texas and northeastern Mexico. Ecology 44:456–466.

Jordan, G. L., and M. L. Maynard. 1970. The San Simon watershed—historical review. Progressive Agriculture in Arizona 22:10–13.

Kearney, T. H., and R. H. Peebles. 1942. Flowering plants and ferns of Arizona. U.S. Department of Agriculture Miscellaneous Publication 423.

Kennedy, C. B., M. M. Karpiscak, and M. C. Parton. 1986. Geographic analysis of agricultural land retirement, Cochise County, Arizona. Pp. 4–6 *in* Proceedings of the Conference on Remote Sensing and Geographic Information Systems in Management, 6–7 November 1986. Arizona Remote Sensing Center, Office of Arid Lands Studies, College of Agriculture, University of Arizona, Tucson.

Kessell, J. L. 1970. Mission of Sorrows: Jesuit Guévavi and the Pimas, 1691–1767. University of Arizona Press, Tucson.

Komarek, E. V. 1968. The nature of lightning fires. Proceedings of the Tall Timbers Fire Ecology Conference 7:5–41.

Lauver, M. E. 1938. A history of the use and management of the forested lands of Arizona 1862–1936. M.S. thesis, University of Arizona, Tucson.

Leopold, A. 1924. Grass, brush, timber, and fire in southern Arizona. Journal of Forestry 22:1–10.

Malin, J. C. 1953. Soil, animal, and plant relations of the grassland, historically reconsidered. Scientific Monthly 76:207–220.

Martin, Paul Schultz. 1975. Vanishings and future of the prairie. Pp. 39–49 *in* R. Kesell (ed.), Grasslands ecology—a symposium. Geoscience and Man, vol. 10. Louisiana State University Press, Baton Rouge.

Martin, P. S. 1979. Prehistory: Mogollon. Pp. 61–74 *in* W. S. Sturtevant and A. Ortiz (eds.), Handbook of North American Indians. Vol. 9: The Southwest. Smithsonian Institution, Washington, D.C.

Martin, S. C. 1975. Ecology and management of southwestern semi-desert grass-shrub ranges: the status of our knowledge. U.S. Department of Agriculture Forest Service Research Paper RM-156.

———. 1983. Responses of semiarid grasses and shrubs to fall burning. Journal of Range Management 36:604–610.

Mattison, R. H. 1946. Early Spanish and Mexican settlements in Arizona. New Mexico Historical Review 21:273–327.

McCarty, K. 1976. Desert documentary: the Spanish years, 1776–1821. Historical Monograph no. 4. Arizona Historical Society, Tucson.

McClaran, M. P., and M. E. Anable. 1992. Spread of introduced Lehmann lovegrass along a grazing intensity gradient. Journal of Applied Ecology 29: 92–98.

Meitl, J. M., P. L. Hathaway, and F. Gregg. 1983. Alternative uses of Arizona lands retired from irrigated agriculture. Cooperative Extension Service, College of Agriculture, University of Arizona, Tucson.

Moorhead, M. 1968. The Apache Frontier: Jacobo Ugarte and Spanish-Indian Relations in Northern New Spain, 1769–1791. University of Oklahoma Press, Norman.

Morrisey, R. J. 1950. The early range cattle industry in Arizona. Agricultural History 24:151–156.

Myrick, D. F. 1975. Railroads of Arizona. Vol. 1: The Southern Roads. Howell-North Books, Berkeley.

Nentvig, J. 1980. Rudo Ensayo: A Description of Sonora and Arizona in 1764. Trans. A. F. Pradeau and R. R. Rasmussen. University of Arizona Press, Tucson.

Officer, J. E. 1987. Hispanic Arizona, 1536–1856. University of Arizona Press, Tucson.

Parker, K. W., and S. C. Martin. 1952. The mesquite problem on southern Arizona ranges. U.S. Department of Agriculture Circular 908.

Peterson, H. V. 1950. The problem of gullying in western valleys. Pp. 407–434 *in* P. D. Trask (ed.), Applied sedimentation. John Wiley & Sons, New York.

Potter, A. F. 1902. Report of the examination of the proposed Santa Rita Forest Reserve. Range conditions in Arizona, 1900–1909, as recorded by various observers in a series of miscellaneous papers. Special Collections Library, University of Arizona, Tucson.

Pyne, S. J. 1982. Fire in America. Princeton University Press, Princeton.

———. 1984. Introduction to Wildland Fire. John Wiley & Sons, New York.

Robert, H. M. 1869. Map of southern Arizona. Bancroft Library, University of California, Berkeley.

Rockfellow, J. A. 1955. Log of an Arizona Trail Blazer. Arizona Silhouettes, Tucson.

Rodgers, W. M. 1965. Historical land occupance of the upper San Pedro River valley since 1870. M.S. thesis, University of Arizona, Tucson.

Sauer, C. O. 1932. The Road to Cibola. Ibero-Americana, vol. 3. University of California Press, Berkeley.

———. 1935. Aboriginal Population of Northwestern Mexico. Ibero-Americana, vol. 10. University of California Press, Berkeley.

———. 1944. A geographic sketch of early man in America. Geographical Review 34:529–573.

Sayles, E. B., and E. Antevs. 1941. The Cochise Culture. Medallion Paper 29. Gila Pueblo, Globe, Ariz.

Schroeder, M. J., and C. C. Buck. 1970. Fire weather: a guide for application of meteorological information to forest fire control operations. U.S. Department of Agriculture Handbook 360.

Serven, J. E. 1965. The military posts on Sonoita Creek. Smoke Signal no. 12. Tucson Corral of the Westerners, Tucson.

Sheridan, T. E. 1986. Los Tucsonenses: The Mexican Community in Tucson, 1854–1941. University of Arizona Press, Tucson.

Smith, G. E. P. 1910. Ground water supply and irrigation in the Rillito Valley. University of Arizona Agricultural Experiment Station Bulletin 64.

Surveyors' Field Notes. 1885–1898. Books 942, 943, 958. U.S. Department of Interior, Bureau of Land Management, Phoenix, Ariz.

Thornber, J. J. 1906. Alfilaria (*Erodium cicutarium*) as a forage plant in Arizona. University of Arizona Agricultural Experiment Station Bulletin 52.

———. 1910. The grazing ranges of Arizona. University of Arizona Agricultural Experiment Station Bulletin 65.

Toumey, J. W. 1891a. I. Notes of some of the range grasses of Arizona. University of Arizona Agricultural Experiment Station Bulletin 2.

———. 1891b. II. Overstocking the range. University of Arizona Agricultural Experiment Station Bulletin 2.

U.S. Department of Commerce, Bureau of the Census. 1864. Eighth Census of the United States, 1860: Population. Washington, D.C.

———. 1883. Tenth Census of the United States, 1880: Population. Washington, D.C.

———. 1902. Twelfth Census of the United States, 1900: Agriculture, vol. 6, pt. 2. Washington, D.C.

———. 1913. Thirteenth Census of the United States, 1910: Agriculture, vol. 6. Washington, D.C.

———. 1922. Fourteenth Census of the United States, 1920: Agriculture, vol. 6, pt. 3. Washington, D.C.

———. 1992. Twentieth Census of the United States, 1990: Population (CP-1-4). Washington, D.C.

U.S. Department of the Interior, Bureau of Land Management. 1978. Final environmental statement—upper Gila–San Simon. Safford, Ariz.

U.S. House of Representatives. 1883. Annual report of the Governor of Arizona. 48th Cong., 1st sess. H. ex. Doc. I. Serial 2191.

———. 1893. Annual report of the Governor of Arizona. 53d Cong., 2d sess. H. ex. Doc. I. Serial 3211.

Voigt, W., Jr. 1976. Public Grazing Lands: Use and Misuse by Industry and Government. Rutgers University Press, New Brunswick, N.J.

Wagoner, J. J. 1951. Development of the cattle industry in southern Arizona, 1870s and 1880s. New Mexico Historical Review 26:204–224.

———. 1952. History of the cattle industry in southern Arizona, 1540–1940. University of Arizona Social Science Bulletin 20. University of Arizona Press, Tucson.

———. 1961. Overstocking of the ranges in southern Arizona during the 1870s and 1880s. Arizoniana 2:23–27.

———. 1975. Early Arizona. University of Arizona Press, Tucson.

Weber, D. J. 1982. The Mexican Frontier, 1821–1846: The American Southwest under Mexico. University of New Mexico Press, Albuquerque.

Whalen, N. M. 1971. Cochise Culture sites in the central San Pedro drainage, Arizona. Ph.D. dissertation, University of Arizona, Tucson.

Willson, R. 1966. Pioneers chopped their hay. *Arizona Republic* (Phoenix), July 3.

Wooton, E. O. 1916. Carrying capacity of grazing ranges in southern Arizona. U.S. Department of Agriculture Bulletin 367.

Wright, H. A. 1980. The role and use of fire in the semidesert grass-shrub type. U.S. Department of Agriculture, Forest Service General Technical Report INT-85.

Wright, H. A., and A. W. Bailey. 1982. Fire Ecology. John Wiley & Sons, New York.

9

Revegetation in the Desert Grassland

Bruce A. Roundy and Sharon H. Biedenbender

The vegetation of the southwestern desert grasslands has changed dramatically since the late 1800s as grasses have decreased and woody plants have increased. The possible causes of this vegetation change were discussed in detail by Cooke and Reeves (1976), Hastings and Turner (1965), Humphrey (1958), Schmutz et al. (1991), and, most recently, by Bahre (1991, this volume), although these authors disagree about the relative importance of livestock grazing, fire, and climatic variation in driving the change. Scientists' and stockmen's early recognition that the grass was disappearing led to an interest in developing appropriate range management practices, including revegetation to restore the grass. Numerous early field trials sought to determine where, when, and how to direct seed or transplant native grasses. Frequent failures, often associated with the lack of consistent precipitation in the grasslands, led to a desire to find more drought-tolerant species. Because the emphasis of range revegetation was on establishing grass for forage production and soil conservation, superior grass species were sought and tested.

After they were introduced in the 1930s, a few nonnative species, especially the African lovegrasses, showed that they seeded and established themselves better than native species. Subsequent revegetation studies sought to determine (1) site potential for revegetation, (2) superior selected native and nonnative cultivars, (3) the best way to prepare seedbeds, and (4) the seeding methods that would optimize establishment success. Highly successful agricultural production, higher energy costs, and increased environmental awareness in the United States after the 1960s led to a deemphasis on production goals for revegetation and a new emphasis on reclamation of disturbed lands and restoration of floral and faunal diversity. These new demands require that revegetation be developed not only as a technology, as in the past, but also as a science, to determine processes of native plant establishment and species coexistence.

This chapter examines past and current goals and approaches to revegetating the desert grassland and the development of seedbed ecology and revegetation science. By presenting a perspective of the past, we hope to clarify future directions and goals.

Vegetation Changes Leading to Early Interest in Revegetation

In the late 1800s ranchers and researchers began noticing with alarm the decline in perennial grasses on rangelands in Arizona (Bahre, this volume). Griffiths, writing in 1901, when Arizona was still a territory and the federal government had not yet assumed control over the use of rangelands, summarized the problem: "The free-range system has led to the ruthless destruction of the native grasses which once covered the magnificent pasture lands of the West, and the time has now come when active measures must be adopted to remedy the evils that have resulted from overstocking and mismanagement" (Griffiths 1901).

It was not the vegetation changes themselves that sparked this early interest in revegetation but their effect on the burgeoning cattle industry. By 1891, cattle numbers in Arizona were at their peak, considered by some to be more than 1.5 million head (Haskett 1935). With the ranges stocked to full capacity, the 1891–1893 drought had disastrous consequences for both vegetation and cattle. In Haskett's words (1935), "In the spring of 1891 the bubble burst; droughty conditions, the first of any consequence, brought on a state of affairs that increased in intensity throughout the next two years. By June, 1892, the grass had practically all disappeared from the ranges, many of the waterholes had failed and cattle losses had been heavy . . . the drought ended in July, 1893. Conservative estimates place the loss of cattle at 50%, and some ranchmen say that it ran as high as 75%."

Subsequent droughts in Arizona after the turn of the century prompted Thornber to document an "extreme shortage of feed, as for example the winter and spring seasons of 1902, 1904, and 1910, when there was little or practically no winter rainfall on the ranges" (Thornber 1911). Thornber chastised the cattle industry for shortsightedness in failing to manage for drought years or provide winter feed, noting that "the complete utilization of our grazing ranges . . . [and] . . . the pronounced tendency on every hand to overstock" made cattle grazing a gamble dependent on timely and adequate precipitation.

Despite the severe losses suffered by cattle ranchers at the turn of the century and the admonitions of early researchers, overstocking and resultant range degradation continued for the next several decades. In 1931, Wilson reported that "tens of millions of acres of range land in the Southwest" were so degraded that natural recovery, further impeded by drought and rodent herbivory, was impossible. In 1940 there was more than a million hectares of deteriorated semidesert rangeland in Arizona, New Mexico, and southwestern Texas (Cassaday and Glendening 1940). According to Flory and Marshall (1942), the decline in range forage production and the loss of topsoil continued at "an alarming rate." They predicted widespread desertification of the Southwest "if drastic and immediate action is not taken to stop further deterioration" (Flory and Marshall 1942). In 1944, Darrow reported that levels of livestock utilization in Cochise County still exceeded the forage capacity of the range.

Although drought was the precipitating factor, early researchers placed the blame for the decline of perennial grasses squarely on overgrazing, which weakened grass root systems, making them unable to withstand drought, set seed for regeneration, or recover when precipitation finally arrived (Flory and Marshall 1942; Griffiths 1901; Parker 1939a,b; Toumey 1891).

As stands of native perennial grasses declined, there was an increase in the incidence of "inferior, worthless, and poisonous plants" (Forsling and Dayton 1931). Parker (1939a) stated that "proper management of livestock offers the most practical means of avoiding death loss from poisonous plants" because "nearly all noxious plants are unpalatable and are eaten only in time of feed shortage," and "when the grasses and other palatable herbs and browse are heavily grazed, natural competition from these is greatly reduced so that there is little to prevent the increase of undesirable and generally unpalatable range plants."

At the turn of the century, when overstocking and droughts were at their most severe, ranchers valued woody plants and cacti as emergency feed that could prevent starvation (Forbes 1895; Thornber 1904; Wooton 1896). For example, Wooton (1896) described mesquite (*Prosopis* spp.) as "a very valuable plant in this locality" for honey, firewood, and stock feed, both the foliage and the seed pods. But attitudes changed when woody species, particularly mesquite, invaded former grasslands. In 1931, mesquite was sometimes "referred to as offering possibilities for range improvement; while again some of the stockmen are inclined to deplore the fact that mesquite is gradually coming in on some of the depleted ranges" (Wilson 1931). By the late 1930s, eradication of woody grassland invaders, primarily mesquite, burroweed (*Isocoma tenuisecta*), and snakeweed (*Gutierrezia sarothrae*), had become a priority (Darrow 1944; Martin 1949; Parker 1939b, 1940).

Opinions varied regarding the cause of the spread of mesquite and other woody species (Bahre, this volume). Cattle disseminated the seeds by consuming mesquite pods, and the overgrazed grasses' reduced ability to compete certainly was a factor (Forsling and Dayton 1931; Glendening and Paulsen 1955). Humphrey (1953, 1958) believed that desert grassland was a subclimax stage maintained by frequent fires that killed the seedlings of woody vegetation. In his view, decreased fire frequency resulting from fire control and the lack of sufficient fuel due to removal of grass by grazing were primarily responsible for the invasion of woody species into former grasslands.

Regardless of the causes, the decrease in grass and the increase in woody and poisonous plant species led to revegetation efforts to stabilize soil and provide livestock forage.

Early Trials and Approaches

New Research Agencies

The Arizona Agricultural Experiment Station was established in Tucson in 1889 in cooperation with the University of Arizona. At that time the cattle industry was at its peak and had not yet suffered the disastrous consequences of overstocking combined with drought. In the Arizona Agricultural Experiment Station's second bulletin, titled "Notes on Some of the Range Grasses of Arizona—Overstocking of the Range," Toumey (1891) observed that the cattle were in poor condition, forced to subsist on shrubs due to the lack of grass.

Only a decade later, Griffiths (1901) documented the effects of overgrazing on Arizona rangelands and established the Arizona Agricultural

Experiment Station Range Reserve Tract in Tucson (now part of Davis Monthan Air Force Base) to conduct revegetation trials. In 1904 he expanded his research to include the Santa Rita Range Reserve, which was then a 14,500-ha grazing exclosure on the foothills of the Santa Rita Mountains about 50 km southeast of Tucson. Later the reserve was expanded to 20,000 ha and became known as the Santa Rita Experimental Range (McClaran, this volume).

Native Perennial Grasses

Griffiths (1907) and other early revegetation researchers seeded valuable native forage grasses such as beardgrasses, or bluestems (*Bothriochloa* [*Andropogon*] spp.); curly mesquite (*Hilaria belangeri*); dropseeds (*Sporobolus* spp.); grama grasses (*Bouteloua* spp.); threeawns (*Aristida* spp.); and tobosa (*Hilaria mutica*). Cultivating on the contour with horse-drawn disks and fine-tooth harrows in various combinations, Griffiths planted his first plots in Tucson in January 1901 (fig. 9.1). His trials over the course of the next 10 years, both in Tucson and on the Santa Rita Range Reserve, failed to show emergence, establishment, or persistence of native perennial grasses. He cited problems with poor germination and obtaining enough native seed for large-scale plantings. He also recognized that perennial grasses became established naturally from seed in the Southwest only sporadically and that successful revegetation could not be accomplished by merely scattering the seeds. But the expense of cultivating huge areas of rangeland was prohibitive. He concluded that successful revegetation attempts would be confined to mountain meadows or cultivated, usually irrigated wheat and ryegrass seedings in the prairie states.

The researchers who followed in Griffiths's footsteps over the course of the next several decades obtained good results in germination tests with native grasses. For example, C. P. Wilson (1928, 1931), at the New Mexico Agricultural Experiment Station, found that "seeds of many New Mexico range forage plants germinate considerably better than generally supposed." In 1937, E. L. Little, of the Southwestern Forest and Range Experiment Station (serving Arizona, New Mexico, and western Texas), summarized the results of germination tests on 22 native grasses since 1923. The germination of most species was quite good, although there was high variation between tests as a result of variable methods, seed ages, accessions (collections from different populations), environmental conditions, and stages of seed maturity. Low germination was typical of species with hard seed coats

Figure 9.1. While working for the Bureau of Plant Industry, David Griffiths conducted early revegetation trials on part of what is now Davis Monthan Air Force Base in Tucson, Arizona. Photograph from Griffiths (1901).

that had not been scarified. Cassaday (1937) used mechanical scarification of seeds to increase the germination of dropseed species from 2–4 percent to 48 percent. Bridges (1941) reported very good germination rates for 121 native and nonnative grasses.

Despite good germination results in the lab, however, the researchers following Griffiths were unable to achieve better emergence or survival success in the field. Forsling and Dayton (1931) summarized the results of experiments conducted on western national forest lands since 1902: "In general, artificial reseeding tests with native and cultivated plants on the range in southern Arizona have thus far been distinct failures, except perhaps where there is summer flooding of the areas." They agreed with Griffiths that cost-effective reseeding was likely only in areas of relatively high precipitation such as mountain meadows.

Wilson (1931) reported that establishment was often poor because young seedlings that germinated during the rainy season died during summer droughts. McGinnies and Arnold (1939) conducted an experiment from 1931 to 1936 on the Santa Rita Experimental Range to determine the relative water requirements of 28 selected range grasses, shrubs, and forbs in relation to weather conditions. They identified the perennial grasses that used water most efficiently during the summer and the cooler seasons. Blue grama (*Bouteloua gracilis*) was highly water-use efficient in summer but not in winter, while tanglehead (*Heteropogon contortus*) was highly water-use efficient in fall, winter, and spring.

Bridges (1941) believed that the failures in the field were caused by unsuitable growing conditions. Reseeding trials were often conducted in areas that had been so severely degraded that the loss of soil, litter, and native seed banks combined to create an inhospitable setting for natural or artificial revegetation. From 1936 to 1941, he conducted seeding trials on the Jornada del Muerto Plain in New Mexico, where average annual precipitation was only 237 mm. He tested 121 species of grasses using various seedbed preparations with and without barley straw mulch. Contour furrowing to catch and hold precipitation was the most successful method for soil preparation. Mulch produced a "marked" increase in germination and survival of most species during both the first and second years. Several native grasses gave promising results, including Rothrock grama (*Bouteloua rothrockii*), sideoats grama (*B. curtipendula*), and sand dropseed (*Sporobolus cryptandrus*). Bridges also had good results with Lehmann lovegrass (*Eragrostis lehmanniana*), a southern African species available at that time only from the Soil Conservation Service for its cooperators but which was expected to become commercially available by 1942.

Continuing studies with native grasses emphasized intensive, expensive treatments such as cultivation, contouring, and mulching. J. T. Cassaday and G. E. Glendening, at the Southwestern Forest and Range Experiment Station, identified the single most limiting factor to range reseeding as the rapid drying of the soil surface layer (Cassaday and Glendening 1940). They described seeding and transplanting methods, contour furrows, and more elaborate water-conserving practices such as crescent basins and check dams, mulch application, and rodent control (fig. 9.2). Glendening (1942) expanded on the mulch experiments on the Santa Rita Experiment Range in 1942, testing 10 native grasses under mulches of barley straw, chopped burroweed stems, and open-mesh gauze fabric after he had observed seedling emergence at or near the soil surface under the mulch. The mulch, by conserving soil moisture and ameliorating soil temperatures, increased emergence by 400–2,000 percent compared with bare ground. These were expensive techniques, however, designed for revegetating small, high-potential areas, which would then serve as seed sources for the eventual revegetation of larger areas.

Flory and Marshall (1942) summarized the Soil Conservation Service's

Figure 9.2. Contour furrowing to catch and concentrate limited precipitation was recommended by Cassaday and Glendening (1940) to enhance soil conservation and revegetate deteriorated desert grasslands. Grass growth has increased on the uphill (left) side of the contour.

revegetation work in Arizona and New Mexico on 30 native and 6 non-native grasses, including harvesting, cleaning, and planting recommendations. During the trials the Soil Conservation Service drilled (mechanically planted) native seeds with awns, hairs, and other appendages using modified equipment. The seeds were planted between 15 June and 30 July to avoid germination in response to the intermittent early-summer rains. Seedlings that germinated during early summer, when precipitation was least reliable, were likely to die. The Soil Conservation Service conducted successful nursery and field trials on the native grasses Arizona cottontop (*Digitaria californica*), blue grama, bush muhly (*Muhlenbergia porteri*), curly mesquite, little bluestem (*Schizachyrium scoparius*), plains bristlegrass (*Setaria macrostachya*), and Rothrock grama. Commercial production of many native species was expected to follow soon afterward.

Shrubs and Cacti

Early revegetation studies were not limited to perennial grasses. There was a great deal of interest in prickly pear and cholla cacti (*Opuntia* spp.) as emergency sources of forage. Although not conducive to weight gain, these succulents could forestall starvation and provide a source of water during drought. Laboratory studies (Thornber 1904) indicated that 75–80 percent of the cactus stems and fruit was water, and that their protein content was about equal to that of blue grama. Work initiated in 1899 by Toumey and continued by Thornber (1904, 1911) on spineless cactus species from Mexico, South America, and Italy proved unsuccessful because the plants were defenseless against rodents and cattle and were sensitive to frost.

Efforts turned to establishing native species whose spines could be singed to make them palatable to cattle, which ate them so voraciously that Thornber recommended singeing only half the cacti in an area to avoid the total destruction of the forage resource. From 1906 to 1909, Thornber planted hundreds of propagules of spiny cacti in several locations in Tucson and on nearby ranges. Mortality was high due to herbivory, low precipitation, poor rooting, sandy soil, or soil underlain with caliche. He recommended that cuttings with at least two or three joints be planted 22–25 cm deep in the winter during years when precipitation was favorable (Thornber 1911). Planting was done in contour furrows created by plowing that were subsequently covered by a second pass of the plow. A third pass created a water catchment to enhance establishment. Thornber estimated that 10 years would be required for the cacti to mature enough for the first harvest, which could then be repeated every 5 years.

Palatable shrubs and perennial grasses were also lost to overgrazing. Fourwing saltbush, or chamiza (*Atriplex canescens*), was a particularly valuable forage during droughts, heavy snowfall, and seasons when grasses were inactive (Griffiths 1901, 1904; Wilson 1928). Chamiza and other shrubs such as winterfat (*Ceratoides lanata*) were included in many germination and seeding experiments along with native and nonnative grasses (Bridges 1941; Griffiths 1901, 1904; Griswold 1936; McGinnies and Arnold 1939; Wilson 1928, 1931). Wilson (1928) studied 11 chamiza accessions and found that despite low germination (9.2 percent) a good stand of a several hundred plants per hectare could be established if planted at a depth of 6–38 mm and protected from herbivory by cattle and native predators (quail, rabbits, mice, and ground squirrels). Wilson determined that chamiza performed better on calcareous soil and that the seedlings were very drought tolerant due to the rapid growth of the taproot.

Miracle Plants, Machines, and Sowing Methods

Plant Trials

Even the earliest revegetation experiments included plants from Australia and Eurasia that had been successfully grown in more mesic areas. Griffiths (1901, 1904) seeded rescuegrass (*Bromus unioloides*) and saltbush (*Atriplex halimoides*) from Australia and several grasses from the northwestern United States, such as annual timothy (*Phleum asperum*), bluebunch wheatgrass (*Agropyron spicatum*), and wild rye (*Elymus* spp.). Some of the latter species germinated following January seeding in response to winter precipitation but died during the ensuing late spring and early summer dry periods. Wilson (1931) received seeds of several Australian grass and shrub species in 1930 for trials in New Mexico. Interestingly, he commented, "It seems hardly probable that any of the [Australian] species are adapted to New Mexico conditions" (Wilson 1931). Forsling and Dayton (1931) expressed the prophetic hope that further research would discover nonnative species preadapted for the Southwest, pointing out that nearly all of the forage plants extensively cultivated in the humid regions of the United States had been introduced from the Old World.

Other researchers (Griffiths 1907; Thornber 1906; Wilson 1931) were interested in seeding filaree, so common on California and Arizona ranges from such an early date that some taxonomists considered it to be native. Thornber found alfilaria, as it was called, to be comparable in nutritional value—including protein content—to alfalfa hay, and ranchers welcomed it as emergency feed for cattle and sheep. It germinated and formed succulent rosettes in the winter, made its greatest growth in the spring, and continued to provide palatable forage into May. Although dense stands excluded native winter annuals, alfilaria did not compete with summer grasses. Thornber recommended that alfilaria be purposefully introduced using sheep, which would not only distribute the seeds but would remove competing vegetation by their close grazing and bury the seeds by trampling. Seeding by more traditional methods proved difficult. Seeds collected in the spring and stored until their natural season of germination in the fall did not germinate well, but seeds sown immediately after spring collection suffered from herbivory and dissipation. Without animals to disperse it, filaree spread very slowly from seeded areas. Thornber suggested collecting seeds by hand due to the "prohibitive price of $1.00 a pound" on the California market.

In 1904, Griffiths described the efforts of rancher Harry L. Heffner of the Empire Cattle Company to reduce gully erosion by planting cuttings of the nonnative Johnson grass (*Sorghum halepense*) at water outlets that flooded one to three times a year. Another nonnative, Bermuda grass (*Cynodon dactylon*), was also tried but was found to require more water for successful establishment. In 1931, Wilson reported that Bermuda grass and Johnson grass were good choices for erosion control but had become serious weeds in many of the warmer irrigated valleys.

The failure of native and introduced species to grow in early revegetation trials at lower elevations led researchers to continue their pursuit of exceptional plants that could establish and produce forage when precipitation was limited (Cox et al. 1982). These efforts led conservationists from the United States, Australia, England, and South Africa to collect and exchange seeds for revegetation from similar rangelands throughout the world (Cox et al. 1988).

By 1932, F. J. Crider, director of the Boyce Thompson Southwestern Arboretum, had received seeds of Boer (*Eragrostis. curvula* var. *conferta*), Lehmann, and weeping (*E. curvula*) lovegrasses from South Africa. He received additional collections of these species in 1934 while he was in charge of the Southwestern Station Erosion Plant Studies in the Division of Plant Exploration and Introduction. In 1935, after being transferred to the Soil Conservation Service, Crider tested these species in irrigated trials at Tucson and selected accessions that matured early and produced seed the first season (Cox et al. 1988; Crider 1945). After the initial introductions, additional accessions of Boer lovegrass, blue panicgrass (*Panicum antidotale*), buffelgrass (*Pennisetum ciliare*), Cochise lovegrass (*Eragrostis lehmanniana* × *E. trichophora*), King Ranch bluestem (*Bothriochloa ischaemum*), Kleingrass (*Panicum coloratum*), weeping lovegrass, Wilman lovegrass (*E. superba*), and others were introduced and field tested. The success of nonnative grasses not only in the desert grassland but also in the big sagebrush (*Artemisia tridentata*) desertscrub in the Great Basin sparked additional research and large-scale revegetation projects throughout the West.

Bridges's (1941) field trials at the Jornada Experimental Range on the lower margin of the desert grassland in New Mexico illustrate the success of nonnative grass species. Of 174 native and introduced grasses, shrubs, and forbs, 2 introduced grasses (Lehmann lovegrass and Boer lovegrass) and 1 native grass (Rothrock grama) were the only species that produced good grass stands in all three trial years. Of the native and Australian shrubs, only the native fourwing saltbush became established in some trials. A few other native grasses such as plains bristlegrass, sideoats grama, Arizona cottontop, and tanglehead established themselves to a limited degree in contour furrows that caught runoff and thus remained moist for a longer time.

Bridges (1941) noted that the native Rothrock grama and the nonnative Boer and Lehmann lovegrasses exhibited characteristics such as high seed production and germinability that facilitated their establishment in open areas. These are the same characteristics that allow species such as Lehmann lovegrass to spread and invade disturbed and undisturbed sites in the desert grassland (Anable et al. 1992).

Such pioneering ability was considered valuable by early researchers such as Crider (1945) and Bridges (1941, 1942), who were greatly concerned about soil conservation and forage for livestock production. Remember that World Wars I and II and the associated needs for food and fiber production occurred within the lifetimes and during part of the professional careers of these researchers. This may explain their emphasis on determining successful plant species and methods of propagation regardless of the provenance of the species used.

After World War II, revegetation research continued to focus on the potential of native and nonnative species to establish and persist on different sites, methods of enhancing establishment of promising species, and determination of the forage value of successful species (Cox et al. 1982). Although some native desert grassland grasses, forbs, and shrubs became established under the right conditions in favorable years, nonnative perennial grasses established and persisted most consistently (Cox et al. 1982; Judd and Judd 1976). In their review of 92 years of southwestern rangeland seedings that included 300 species and 400 planting sites, Cox et al. (1982) concluded that only 11 grasses could be expected to successfully establish in one out of two or three years, and then only in specific geographical areas. Of these 11 grasses, only sideoats grama and green sprangletop (*Leptochloa dubia*) were native species, and they were recommended for seeding higher elevations (1,385–1,850 m) within the Chihuahuan interior plateau. Even on Sonoran interior upland, some of the best grassland in southern Arizona, only 3 grasses were considered by Cox et al. (1982) to be able to consistently establish, and these were all nonnative lovegrasses: Lehmann, Boer (A-84 and Catalina), and Cochise.

Pelleted Seed

After the introduction and obvious success of nonnative grasses, revegetation research shifted its focus away from establishing native species and toward evaluating improved or selected nonnative accessions and devising ways to enhance their establishment on degraded desert grassland and shrubland sites to control erosion. The research of Gilbert L. Jordan is a classic case study of revegetation research in the 1960s and 1970s.

Jordan began his revegetation research in Arizona on the use of pelleted seeds. The pelleted seeding studies, conducted from the 1940s to the 1960s, sought an easy and inexpensive way to reseed degraded western rangelands. The idea of broadcasting pelleted seeds was conceived by Lytle Adams, a retired dentist (Jordan 1967), who first thought of the idea in southern California when he noticed small cactus plants emerging from rabbit droppings. In 1945 he interested Congress in the idea of aerially distributing pelleted seed. The idea was that seed could be compressed into earthen pellets with fertilizers, fungicides, and insect and rodent repellents. The weight of the pellets would promote their even distribution and penetration into the soil (Jordan 1967). This idea was inexpensive and appealing, especially when compared with the alternative methods of broadcasting seeds alone, in which success was usually limited by lack of soil coverage, and mechanical seedbed preparation and sowing on the ground, which was slow and expensive over a large area. Dr. and Mrs. Adams actively promoted the pelleted seeding program at the local and national levels. Mrs. Adams made hats from Lehmann lovegrass and devil's claw (*Proboscidea* spp.) and presented them to such prominent ladies as Mrs. Morris Udall and Jacqueline Kennedy.

Pelleted seeding research was conducted in many western states from 1945 into the 1960s (Hull et al. 1963; Jordan 1967). Jordan's (1967) findings on a Chihuahuan desertscrub site were similar to the results of other studies: pelleted seeds established to a similar or lesser extent than aerially broadcast nonpelleted seed. Mechanical seedbed preparation and sowing on the ground produced successful stands of lovegrass when pelleted seeding did not, even in years with favorable precipitation. Pelleted seeding failed primarily because of inadequate soil coverage but also partly because the pelleting process reduced germination (Jordan 1967). When Adams's advocate in Congress, Representative Ben F. Jensen of Iowa, learned that pelleted seed research money had been used to make comparisons with other methods, he accused the University of Arizona of sabotaging the program. The pelleted seeding story is interesting for two reasons: it shows the interest and concern at the local and national levels in revegetating the range, and it demonstrates the need for careful comparative research to test apparently easy solutions.

Seedbed Preparation

Proper seedbed preparation increases water availability to seeds and seedlings by controlling competing species, covering seeds at a shallow depth, and creating depressions to catch and hold water from intense thunderstorms (Roundy and Call 1988). Scientists have sought to develop seedbed preparation equipment that would improve the chances for successful revegetation. On semiarid grasslands and shrublands, this means increasing the length of time that seeds and seedlings are exposed to water. The most common ways of doing this involve removing competing species whose extensive roots use water, creating pits or furrows to catch and hold water, packing and firming the seedbed to increase seed-soil contact, sowing seeds shallowly, and applying mulch to shade the soil surface and reduce soil water evaporation. The effectiveness of seedbed modifications and sowing methods depends on soil texture and the precipitation pattern during seedling establishment (Cox et al. 1982). Ideally, seedbed preparation studies should be conducted on a number of different soils over a period of years and then compared with some standard practice.

After the pelleted seeding studies, Jordan conducted revegetation research from 1963 to 1973 on species, seedbed preparation, and sowing methods on former desert grassland and desert shrubland sites in the San Simon and Gila Valleys, Arizona, the Date Creek Mountains near Congress, Arizona, and on the Arizona Strip (fig. 9.3). His primary objective in southeastern Arizona was to find successful revegetation techniques for rehabilitating the badly eroded San Simon watershed (Bahre, this volume; Jordan and Maynard 1970a,b). The seeding sites, located in an area transitional between the Sonoran and Chihuahuan Deserts, averaged about 250 mm annual precipitation. Jordan's studies, funded by the Department of the Interior Bureau of Land Management, represented an extensive approach to revegetation research. He tested chemical and mechanical means of brush control and seeding techniques: seeding in fall, early and late spring, or summer; drilling versus broadcasting seeds; seeding at different rates; and treating seeds with fungicides. He tested at least 15 species of native grasses and six nonnative grasses, as well as a number of nonnative experimental

Figure 9.3. Gilbert L. Jordan conducted extensive brush control and revegetation studies in the San Simon Valley, Arizona, in the 1960s and 1970s. Grass establishment was most successful when seeds were broadcast in March on seedbeds highly disturbed by root plowing, like those pictured here.

selections developed by L. Neal Wright of the United States Department of Agriculture's Agricultural Research Service. At the peak of his research activity, Jordan was annually seeding dozens of plots on five different sites, as well as evaluating density and forage yield on hundreds of plots from previous years. He did this for 10 years, often with a small field crew that included student employees, and faithfully produced annual reports documenting his progress for the Bureau of Land Management.

In an effort to efficiently modify seedbeds to increase the probability of establishment, Jordan developed a disk pitter that could be attached behind a root plow and permitted him to plow, pit, and broadcast seed in one operation (fig. 9.4A). He used the transmission from a small car as a gearing mechanism to drive a modified Nisbet shaft-type seeder to allow seeding of tiny lovegrass seeds at controlled rates with a rangeland drill (fig. 9.4B). Jordan's ingenuity in equipment development is typical of the revegetation

Figure 9.4. Equipment developed and modified by Gilbert L. Jordan to reseed the desert grassland. A. A disk pitter attached behind a root plow permitted plowing, pitting, and broadcast seeding in one operation. B. Nisbet shaft-type seeders (cannister-like devices

scientists who developed much of our current rangeland seeding technology. The rough terrain and the huge extent of the rangelands in need of revegetation required scientists to develop heavy equipment such as the brushland plow and the rangeland drill to do the job (Young and McKenzie 1982). The capital expense of this heavy equipment was and continues to be a major economic constraint on revegetation projects (Young 1990).

The results from this extensive, comparatively long-term research program were incorporated into Jordan's (1981) revision of a guidebook on seeding southern Arizona rangelands first written by Anderson et al. (1953, rev. 1957). That guidebook continues to be used to plan Arizona range seeding.

Jordan's conclusions regarding revegetation of brush-invaded sites on the San Simon watershed were of practical value. He found rootplowing (fig. 9.3) to be the most effective means of controlling resprouting of woody species and for preparing seedbeds. When done in March with some winter soil moisture present to help maintain soil structure, rootplowing sub-

to hold the seed with a rotating shaft below to catch and expel seed) allowed controlled seeding rates of small-seeded species with a rangeland drill. Broadcast seeding (A) of small-seeded species was as successful as drilling (B) in the desert grassland.

sequently improved infiltration during intense summer rainstorms. Broadcasting small-seeded species within two weeks of rootplowing resulted in adequate seed coverage and was as successful as drilling the seed into the soil. Pitting the soil (fig. 9.5) increased grass density and yield and prevented seeding failure about once in every five seedings.

Most of the native grasses tested could successfully establish when rainfall was well above average, which happens about once in 10 years. Plains bristlegrass was the most successful native grass, but even it required above-normal precipitation to establish. In contrast, most of the nonnative grasses tested were considered promising for revegetation. Accession A-68 of Lehmann lovegrass was the most dependable. It could emerge after a 20-mm rainstorm, would survive if more rain occurred within five days after emergence, and was predicted to persist on sites receiving 127 mm of summer precipitation and 64 mm of winter precipitation. None of the experimental accessions of Lehmann lovegrass established as dependably as A-68, and

Figure 9.5. Lovegrasses (*Eragrostis* spp.) established in soil pits that catch rainwater near Bowie, Arizona.

Wilman lovegrass was not recommended at all because it could not tolerate drought. Catalina and A-84 Boer lovegrasses and Cochise, or Atherstone, lovegrass were considered slightly less drought tolerant than A-68 Lehmann lovegrass. Blue panic was slower to establish than the lovegrasses, but it was recommended for seeding in washes or sites receiving runoff water.

Subsequent studies of grass persistence after 10 years on Jordan's seeded plots at the San Simon site found significant reductions in density and yield of the seeded species (Cox and Jordan 1983). A decrease in yield definitely makes revegetation on such sites less cost-effective if expected gain in livestock production is the goal. However, seedings initially established at high densities on this site retained a fairly high grass cover compared with adjacent unseeded areas, and in many places seeded grasses had spread to adjacent small arroyos. Therefore, revegetation on these sites might be justified in terms of watershed rehabilitation.

The once-dense stands of seeded grasses on Jordan's plots near Bowie and Pima, Arizona, now mainly support burroweed and snakeweed

(Roundy and Jordan 1988). It should be noted that the winter precipitation decreases and winter temperatures become cooler going east from the San Simon Valley and into New Mexico. In these marginal sites with erratic precipitation, nonnative grasses may establish more dependably than native grasses but may not persist in grassland densities even in the absence of grazing, while native grasses such as bush muhly are adapted to the prevailing conditions and can persist as mature plants. Cox et al. (1982) reported evaluations of Soil Conservation Service seedings in New Mexico conducted from 1961 to 1979 in which native grasses such as blue grama, galleta grass (*Hilaria jamesii*), and tobosa persisted longer than nonnative lovegrasses. This underscores the importance of continued research and development of native plant materials to improve their establishment.

Seedbed preparation was also important in the development of the arid land seeder by G. H. Abernathy and C. H. Herbel (1973). That device follows a rootplow and picks up brush, forms pits, plants the seed, firms the seedbed, and mulches the seeded area with brush. The mulch decreases soil temperatures and increases the time of soil water availability after seeding (Herbel 1972). This technology produced good stands of native and introduced grasses in 10 of 23 seeding trials in southern New Mexico (Herbel et al. 1973). Black grama (*Bouteloua eriopoda*), blue panicgrass, Boer lovegrass, fourwing saltbush, Lehmann lovegrass, sideoats grama, and yellow bluestem established on sandy to loamy sites; and alkali sacaton (*Sporobolus airoides*), sideoats grama, and yellow bluestem established best on heavier-textured soils. Seeding failures were attributed to erosive or crusting soils and droughty conditions. The results of this study illustrate the site and precipitation-year dependency of range seeding success.

The land imprinter developed by R. M. Dixon (1978) is a heavy, water-filled cylinder with metal ridges that modifies seedbeds by imprinting rain-catching furrows in the soil (Dixon and Simanton 1980). Theoretically, imprinting enhances soil surface roughness and promotes infiltration by allowing water to replace air in macropores that are open to the soil surface (Dixon 1978). The imprinter has been highly promoted as a tool to greatly improve revegetation of semiarid rangelands in the Southwest and throughout the world. Imprinting helps bury seeds (Winkel et al. 1991a), firms seedbeds, reduces wind erosion, increases seed contact with the soil, and increases seedling emergence on some soils (Clary 1989; Haferkamp et al. 1987); but it is not effective for controlling sprouting brush (Cox et al. 1986; Larson 1980). Imprinted furrows on sandy soils may slough and bury broadcast seeds too deep for emergence.

On a desert grassland site in the Altar Valley, Arizona, soil disturbance caused by once-heavy cattle trampling or land imprinting increased surface soil water availability by one to two days during seedling establishment. Grass seedling emergence was greater on areas treated by imprinting and cattle trampling than on undisturbed soils during a moderately wet summer rainy season (Roundy et al. 1992b; Winkel and Roundy 1991). On the sandy loam soil of the plot used in this study, seedling establishment was successful without soil disturbance during a wet summer and was a failure during a dry summer, despite disturbance by imprinting or cattle trampling (Winkel and Roundy 1991). Thus, whether or not seedbed modifications increase water availability long enough to make a difference in seedling establishment in the desert grassland depends on the timing, frequency, and amount of summer rainfall.

Several seedbed preparation treatments were compared on a variety of sites to develop recommendations for their use (Cox et al. 1986). Cox et al. studied establishment on four sites in Arizona and Mexico in relation to seven plant control/seedbed preparation methods. Grasses established and persisted best on plots that were plowed and contour furrowed. Grasses on railed (brush removed by pulling a long piece of railroad track perpendicular to the tractor) and imprinted plots died after three or four years, presumably because of inadequate brush control.

Barrow (1992) proposed a more passive revegetation technique that focuses on small channels that flood ephemerally. Floodwaters tip a container, which dumps seeds into the stream. Seeds, soil, and organic debris settle out as the flow is slowed by natural features or artificial dikes. The flow channels retain moisture longer than the adjacent uplands do and may permit better seedling establishment. Once established, vegetation would further decrease flows, control erosion, and allow runoff water to infiltrate. The hydrology and establishment ecology of this approach warrant further development. The use of animals to disperse and establish desirable species also warrants additional investigation (Barrow and Havstad 1992).

Revegetation Science and Desert Grassland Revegetation

Traditional and New Research Questions

Traditional revegetation research focused on determining appropriate species and techniques for successful plant establishment on specific sites. This approach, as exemplified by Jordan's studies, involved testing numerous species using different techniques on various sites over a number of years.

But determining reliable site recommendations for specific plant species can be difficult due to the precipitation dynamics of semiarid areas. The first criterion for success, obviously, is seedling emergence, but some species will emerge and establish in unusually wet years but not in average or dry years. Too often, long-term persistence is not evaluated, and this can lead to inappropriate recommendations. For example, Cox et al. (1982) noted that Lehmann lovegrass and Wilman lovegrass are recommended for southern New Mexico and southeastern Arizona. Although these species do establish there during wet summers, they persist only on sites or in years with mild, wet winters. Successful persistence is a function of the genetic potential of a population to live, grow, and reproduce within the physical and biological environment on a particular site. Since the interactions between plant populations and the environment are difficult to measure even on one site, the emphasis has been on *what* persists rather than *how* or *why* certain plants establish and persist.

Early researchers (Griffiths 1901; Kennedy and Doten 1901) assumed that native species would be more likely to persist over their traditional range than nonnative species, and the same assumption is made today. Soil Conservation Service range site guides (Shiflet 1973) list native species that are believed to have been present before Europeans arrived and disturbed the area. This approach to species selection assumes that the historical species distribution and composition are known and that environmental conditions for establishment and persistence are unchanged. In fact, however, no one knows how much of the modern desert shrubland was once grassland (Bahre 1991; Burgess, this volume; Gardner 1951), and the decline of grass and associated environmental changes away from the pre-European grassland may have rendered certain sites less favorable to establishment and persistence of native species. For example, the spread of velvet mesquite (*Prosopis velutina*) has resulted in the concentration of nutrients under the mesquite canopies (Tiedemann and Klemmedson 1973), while under grassland conditions these nutrients were probably more dispersed. Some native grasses, such as Arizona cottontop, bush muhly, and plains bristlegrass, are actually more abundant under velvet mesquite canopies (Hague et al. 1991; Livingston 1992) than they are in more open areas. These native grasses may require greater soil fertility to dominate than currently exists in the spaces between mesquite trees.

The range of adaptation of nonnative species cannot be inferred from descriptions of the sites and plant communities they now inhabit as is done for native species. It is not necessarily clear which nonnative species have the potential to be beneficial or to become weedy. Obviously it is impossible to test every nonnative species for revegetation on every site. This dilemma led Cox et al. (1988) to determine the environmental characteristics (soils, temperature, and precipitation) of nonnative grasses in their original environments as well as in their adopted ones. Such information is helpful for determining the potential spread and usefulness of these species.

Revegetation recommendations are often based on the ability of species to consistently emerge as seedlings from seeds in the soil. But longer-lived semiarid perennials do not reestablish from seed every year. On the contrary, they persist with episodic establishment only in unusually wet years. Therefore it is not surprising that the shorter-lived Lehmann lovegrass, with its high seed production and establishment ability, has been more consistently successful on degraded sites than native grasses.

Past revegetation objectives emphasized the economical establishment of grass for soil cover and forage, and only the most consistently successful species were recommended for revegetation. Most often, nonnative species were recommended because they met those narrow objectives, and Jordan's (1981) handbook recommending sites and seeding techniques for the desert grassland emphasizes the use of nonnative species for that reason. Since past revegetation objectives emphasized the establishment of grass for soil cover and forage with the least expense, only the most consistently successful species were recommended.

Many native grasses produce viable and germinable seed (Biedenbender 1994; Bridges 1941; Little 1937; Livingston 1992; Wilson 1931) and seem to have potential for use in revegetation. Unfortunately, the revegetation research of the past is inadequate to meet the new goals of restoring native plant diversity (Call and Roundy 1991). In that context a number of questions arise. Why do some plants establish better than others? What biological characteristics are necessary for establishment under prevailing weather conditions? Can we select or breed native species to improve establishment and persistence? Can we establish mixtures of species that persist? How much do seedbed modifications affect the seedbed environment and subsequent plant establishment?

These questions are addressed by relating species' characteristics to the establishment environments (seedbed ecology) and by improved plant development. Neal Wright was a pioneer in grass improvement studies in the Southwest (N. Wright 1975; N. Wright and Streetman 1960). Improved plant development for semiarid lands involves characterizing the environmental conditions of the establishment environment, defining the selection criteria for successful adaptations for establishment, and developing appropriate screening techniques (D. A. Johnson et al. 1981).

Selecting for Drought Tolerance

The adaptations employed by desert plants that allow them to tolerate drought are fairly obvious and well known (Ehleringer 1985). However, differences in drought tolerance and their physiological or morphological basis among similar life forms, different genotypes within a species, are difficult to determine. Plant improvement specialists have been unable to identify obvious physiological and morphological characteristics that can be consistently used to develop selection criteria for improved plant establishment (D. A. Johnson et al. 1981; N. Wright and Streetman 1960). Wright and Streetman (1960) extensively reviewed the literature on drought tolerance and concluded that exposure of warm-season grass seedlings to simulated field environmental conditions would be the most efficient screening method for drought tolerance. They used a growth chamber programmed to simulate the temperature and humidity conditions of the first 10 days of August in the Southwest (N. Wright 1961, 1964) and subjected the seedlings to different watering regimes to determine their survival in relation to the length of the stress periods. Lovegrass and the panicgrasses ranked in the chamber studies just as they had been ranked for drought tolerance in rangeland seeding trials (N. Wright 1964). Lehmann lovegrass was more drought tolerant than Boer lovegrass, which was more tolerant than plains lovegrass (*Eragrostis intermedia*). All the lovegrasses survived longer under stress than the panicgrasses.

The researchers subsequently used the growth chamber to compare seedling drought tolerance of Boer and Lehmann lovegrass accessions (L. N. Wright and Brauen 1971; L. N. Wright and Jordan 1970). Wright and Brauen (1971) compared seedling survival of 37 accessions of Lehmann lovegrass after 19 days of withholding water in the programmed chamber. They identified a number of accessions that survived drought better than the standard A-68 accession; and they also noted a positive correlation between seed dormancy, seed weight, and seedling drought tolerance. However, Jordan's field comparisons of 4 of Wright's most tolerant accessions did not establish or persist any better than the standard A-68 (Cox and Jordan 1983).

Plant Response to Environmental Conditions

It has been difficult to determine which biological characteristics are responsible for establishment success in different environments, partly because it is difficult to measure and characterize the seed and seedling environment at a biologically significant scale (Call and Roundy 1991). Semiarid seedbed temperatures and water potentials may be highly dynamic in time and variable in space (Burgess, this volume). The continuous measurement necessary to characterize these seedbeds was quite difficult in the past (Evans and Young 1987) but is much easier now that electronic microloggers are available. Soil temperatures at the scale of a seed can easily be measured electronically with inexpensive fine-wire thermocouples. On the other hand, soil water sensors are generally at least 2 cm wide. Small changes in soil water conditions near seeds on the soil surface or buried at a shallow depth greatly affect germination but are difficult to detect with the sensors currently available (Harper et al. 1965; Winkel et al. 1991b).

On Intermountain cold desert areas, where plants germinate and establish in the spring on soil moisture accumulated from winter storms, the success of cheatgrass has been attributed to its ability to germinate and develop roots at cool temperatures (Harris and Wilson 1970). Successful cool-season perennial grass competitors would also need to germinate and develop roots during the winter and early spring to avoid preemption of soil moisture by cheatgrass and other undesirable species (Aguirre and Johnson 1991).

Rapid root growth is not necessarily associated with the ability of warm-season grasses to establish during the summer rainy season in the Southwest desert grassland. In fact, Cochise and Lehmann lovegrasses, which are more easily established in revegetation projects, exhibit less and slower seminal root growth than sideoats grama and blue panicgrass (Simanton and Jordan 1986). Warm-season grasses elongate their subcoleoptile internodes to produce adventitious roots at the soil surface (fig. 9.6; Hyder et al. 1971; Winkel 1990). They generally require 9 to 11 days of continuously available soil moisture to develop adventitious roots (Winkel 1990). Adventitious roots are much more effective than seminal roots in supplying water to seedlings (Tischler and Voigt 1987; Vander Sluijs and Hyder 1974). Prolonged available moisture after germination or the ability to tolerate drought until consistent rainfall occurs is apparently necessary for warm-season grass species to successfully develop adventitious roots and establish.

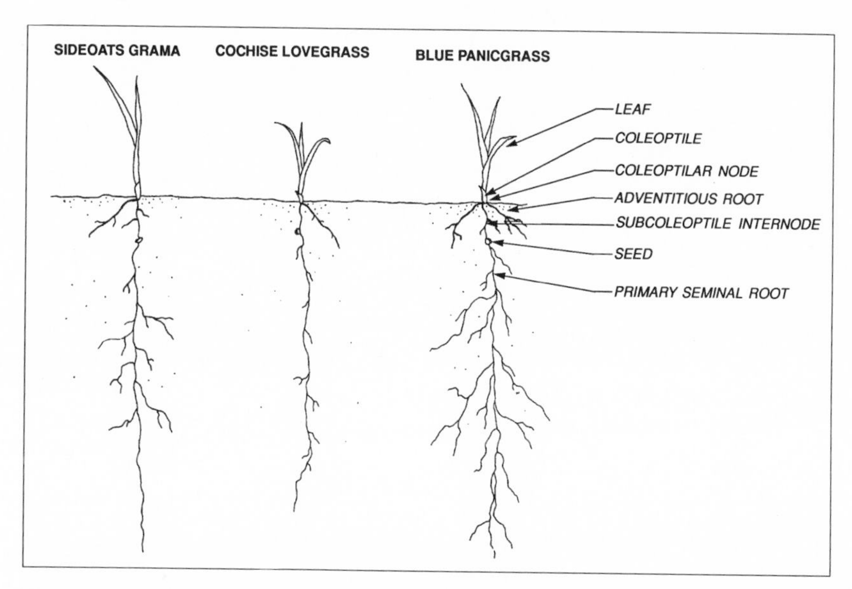

Figure 9.6. Root morphology of warm-season grasses frequently used in desert grassland revegetation (Winkel 1990).

Laboratory root-growth studies (Winkel 1990) and field moisture measurements (Roundy et al. 1992b) suggest that if germinating rains are not followed by more rains, the soil drying front (the depth at which the soil is dry) will proceed faster than the seminal root growth, and the seedlings will desiccate (fig. 9.7). Although the seminal roots of sideoats grama grow faster than those of Cochise lovegrass, they are not fast enough to stay ahead of the soil drying front in the absence of rain.

The reason why nonnative lovegrasses establish better than most native grasses may relate to their germination rates. Frasier et al. (1985, 1987) showed that sideoats grama may germinate rapidly during a short wet period and thereby be vulnerable to seedling desiccation during the subsequent dry period. In contrast, the nonnative lovegrasses may be slower to germinate under a short wet period (Frasier et al. 1985, 1987). Their requirement of a longer wet period to germinate makes them more likely to survive. Abbott et al. (1995) found that native grasses germinated after the initial intermittent summer rains while Lehmann lovegrass did not. The success of Lehmann lovegrass may be associated with this delay in germination following the initial summer rainstorms in July because this avoids the dry period that usually follows; thus germination occurs when more consistent rainfall is present in late July and August. On the other hand, Weaver and

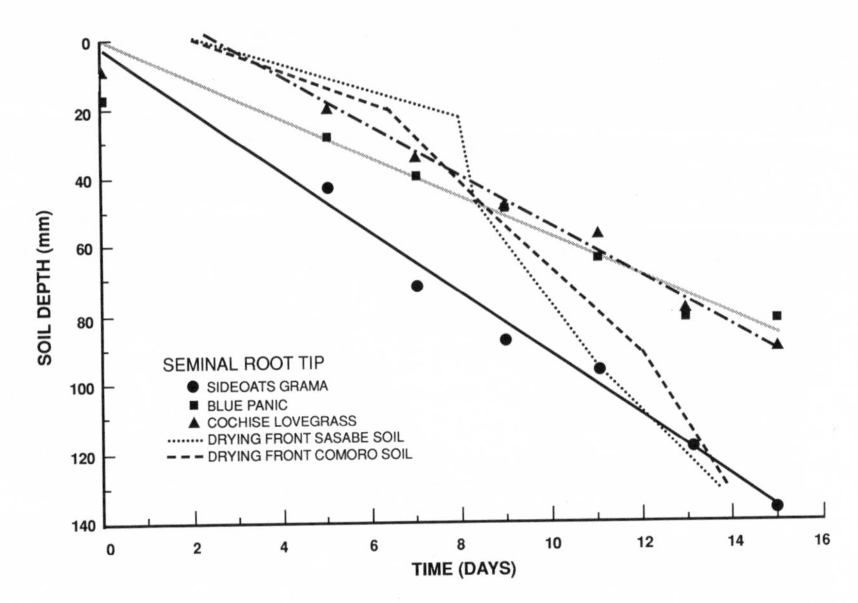

Figure 9.7. Maximum seminal root depth as a function of time since seed wetting for three warm-season grasses (Roundy et al. 1993; Winkel 1990). The dashed line and the dotted line represent the time that soil matric potentials decrease to below -1.5 MPa at a given depth after the end of two days of intermittent rain on two typical sandy loam soils during the summer rainy season in southern Arizona. If no more rain falls, the soil drying front may exceed the rate of seminal root elongation.

Jordan (1985) thought rapid germination rates would increase establishment and looked for methods of pretreating Lehmann lovegrass seed to decrease the time required for germination.

Seedling Establishment Models

There is a need to relate meteorological conditions and actual field temperature and moisture conditions to biological establishment responses. The many different possible meteorological scenarios and other variables make this a problem best addressed by modeling seedling establishment. Such models have not yet been developed, but they would have three basic parts: (1) a meteorological model to determine and set weather conditions of known probabilities, (2) an environmental model to estimate seedbed water and temperature conditions as a function of soil characteristics and meteorological inputs, and (3) a biological response model to estimate establishment responses such as germination, root and shoot growth, and seedling survival in relation to temperature and moisture dynamics and thresholds.

The critical biological responses and thresholds still need to be defined—for example, the effects of alternate wetting and drying on seed germination (Griswold 1936). Alternating wet and dry periods can either stimulate or reduce germination, depending on the species (Hegarty 1978). Large diurnal temperature fluctuations in dry seedbeds in May and June before the summer rains may have positive or negative effects on germination, and the effects need to be determined for different species (Biedenbender 1994). The ability of seedlings to tolerate desiccation at different stages of growth and for different periods needs to be determined in relation to water potential in the root zone. The current interest in native species offers research opportunities to determine their establishment requirements. Once these parameters have been determined, revegetation will become a much more predictable science.

Seedbed ecology studies may also help us determine how to replace unwanted species with more desirable species. For example, we know that emergence of Lehmann lovegrass seedlings is limited under litter or a dense grass canopy, probably because the seeds require a certain light and/or temperature for germination (Roundy et al. 1992a,c; Sumrall et al. 1991). Gravel and litter mulches decreased Lehmann lovegrass seedling density and increased establishment and survival of Arizona cottontop during a wet year (Livingston 1992). This suggests that treatments such as spraying with herbicide and mowing could be used to replace a stand of Lehmann lovegrass with native grasses. Mature lovegrass would be killed by the herbicide, and the dead canopy could be mowed to produce a mulch. Biedenbender (1994) found that such a treatment reduced the density of Lehmann lovegrass seedlings and improved seedling establishment of four native species.

Revegetation or Restoration—What Do We Want?

The successful accomplishment of revegetation goals for the desert grassland depends on site potential, management practices, and economics. Goals may be described in terms of outputs from the ecosystem or even ecosystem functions (Jackson 1992). In the past, the desert grassland has been managed mainly to produce fiber and protect the watershed (erosion and runoff control). Productive grassland, regardless of its species composition, has traditionally been the plant community pursued to accomplish these goals (Burgess, this volume).

Besides the ongoing goals of watershed protection and production of forage for domestic livestock, current goals may include increasing or maintaining endangered species populations, increasing biodiversity or the variety of life and its processes (West 1993), creating mixed plant communities that favor certain wildlife species, and just establishing native grasslands for their intrinsic value. Most of these goals involve revegetating with native species.

Jackson (1992) discussed goals relative to altered lands and the ethics and *re-* words—revegetation, rehabilitation, reclamation, and restoration—associated with treating them for utilitarian purposes or for their natural intrinsic value. Box (1978) defined some of these terms in an attempt to clarify goals for treating drastically disturbed lands. *Rehabilitation* involves returning a site to a form and productivity that conforms to its predisturbance use. *Reclamation* returns a site to a condition that supports a composition and density of organisms similar to the one present before the disturbance. *Restoration* returns a site to its exact predisturbance condition.

Restoration as a practice must be based on an understanding of the ecology of the system involved in order to re-create its structure, functions, diversity, and dynamics (Society of Ecological Restoration 1991). Jackson (1992) contrasted the small-scale, intensive effort of prairie restoration in the Midwest with the large-scale, agronomic approach to rangeland revegetation in the West. The latter was funded by a production- and soil conservation–conscious society faced with rehabilitating a tremendous amount of land as efficiently as possible (Young 1990). The goals of that effort were simple compared with those of some of the restoration efforts being discussed today.

True restoration is difficult ecologically, expensive, and hard to justify on a large scale in areas where natural resources and land must support people. Consequently, it is usually done on a very small scale with research and educational values in mind. Given the long-term dynamics of plant populations and communities, it is not always clear just which species composition should be restored (Diamond 1987). H. B. Johnson and Mayeux (1992) demonstrated that, contrary to popular assumptions, the species composition of an unperturbed ecosystem may be dynamic without greatly affecting the ecosystem's function. Consequently, goals centered on ecosystem function and desired outputs may be much more realistic than those centered on some preconceived notion of pristine species composition. Without the perspective of ecosystem function and natural dynamics, frivolous goals may be proposed that have little economic or ecological justification.

In the past, lack of seed availability discouraged the use of native seed for revegetation projects. Seed appendages, low seed production and viability, and natural dormancy have often made the use of native species risky and expensive (Schwendiman and Sackman 1940). New machines like the flail-vac (Dewald 1991) and the brush machine to harvest and clean trashy seeds as well as devices to sow them (Wiedemann 1991) are necessary for development of a native seed industry. The Soil Conservation Service's Plant Material Center in Tucson will play an important role in selecting successful native plant materials and developing and demonstrating native seed production technology for the desert grassland (U.S. Department of Agriculture 1991).

The recognition of the value of vegetation diversity for associated faunal diversity has created an interest in establishing mixed native plant communities. Of special concern is the loss of biodiversity on lands invaded and dominated by nonnative plants, such as Lehmann and other southern African lovegrasses in the desert grassland. Originally seeded on 69,000 ha, this grass has spread to an additional 76,000 ha (Cox and Ruyle 1986). Although it has filled an important role in conserving soil and providing forage, it is seen as a threat to biodiversity. Bock et al. (1986) found lower floral and faunal diversities in a nonnative lovegrass stand than in a nearby mixed native species grass stand.

Nonnative lovegrasses have primarily been seeded on and have invaded rangelands lacking native perennial grasses and dominated by mesquite, snakeweed, and burroweed. Lehmann lovegrass does not usually replace native grasslands in good condition, but there is evidence that it is continuing to spread under these good conditions (Anable et al. 1992) and can replace native grasses damaged by drought (Robinette 1992) or fire (Cable 1965, 1971). Its prolific seed production gives it an advantage over native perennial grasses in areas where disturbances have reduced mature plant densities and where vegetation dominance is dictated by the ability to reestablish from seed. Its ability to reseed itself after fire (Cable 1965; Sumrall et al. 1991) may allow it to dominate desert grasslands, where it can provide enough fine fuels to permit frequent fires (Anable et al. 1992). This produces the ironic situation that native grasslands once maintained by frequent fires may become nonnative grasslands through the agency of fire.

Domestic livestock grazing may reduce this potential increase in fire frequency, and removal of livestock could greatly increase fire frequency—and dominance by the nonnatives (Anable et al. 1992). Research on the competitive ability and population dynamics of native and nonnative plants is necessary to determine the potential for restoring native biodiversity in the desert grassland.

Revegetation to improve habitat for threatened and endangered animal species and other wildlife species may receive increased emphasis in the future. For example, restoration of the masked bobwhite quail (*Colinus virginianus ridgwayi*) at the Buenos Aires Wildlife Refuge may be encouraged by establishing legumes such as whiteball acacia (*Acacia angustissima*), which produces persistent seeds for critical winter food. There is substantial potential to improve wildlife habitat using standard methods of vegetation manipulation (Ruyle and Roundy 1990) and revegetation with legumes, shrubs, and forbs to fulfill food and cover requirements (Roundy, in press).

Summary

The desert grassland has undergone dramatic changes in vegetation composition. Once-dominant native perennial grasses have been replaced by native woody species such as mesquite and snakeweed. Nonnative lovegrasses, successfully seeded on degraded desert grasslands to provide forage and stabilize the soil, continue to spread. On many sites, what was once a native perennial grassland is now mesquite shrubland with a half shrub and/or nonnative grass understory.

Revegetation research and projects have traditionally focused on reestablishing grass for soil conservation and livestock forage. Although many native grasses produce highly germinable seed, they have not established as well as nonnative grasses. Revegetation research since the early 1900s has shown the necessity of modifying the seedbed to catch and retain precipitation to increase establishment success; however, these treatments do not ensure success. Instead, revegetation success is largely determined by the pattern of summer precipitation in a given year. Revegetation technology presented in numerous guidebooks written over the years helps increase chances of establishment and is centered on appropriate site potential and seeding methods.

Contemporary goals for the desert grassland may include restoring native plant diversity to favor specific wildlife species, to increase wildlife diversity in general, or to generally increase biodiversity. It is not clear who will pay for revegetation with native plants, since revegetation continues to be an intensive and expensive practice. True grassland restoration will probably be confined to small areas and will serve educational and research purposes.

Studies of seedbed environmental dynamics and germination and seedling responses of native plants will help determine site potential and appropriate seeding techniques for native species. Better determination of successful establishment characteristics may permit us to select for native plant genotypes that are able to establish as well as or better than the successful nonnative grasses. Evaluation of success requires time for seeded species to demonstrate their ability to persist and reproduce under the environmental and management conditions of the revegetation site. Development of new equipment to harvest, clean, and sow native seeds with natural appendages will help facilitate expansion of a native seed industry. Revegetation in the desert grassland will increasingly involve native species. However, most revegetation efforts will probably be directed at small-scale disturbances, demonstration areas, or critical wildlife habitats rather than expansive areas.

Literature Cited

Abbott, L. B., B. A. Roundy, and S. H. Biedenbender, 1995. Seed fate of warm-season perennial grasses. Pp. 37–43 *in* Proceedings of the Wildland Shrub and Arid Land Restoration Symposium, 19–21 October 1993, Las Vegas. U.S. Department of Agriculture, Forest Service General Technical Report INT-315.

Abernathy, G. H., and C. H. Herbel. 1973. Brush eradicating, basin pitting, and seeding machine for arid to semiarid rangeland. Journal of Range Management 26:189–192.

Aguirre, L., and D. A. Johnson. 1991. Influence of temperature and cheatgrass competition on seedling development of two bunchgrasses. Journal of Range Management 44:347–354.

Anable, M. E., M. P. McClaran, and G. B. Ruyle. 1992. Spread of introduced Lehmann lovegrass (*Eragrostis lehmanniana* Nees.) in southern Arizona, USA. Biological Conservation 61:181–188.

Anderson, D., L. P. Hamilton, H. G. Reynolds, and R. R. Humphrey. 1953.

Reseeding desert grassland ranges in southern Arizona. University of Arizona Agricultural Experiment Station Bulletin 249.

Anderson, D., L. P. Hamilton, H. G. Reynolds, and R. R. Humphrey. 1957. Reseeding grassland ranges in southern Arizona. University of Arizona Agricultural Experiment Station Bulletin 249, revised.

Bahre, C. J. 1991. A Legacy of Change: Historic Impact on Vegetation in the Arizona Borderlands. University of Arizona Press, Tucson.

Barrow, J. R. 1992. Use of floodwater to disperse grass and shrub seeds on native arid lands. Pp. 167–169 *in* W. P. Clary, E. D. McArthur, D. Bedunah, and C. L. Wambolt (comps.), Symposium on ecology and management of riparian shrub communities. U.S. Department of Agriculture, Forest Service General Technical Report INT-289.

Barrow, J. R., and K. M. Havstad. 1992. Recovery and germination of gelatin-encapsulated seeds fed to cattle. Journal of Arid Environments 22: 395–399.

Biedenbender, S. H. 1994. Germination requirements of Arizona native perennial grasses and their establishment in existing stands of *Eragrostis lehmanniana* Nees. M.S. thesis, University of Arizona, Tucson.

Bock, C. E., J. H. Bock, K. L. Jepson, and J. C. Ortega. 1986. Ecological effects of planting African lovegrasses in Arizona. National Geographic Research 2:456–463.

Box, T. W. 1978. The significance and responsibility of rehabilitating drastically disturbed land. Pp. 1–10 *in* F. W. Schaller and P. Sutton (eds.), Reclamation of drastically disturbed lands. American Society of Agronomy, Madison, Wisc.

Bridges, J. O. 1941. Reseeding trials on arid range. New Mexico State University Agricultural Experiment Station Bulletin 278.

———. 1942. Reseeding practices for New Mexico ranges. New Mexico State University Agricultural Experiment Station Bulletin 291.

Cable, D. R. 1965. Damage to mesquite, Lehmann lovegrass, and black grama by a hot June fire. Journal of Range Management 18:326–329.

———. 1971. Lehmann lovegrass on the Santa Rita Experimental Range, 1937–1968. Journal of Range Management 24:17–21.

Call, C. A., and B. A. Roundy. 1991. Perspectives and processes in revegetation of arid and semiarid rangelands. Journal of Range Management 44:543–549.

Cassaday, J. T. 1937. Scarification of hard-coated grass seeds to increase germination. U.S. Department of Agriculture, Forest Service, Southwestern Forest and Range Experiment Station Note 13.

Cassaday, J. T., and G. E. Glendening. 1940. Revegetating semidesert rangelands in the Southwest. U.S. Department of Agriculture, Forest Service, Southwestern Forest and Range Experiment Station, Forestry Publication 8.

Clary, W. P. 1989. Revegetation by land imprinter and rangeland drill. U.S. Department of Agriculture, Forest Service Research Paper INT-397.

Cooke, R. U., and R. W. Reeves. 1976. Arroyos and Environmental Change in the American Southwest. Clarendon Press, Oxford.

Cox, J. R., and G. L. Jordan. 1983. Density and production of seeded range grasses in southeastern Arizona (1970–1982). Journal of Range Management 36: 649–652.

Cox, J. R., M. H. Martin R., F. A. Ibarra F., J. H. Fourie, N. F. G. Rethman, and D. G. Wilcox. 1988. The influence of climate and soils on the distribution of four African grasses. Journal of Range Management 41:127–139.

Cox, J. R., M. H. Martin R., F. A. Ibarra F., and H. L. Morton. 1986. Establishment of range grasses on various seedbeds at creosotebush (*Larrea tridentata*) sites in Arizona, U.S.A., and Chihuahua, Mexico. Journal of Range Management 39:540–546.

Cox, J. R., H. L. Morton, T. N. Johnsen, Jr., G. L. Jordan, S. C. Martin, and L. C. Fierro. 1982. Vegetation restoration in the Chihuahuan and Sonoran Deserts of North America. U.S. Department of Agriculture, Agricultural Research Service, Agricultural Reviews and Manuals, ARM-W-28.

Cox, J. R., and G. B. Ruyle. 1986. Influence of climatic and edaphic factors on the distribution of *Eragrostis lehmanniana* Nees. in Arizona, USA. Journal of the Grassland Society of South Africa 3:25–29.

Crafts, E. C., and L. A. Wall. 1938. Tentative range utilization standards. U.S. Department of Agriculture, Forest Service, Southwestern Forest and Range Experiment Station Note 25.

Crider, F. J. 1945. Three introduced lovegrasses for soil conservation. U.S. Department of Agriculture Circular 730.

Darrow, R. A. 1944. Arizona range resources and their utilization. I. Cochise County. University of Arizona Agricultural Experiment Station Technical Bulletin 103:311–366.

Dewald, C. L. 1991. Woodward fail-vac seed stripper-update. P. 48 *in* Rangeland Technology Equipment Council, 1990 Annual Report. U.S. Department of Agriculture, Forest Service Technology and Development Center, Missoula, Mont.

Diamond, J. 1987. Reflections on goals and on the relationship between theory and practice. Pp. 329–336 *in* W. R. Jordan III, M. E. Gilpin, and J. D. Aber (eds.), Restoration ecology. Cambridge University Press, Cambridge, England.

Dixon, R. M. 1978. Water infiltration control on rangelands: principles and practices. Pp. 322–325 *in* D. N. Hyder (ed.), Proceedings of the First International Rangeland Congress. Society for Range Management, Denver.

Dixon, R. M., and J. R. Simanton. 1980. Land imprinting for better watershed management. Pp. 809–826 *in* Symposium on Watershed Management, vol. 2. American Society of Civil Engineers, Boise, Ida.

Ehleringer, J. 1985. Annuals and perennials of warm deserts. Pp. 161–180 *in* B. F. Chabot and H. A. Mooney (eds.), Physiological ecology of North American plant communities. Chapman & Hall, New York.

Evans, R. A., and J. A. Young. 1987. Seedbed microenvironment, seedling recruitment, and plant establishment on rangelands. Pp. 212–224 *in* G. W. Frasier and R. A. Evans (eds.), Proceedings of symposium on Seed and Seedbed Ecology of Rangeland Plants. U.S. Department of Agriculture, Agricultural Research Service, National Technical Information Service.

Flory, E. L., and C. G. Marshall. 1942. Regrassing for soil protection in the Southwest. U.S. Department of Agriculture Farmers' Bulletin 1913.

Forbes, R. H. 1895. The mesquite tree: its products and uses. University of Arizona Agricultural Experiment Station Bulletin 13:12–26.

Forsling, C. L., and W. A. Dayton. 1931. Artificial reseeding on western mountain rangelands. U.S. Department of Agriculture Circular 178.

Frasier, G. W., J. R. Cox, and D. A. Woolhiser. 1985. Emergence and survival responses of seven grasses for six wet-dry sequences. Journal of Range Management 38:372–377.

———. 1987. Wet-dry cycle effects on warm-season grass seedling establishment. Journal of Range Management 40:2–6.

Gardner, J. L. 1951. Vegetation of the creosote area of the Rio Grande valley in New Mexico. Ecological Monographs 21:379–403.

Glendening, G. E. 1942. Germination and emergence of some native grasses in relation to litter cover and soil moisture. Journal of the American Society of Agronomy 34:797–804.

Glendening, G. E., and H. A. Paulsen, Jr. 1955. Reproduction and establishment of velvet mesquite as related to invasion of semidesert grasslands. U.S. Department of Agriculture Technical Bulletin 1127.

Griffiths, D. A. 1901. Range improvement in Arizona. U.S. Department of Agriculture, Bureau of Plant Industry Bulletin 4.

———. 1904. Range investigations in Arizona. U.S. Department of Agriculture, Bureau of Plant Industry Bulletin 67.

———. 1907. The reseeding of depleted range and native pasture. U.S. Department of Agriculture, Bureau of Plant Industry Bulletin 117.

Griswold, S. M. 1936. Effect of alternate moistening and drying on germination of seeds of western range plants. Botanical Gazette 98:243–269.

Haferkamp, M. R., D. C. Ganskopp, R. F. Miller, and F. A. Sneva. 1987. Drilling versus imprinting for establishing crested wheatgrass in the sagebrush bunchgrass steppe. Journal of Range Management 40:524–530.

Hague, Z., A. Young, K. C. McDaniel, and R. D. Pieper. 1991. Two-phase pattern in mesquite-herbland vegetation in southern New Mexico. Southwestern Naturalist 36:54–59.

Harper, J. L., J. T. Williams, and G. R. Sagar. 1965. The behavior of seeds in soil. I. The heterogeneity of soil surfaces and its role in determining establishment of plants from seed. Journal of Ecology 53:273–286.

Harris, G. W., and A. M. Wilson. 1970. Competition for moisture among seedlings of annual and perennial grasses as influenced by root elongation at low temperature. Ecology 51:530–534.

Haskett, B. 1935. Early history of the cattle industry in Arizona. Arizona Historical Review 6:3–42.

Hastings, J. R., and R. M. Turner. 1965. The Changing Mile: An Ecological Study of Vegetation Change with Time in the Lower Mile of an Arid and Semiarid Region. University of Arizona Press, Tucson.

Hegarty, T. W. 1978. The physiology of seed hydration and dehydration, and the relation between water stress and the control of germination: a review. Plant, Cell, and Environment 1:101–119.

Herbel, C. H. 1972. Environmental modification for seedling establishment. Pp. 101–114 *in* V. B. Youngner and C. M. McKell (eds.), The biology and utilization of grasses. Academic Press, New York.

Herbel, C. H., G. H. Abernathy, C. C. Yarbrough, and D. K. Gardner. 1973. Rootplowing and seeding arid rangelands in the Southwest. Journal of Range Management 26:193–200.

Hull, A. C., Jr., R. C. Holmgren, W. H. Berry, and J. A. Wagner. 1963. Pelleted seeding on western rangelands. U.S. Department of Agriculture and U.S. Department of Interior Miscellaneous Publication 922.

Humphrey, R. R. 1953. The desert grassland, past and present. Journal of Range Management 6:159–164.

———. 1958. The desert grassland. Botanical Review 24:193–252.

Hyder, D. N., A. C. Everson, and R. E. Bement. 1971. Seedling morphology and seeding failures with blue grama. Journal of Range Management 24: 287–292.

Jackson, L. L. 1992. The role of ecological restoration in conservation biology. Pp. 433–451 *in* P. L. Fiedler and S. K. Jain (eds.), Conservation biology. Chapman & Hall, New York.

Johnson, D. A., M. D. Rumbaugh, and R. H. Asay. 1981. Plant improvement for semi-arid rangelands: possibilities for drought resistance and nitrogen fixation. Plant and Soil 58:279–303.

Johnson, H. B., and H. S. Mayeux. 1992. Viewpoint: A view on species additions and deletions and the balance of nature. Journal of Range Management 45:322–333.

Jordan, G. L. 1967. An evaluation of pelleted seeds for seeding Arizona rangelands.

University of Arizona Agricultural Experiment Station Technical Bulletin 183.

———. 1981. Range seeding and brush management on Arizona rangelands. University of Arizona Agricultural Experiment Station T81121.

Jordan, G. L., and M. L. Maynard. 1970a. The San Simon Watershed; historical review. Progressive Agriculture in Arizona 4:10–13.

———. 1970b. The San Simon Watershed; revegetation. Progressive Agriculture in Arizona 4:4–7.

Judd, I. B., and L. W. Judd. 1976. Plant survival in the arid Southwest 30 years after seeding. Journal of Range Management 29:248–251.

Kennedy, P. B., and S. B. Doten. 1901. A preliminary report on the summer ranges of western sheep. Nevada State University Agricultural Experiment Station Bulletin 51.

Larson, J. E. 1980. Revegetation equipment catalog. U.S. Department of Agriculture, Forest Service, Equipment Development Center, Missoula, Mont.

Little, E. L. 1937. Viability of seeds of southern New Mexico range grasses. U.S. Department of Agriculture, Forest Service, Southwestern Forest and Range Experiment Station Note 6.

Livingston, M. L. 1992. Factors influencing germination and establishment of Arizona cottontop, bush muhly and plains lovegrass in southern Arizona. Ph.D. dissertation, University of Arizona, Tucson.

Martin, S. C. 1949. Controlling mesquite with diesel oil. U.S. Department of Agriculture, Forest Service, Southwestern Forest and Range Experiment Station Note 115.

McGinnies, W. G., and J. F. Arnold. 1939. Relative water requirements of Arizona range plants. University of Arizona Agricultural Experiment Station Technical Bulletin 80.

Parker, K. W. 1939a. Management of livestock to avoid losses from poisonous plants. U.S. Department of Agriculture, Forest Service, Southwestern Forest and Range Experiment Station Note 70.

———. 1939b. The control of snakeweed in the Southwest. U.S. Department of Agriculture, Forest Service, Southwestern Forest and Range Experiment Station Note 77.

———. 1940. The use of arsenic in the control of mesquite in the Southwest. U.S. Department of Agriculture, Forest Service, Southwestern Forest and Range Experiment Station Note 88.

Robinette, D. 1992. Lehmann lovegrass and drought in southern Arizona. Rangelands 14:100–103.

Roundy, B. A., in press. Revegetation of rangelands for wildlife. *In* P. R. Krausman (ed.), Rangeland wildlife. Society for Range Management, Denver.

Roundy, B. A., and C. A. Call. 1988. Revegetation of arid and semiarid rangelands. Pp. 607–635 *in* P. T. Tueller (ed.), Vegetation science applications for rangeland analysis and management. Kluwer Academic Publishers, Boston.

Roundy, B. A., and G. L. Jordan. 1988. Vegetation changes in relation to livestock exclusion and rootplowing in southeastern Arizona. Southwestern Naturalist 33:425–436.

Roundy, B. A., R. B. Taylorson, and L. B. Sumrall. 1992a. Germination responses of Lehmann lovegrass to light. Journal of Range Management 45:81–84.

Roundy, B. A., V. K. Winkel, J. R. Cox, A. K. Dobrenz, and H. Tewolde. 1993. Sowing depth and soil water effects on seedling emergence and root morphology of three warm season grasses. Agronomy Journal 85:975–982.

Roundy, B. A., V. K. Winkel, H. Khalifa, and A. D. Matthias. 1992b. Soil water availability and temperature dynamics after one-time heavy cattle tramping and land imprinting. Arid Soil Research and Rehabilitation 6: 53–69.

Roundy, B. A., J. A. Young, L. B. Sumrall, and M. Livingston. 1992c. Laboratory germination responses of 3 lovegrasses to temperature in relation to seedbed temperatures. Journal of Range Management 39:63–67.

Ruyle, G. B., and B. A. Roundy. 1990. Effects of vegetation manipulation on wildlife on semidesert rangelands. Pp. 230–238 *in* P. R. Krausman and N. S. Smith (eds.), Managing wildlife in the Southwest. Arizona Chapter of the Wildlife Society, Phoenix.

Schmutz, E. M., B. N. Freeman, and R. E. Reed. 1974. Livestock-Poisoning Plants of Arizona. University of Arizona Press, Tucson.

Schmutz, E. M., E. L. Smith, P. R. Ogden, M. L. Cox, J. O. Klemmedson, J. J. Norris, and L. C. Fierro. 1991. Desert grassland. Pp. 337–362 *in* R. T. Coupland (ed.), Natural grasslands, introduction and Western Hemisphere. Ecosystems of the World 8A. Elsevier, Amsterdam.

Schwendiman, J. L., and R. F. Sackman. 1940. Processing seed of grasses and other plants to remove awns and appendages. U.S. Department of Agriculture Circular 558.

Shiflet, T. N. 1973. Range sites and soils in the United States. Pp. 26–33 *in* Arid shrublands. Proceedings of the Third Workshop of the United States/ Australian Rangelands Panel, 26 March–5 April 1973, Tucson. Society for Range Management, Denver.

Simanton, J. R., and G. L. Jordan. 1986. Early root and shoot elongation of selected warm-season perennial grasses. Journal of Range Management 39: 63–67.

Society of Ecological Restoration. 1991. Program and abstracts, Third annual conference, Orlando, Fla., 18–23 May.

Sumrall, L. B., B. A. Roundy, J. R. Cox, and V. K. Winkel. 1991. Influence of canopy removal by burning or clipping on emergence of *Eragrostis lehmanniana* seedlings. International Journal of Wildland Fire 1:35–40.

Thornber, J. J. 1904. Singed cacti as a forage. University of Arizona Agricultural Experiment Station Bulletin 51:546–549.

———. 1906. Alfilaria. *Erodium cicutarium* as a forage plant in Arizona. University of Arizona Agricultural Experiment Station Bulletin 51:23–58.

———. 1911. Native cacti as emergency forage plants. University of Arizona Agricultural Experiment Station Bulletin 67:457–508.

Tiedemann, A. R., and J. O. Klemmedson. 1973. Nutrient availability in desert grassland soils under mesquite (*Prosopis juliflora*) trees and adjacent open areas. Soil Science Society of America Proceedings 37:107–111.

Tischler, C. R., and P. W. Voigt. 1987. Seedling morphology and anatomy of rangeland plant species. Pp. 5–13 *in* G. W. Frasier and R. A. Evans (eds.), Proceedings of symposium on Seed and Seedbed Ecology of Rangeland Plants. U.S. Department of Agriculture, Agricultural Research Service, National Technical Information Service, Springfield, Va.

Toumey, J. W. 1891. I. Notes on some of the range grasses of Arizona. II. Overstocking of the range. University of Arizona Agricultural Experiment Station Bulletin 2.

U.S. Department of Agriculture, Soil Conservation Service. 1991. Technical report of the Tucson Plant Materials Center, 1991. Tucson Plant Materials Center.

Vander Sluijs, P. H., and D. N. Hyder. 1974. Growth and longevity of blue grama seedlings restricted to seminal roots. Journal of Range Management 27:117–119.

Weaver, L. C., and G. L. Jordan. 1985. Effects of selected seed treatment on germination rates of five range plants. Journal of Range Management 38:415–418.

West, N. E. 1993. Biodiversity of rangelands. Journal of Range Management 46:2–13.

Wiedemann, H. T. 1991. Seeding chaffy grass seed and grass seed mixtures. P. 71 *in* Rangeland Technology Council, 1990 Annual report. U.S. Department of Agriculture, Forest Service Technology Development Center, Missoula, Mont.

Wilson, C. P. 1928. Factors affecting the germination and growth of chamize (*Atriplex canescens*). New Mexico State University Agricultural Experiment Station Bulletin 169.

———. 1931. The artificial reseeding of New Mexico ranges. New Mexico State University Agricultural Experiment Station Bulletin 189.

Winkel, V. K. 1990. Effects of seedbed modification, sowing depth, and soil water on emergence of warm-season grasses. Ph.D. dissertation, University of Arizona, Tucson.

Winkel, V. K., and B. A. Roundy. 1991. Effects of cattle trampling and mechanical seedbed preparation on grass seedling emergence. Journal of Range Management 44:171–175.

Winkel, V. K., B. A. Roundy, and D. K. Blough. 1991a. Effects of seedbed preparation and cattle trampling on burial of grass seeds. Journal of Range Management 44:176–180.

Winkel, V. K., B. A. Roundy, and J. R. Cox. 1991b. Influence of seedbed microsite characteristics on grass seedling emergence. Journal of Range Management 44:210–214.

Wooton, E. O. 1896. Some New Mexico forage plants. New Mexico State University Agricultural Experiment Station Bulletin 18.

Wright, H. A., and A. W. Bailey. 1982. Fire Ecology: United States and Southern Canada. John Wiley & Sons, New York.

Wright, L. N., and S. E. Brauen. 1971. Artificial selection for seedling drought tolerance and association of plant characteristics of Lehmann lovegrass. Crop Science 11:324–326.

Wright, L. N., and G. L. Jordan. 1970. Artificial selection for seedling drought tolerance in Boer lovegrass (*Eragrostis curvula*) Nees. Crop Science 10: 99–102.

Wright, N. 1961. A program-controlled environmental plant-growth chamber. University of Arizona Agricultural Experiment Station Technical Bulletin 148.

———. 1964. Drought tolerance—program-controlled environmental evaluation among range grass genera and species. Crop Science 4:472–474.

———. 1975. Improving range grasses for germination and seedling establishment under stress environments. Pp. 3–22 *in* P. S. Campbell and C. H. Herbel (eds.), Improved range plants. Range Symposium Series no. 1. Society for Range Management, Denver.

Wright, N., and L. J. Streetman. 1960. Grass improvement for the Southwest relative to drought evaluation. University of Arizona Agricultural Experiment Station Technical Bulletin 143.

Young, J. A. 1990. Revegetation technology. Pp. 92–102 *in* C. A. Call and B. A. Roundy (comps.), Proceedings of symposium on Perspectives and Processes in Rangeland Revegetation, 43d Annual Meeting of the Society for Range Management, 11–16 February 1990, Reno, Nev.

Young, J. A., and D. McKenzie. 1982. Rangeland drill. Rangelands 4:108–113.

Appendix 1

Common and Scientific Plant Names

Acacia	*Acacia* spp.
African sumac	*Rhus lancea*
Alfilaria	*Erodium cicutarium*
Algerita	*Berberis haematocarpa*
Allthorn	*Koeberlinia spinosa*
Annual broomweed	*Xanthocephalum dracunculoides*
*Archaic grasses	*Berriochloa, Graminidite, Graminites, Graminocarpon, Monoporites, Poacites, *Stipideum*
Arizona cottontop	*Digitaria californica*
Arizona rosewood	*Vauquelinia californica*
Arizona white oak	*Quercus arizonica*
Barley grass	*Horduem* spp.
Beans	*Phaseolus vulgaris*
Beardgrass	*Bothriochloa*

* = extinct taxon

Beargrass	*Nolina* spp.
Bentgrass	*Agrostis* spp.
Bermuda grass	*Cynodon dactylon*
Big bluestem	*Andropogon gerardii*
Big galleta	*Hilaria rigida*
Big sacaton	*Sporobolous wrightii*
Big sagebrush	*Artemisia tridentata*
Bimble box	*Eucalyptus populnea*
Birch	*Betula* spp.
Black grama	*Bouteloua eriopoda*
Blind prickly pear	*Opuntia rufida*
Bluebunch wheatgrass	*Agropyron spicatum*
Blue grama	*Bouteloua gracilis*
Blue paloverde	*Cercidium floridum*
Blue panicgrass	*Panicum antidotale*
Bluestem	*Andropogon* spp., *Bothriochloa* spp.
Boer lovegrass	*Eragrostis curvula* var. *conferta*
Borages (family)	Boraginaceae
*Bristlecone pine ancestor	**Pinus crossii*
Bristlegrass	*Setaria* spp.
Brittlebush	*Encelia farinosa*
Broom dalea	*Dalea scoparia*
Browntop panicgrass	*Brachiaria fasciculata*
Buffalo grass	*Buchloë dactyloides*
Buffelgrass	*Pennisetum ciliare* (*Cenchrus ciliaris*)
Burrograss	*Scleropogon brevifolius*
Burroweed	*Isocoma tenuisecta*
Bush muhly	*Muhlenbergia porteri*
California juniper	*Juniperus californicus*
Camphor weed	*Heterotheca subaxillaris*
Cane beardgrass	*Bothriochloa barbinodis*
Cane cholla	*Opuntia spinosior*
Carrizo	*Phragmites communis*
Catclaw	*Acacia greggii*
Chainfruit cholla	*Opuntia fulgida*

Chamiza	*Atriplex canescens*
Cheatgrass	*Bromus tectorum*
Chino grass	*Bouteloua ramosa*
Cholla	*Opuntia* spp.
Christmas cholla	*Opuntia leptocaulis*
Cochise lovegrass	*Eragrostis lehmanniana* × *E. trichophora*
Cotta grass	*Cottea pappophoroides*
Cotton	*Gossypium* spp.
Creosotebush	*Larrea divaricata* (*L. tridentata*)
Crested wheatgrass	*Agropyron cristatum*
Cudweed	*Gnaphalium* spp.
Curly mesquite	*Hilaria belangeri*
Cycads	*Dion, Zamia*
Desert hackberry	*Celtis pallida*
Desert vine	*Janusia gracilis*
Devil's claw	*Proboscidea* spp.
Douglas fir	*Pseudotsuga menziesii*
Dropseed	*Sporobolus* spp.
Dune mesquite	*Prosopis torreyana*
Emory oak	*Quercus emoryi*
English walnut	*Juglans regia*
Fairy duster	*Calliandra eriophylla*
Feather dalea	*Dalea formosa*
Feather fingergrass	*Chloris virgata*
Fescue	*Festuca* spp.
Filaree	*Erodium cicutarium*
Fingergrass	*Chloris* spp.
Fishhook barrel cactus	*Ferocactus wislizenii*
Flameflower	*Talinum aurantiacum*
Fluffgrass	*Erioneuron pulchellum*
Foothills paloverde	*Cercidium microphyllum*
Fourwing saltbush	*Atriplex canescens*
Fremont cottonwood	*Populus fremontii*
Fuzzynode	*Dichanthium annulatum*
Galleta	*Hilaria jamesii*

Ghaf	*Prosopis spicigera*
Giant cane	*Arundo donax*
Goosefoot	*Chenopodium* spp.
Goosefoot (family)	Chenopodiaceae
Grama grass	*Bouteloua* spp.
Grasses (family)	Gramineae (Poaceae)
Great Basin sagebrush	*Atemisia tridentata* and relatives
Green sprangletop	*Leptochloa dubia*
Hall panicgrass	*Panicum hallii*
Honey mesquite	*Prosopis glandulosa*
Hop bush	*Dodonaea viscosa* var. *angustifolia*
Huisache	*Acacia schaffneri*
Indian ricegrass	*Oryzopsis hymenoides*
Indian wheat	*Plantago fastigiata, P. patagonica*
Jack pine	*Pinus banksiana*
Johnson grass	*Sorghum halepense*
Joint fir	*Ephedra trifurca*
Joshua tree	*Yucca brevifolia*
Juniper	*Juniperus* spp.
King Ranch bluestem	*Bothriochloa ischaemum*
Kleberg bluestem	*Dichanthium annulatum*
Kleingrass	*Panicum coloratum*
Lechuguilla	*Agave lechuguilla*
Legumes (family)	Leguminosae (Fabaceae)
Lehmann lovegrass	*Eragrostis lehmanniana*
Little bluestem	*Schizachyrium scoparium*
Littleleaf sumac	*Rhus microphylla*
Lovegrasses	*Eragrostis* spp.
Maize	*Zea mays*
Mariola	*Parthenium incanum*
Melicgrass	*Melica* spp.
Mesa dropseed	*Sporobolus flexuosus*
Mesa muhly	*Muhlenbergia monticola*
Mesquite	*Prosopis* spp.
Mitchell grasses	*Astrebla* spp.

Mustards (family)	Cruciferae (Brassicaceae)
Needlegrasses	*Stipa* spp.
Negrito	*Lasiacus ruscifolia*
New Mexico feathergrass	*Stipa neomexicana*
New Mexico muhly	*Muhlenbergia pauciflora*
Nopal cenizo	*Opuntia durangensis*
Oak	*Quercus* spp.
Ocotillo	*Fouquieria splendens*
Oneseed juniper	*Juniperus monosperma*
Orchids (family)	Orchidaceae
Paloverde	*Cercidium* spp.
Pancake prickly pear	*Opuntia chlorotica*
Panicgrass	*Panicum* spp.
Panicillo caricillo	*Panicum trichoides*
Pappusgrass	*Pappophorum vaginatum*
Peach	*Prunus persica*
Pentzia	*Pentzia incana*
Pigweed	*Amaranthus palmeri, Amaranthus* spp.
Pinyon pine	*Pinus discolor, P. edulis, P. monophylla, P. remota*
Pinyon ricegrass	*Piptochaetium fimbriatum*
Plains bristlegrass	*Setaria leucopila, S. macrostachya*
Plains lovegrass	*Eragrostis intermedia*
Plains prickly pear	*Opuntia polyacantha*
Ponderosa pine	*Pinus ponderosa*
Prickly pear	*Opuntia* spp.
Primitive fern	*Anemia*
Primitive grass	*Pharus cornutus*
Purgeroot	*Jatropha macrorhiza*
Purple threeawn	*Aristida purpurea*
Rabbit brush	*Chrysothamnus* spp.
Ragweed	*Ambrosia* spp.
Range ratany	*Krameria erecta*
Red brome	*Bromus rubens*
Red threeawn	*Aristida longiseta*
Rescuegrass	*Bromus unioloides*

Resinbush	*Euryops multifidus*
Rocky Mountain juniper	*Juniperus scopulorum*
Rothrock grama	*Bouteloua rothrockii*
Saguaro	*Carnegiea gigantea*
Saltbush	*Atriplex halimoides, Atriplex* spp.
Sand dropseed	*Sporobolus cryptandrus*
Sand sage	*Artemisia filifolia*
Santa Rita threeawn	*Aristida glabrata*
Shortleaf tridens	*Erioneuron grandiflorum*
Shrub live oak	*Quercus turbinella*
Shrub oak	*Quercus pungens*
Sideoats grama	*Bouteloua curtipendula*
Sidr	*Ziziphus nummularia*
Singleleaf pinyon	*Pinus monophylla*
Six-weeks needle grama	*Bouteloua aristidoides*
Six-weeks threeawn	*Aristida adscensionis*
Slender grama	*Bouteloua repens*
Slim tridens	*Tridens muticus*
Snakeweed	*Gutierrezia sarothrae, Gutierrezia* spp.
Soaptree yucca	*Yucca elata*
Sotol	*Dasylirion wheeleri, Dasylirion* spp.
Spider grass	*Aristida ternipes* var. *ternipes*
Spike pappusgrass	*Enneapogon desvauxii*
Spruce	*Picea* spp.
Sprucetop grama	*Bouteloua chondrosioides*
Squash	*Cucurbita* spp.
Squirreltail grass	*Elymus elymoides* (*Sitanion hystrix*)
Sunflower	*Helianthus anuus*
Sunflower (family)	Compositae (Asteraceae)
Tanglehead	*Heteropogon contortus*
Tarbush	*Flourensia cernua*
Texas filaree	*Erodium texanum*
Threeawns	*Aristida* spp.
Thumor	*Acacia senegal*
Timothy	*Phleum asperum*

Tobacco	*Nicotiana* spp.
Tobosa	*Hilaria mutica*
Torrey yucca	*Yucca torreyi*
Triangle leaf bursage	*Ambrosia deltoidea*
Tumbleweed	*Salsola australis*
Tuna blanca	*Opuntia megacantha*
Utah juniper	*Juniperus osteosperma*
Variable prickly pear	*Opuntia phaeacantha*
Velvet ash	*Fraxinus pennsylvanica* var. *velutina*
Velvet mesquite	*Prosopis velutina*
Velvet pod mimosa	*Mimosa dysocarpa*
Vine mesquite	*Panicum obtusum*
Wait-a-minute bush	*Mimosa biuncifera*
Weeping lovegrass	*Eragrostis curvula*
Western wheatgrass	*Agropyron smithii*
Wheatgrasses	*Agropyron* spp.
Whiteball acacia	*Acacia angustissima*
Whitethorn acacia	*Acacia constricta*
Wild rose	*Rosa stellata*
Wild rye	*Elymus* spp.
Wilman lovegrass	*Eragrostis superba*
Winterfat	*Ceratoides lanata*
Wolftail	*Lycurus setosus* (*L. phleoides*)
Wright buckwheat	*Eriogonum wrightii*
Wright threeawn	*Aristida purpurea* var. *wrightii*
Yellow bluestem	*Bothriochloa ischaemum*
Yucca	*Yucca* spp.
Zacate barbón	*Oplismenus hirtellus*

Appendix 2

Common and Scientific Invertebrate Names

Ant lions (order, family)	Neuroptera, Myrmeleontidae
Back swimmers (family)	Notonectidae
Bandwinged grasshopper	*Trimerotropis pallidipennis*
Bee flies (family)	Bombyliidae
Bees (order)	Hymenoptera
Beetles (order)	Coleoptera
Black harvester ant	*Pogonomyrmex rugosus*
Bloodworm flies (family)	Chironomidae
Butterflies (order)	Lepidoptera
Cactus bug	*Chelinidea vittiger*
Camel cricket	*Ceuthophilus* sp.
Camel crickets (family)	Gryllacrididae
Centipedes (order)	Chilopoda
Clam shrimp	*Eulimnadia texana*
Clam shrimp (order)	Conchostraca
Click beetles (family)	Elateridae
Cockroaches (order)	Blattodea

Common cricket	*Gryllus* sp.
Common crickets (family)	Gryllidae
Crazy ant	*Conomyrma insana*
Crickets (order)	Orthoptera
Damselflies (order)	Odonata
Darkling beetle	*Araeoschizus decipiens, Eleodes* spp.
Darkling beetles (family)	Tenebrionidae
Desert cockroach	*Arenivaga* spp.
Desert cockroaches (family)	Polyphagidae
Desert fire ant	*Solenopsis xyloni*
Desert harvester ant	*Pogonomyrmex desertorum*
Desert long-legged ant	*Aphaenogaster cockerelli*
Desert millipede	*Orthoporus ornatus*
Desert millipede (family)	Spirostrepidae
Dragonflies (order)	Odonata
Dung beetles	*Canthon puncticollis*, *Canthon* spp., *Copris* spp., *Onthophagus* spp., *Phanaeus* spp.
Dung beetles (family)	Scarabaeidae
Fairy shrimp	*Streptocephalus texanus*, *Thamnocephalus platyurus*
Fairy shrimp (order)	Anostraca
Fire ant	*Solenopsis krockowi*
Flies (order)	Diptera
Giant desert centipede	*Scolopendra heros*
Gladstone spurthroat grasshopper	*Melanoplus gladstoni*
Grasshoppers	*Cibolacris parviceps*, *Derotmema haydenii*, *Hippopedon gracilipes*, *Opeia obscura*, *Orphulella pelidna*, *Parapomala pallida*, *Phrynotettix robustus*, *Psoloessa delicatula*, *Trimerotropis strenua*
Grasshoppers (order)	Orthoptera
Ground beetles	*Pasimachus duplicatus*, *Tetragonoderus pallidus*
Ground beetles (family)	Carabidae
Hairless honeypot ant	*Myrmecocystus depilis*

Harvester ants	*Pogonomyrmex texanum*, *Pogonomyrmex* spp.
Lace bugs	*Corythucha morrilli*, *C. venusta*, *Gargaphia opacula*
Lace bugs (family)	Tingidae
Leafhoppers (family)	Cicadellidae
Millipedes (order)	Diplopoda
Mosquito	*Aedes* spp.
Moths (order)	Lepidoptera
Pinacate beetles	*Eleodes extricatus*, *E. gracilis*, *E. hispilabrus*, *Eleodes*
Predaceous diving beetles (family)	Dytiscidae
Pseudoscorpions (order)	Pseudoscorpiones
Red piss ant	*Iridomyrmex pruinosum*
Robber flies	*Ablautus flavipes*, *A. rufotibialis*, *Cerotainiops abdominalis*, *C. lucyae*, *Dicropaltatum mesae*, *Efferia apache*, *E. benedicti*, *E. bicolor*, *E. cressoni*, *E. helenae*, *E. kelloggi*, *E. luna*, *E. mortensoni*, *E. ordwayae*, *E. pallidula*, *E. pilosa*, *E. subarida*, *E. tuberculata*, *E. tucsoni*, *E. varipes*, *Heteropogon cazieri*, *H. johnsoni*, *H. patruelis*, *H. wilcoxi*, *Leptogaster hesperis*, *L. patula*, *Machimus* nr. *erythocnemius*, *Mallophora fautrix*, *Megaphorus lascrucensis*, *M. prudens*, *M. pulchrus*, *Metapogon punctipennis*, *Omninablautus arenosus*, *Ospriocerus abdominals*, *Polacantha composita*, *Proctacanthella leucopogon*, *Proctacanthus milberti*, *P. nearno*, *P. nigrofemoratus*, *Promachus giganteus*, *P. nigrialbus*, *Psilocurus* sp., *Regasilus blantoni*, *Saropogon coquillettii*, *Scleropogon duncani*, *S. indistinctus*, *S. picticornis*, *Stichopogon fragilis*, *Wilcoxia* nr. *martinorum*
Robber flies (order, family)	Diptera, Asilidae
Rotifers	*Rotifera*
Scarab beetles	*Diplotaxis* sp., *Trox* sp.
Scarab beetles (family)	Scarabaeidae
Scorpions (order)	Scorpiones

Seed bugs (family)	Lygaeidae
Seed-harvester ants	*Pheidole militicida*, *P. xerophila*
Soil mites (families)	Cryptostigmatidae, Prostigmatidae
Spiders (order)	Araneae
Stink beetles	*Eleodes extricans*, *E. gracilis*, *E. hispilabrus*, *Eleodes*
Subterranean termites	*Gnathamitermes perplexus*, *Heterotermes aureus*
Sun spiders (order, family)	Solifugida, Ermobatidae
Tadpole shrimp	*Triops longicaudatus*
Tadpole shrimp (order)	Notostraca
Termites (order)	Isoptera
Thrips	*Chirothrips falsus*, *C. simplex*, *Haplothrips haplophilus*
Thrips (order)	Thysanoptera
True bugs (order)	Hemiptera
Vinegaroons (order, family)	Uropygi, Thelyphonidae
Wasps (order)	Hymenoptera
Water flea	*Moina wierzeiskii*
Water fleas (order)	Cladocera
Whip scorpions (order, family)	Uropygi, Thelyphonidae
Yucca moth	*Tegeticula yuccasella*
Yucca moth (family)	Prodoxidae

Appendix 3

Common and Scientific Vertebrate Names

*American lion	**Felis atrox*
*Antelope	*Antilocapra americana,* **Capromeryx* spp.,**Stockoceros onusoagris*
Arizona coral snake	*Micruroides euryxanthus*
Arizona cotton rat	*Sigmodon arizonae*
Arizona pocket mouse	*Perognathus amplus*
Arizona prairie dog	*Cynomys ludovicianus arizonensis*
Badger	*Taxidea taxus*
Bailey pocket mouse	*Chaetodipus baileyi*
Baird's sparrow	*Ammodramus bairdii*
Bannertail kangaroo rat	*Dipodomys spectabilis*
Bighorn sheep	*Ovis canadensis*
Bison	*Bison bison, *Bison antiguus, *B. latifrons*
Black bear	*Ursus americanus*

* = extinct taxon

Black-tailed prairie dog	*Cynomys ludovicianus*
Black-throated sparrow	*Amphispiza bilineata*
Botteri's sparrow	*Aimophila botterii*
Brown vine snake	*Oxybelis aeneus*
Brush mouse	*Peromyscus boylei*
Bullfrog	*Rana catesbeiana*
Bullsnake	*Pituophis melanoleucus sayi*
Burrowing owl	*Athene cunicularia*
Cactus mouse	*Peromyscus eremicus*
California condor	*Gymnogyps californianus*
*Camel	**Camelops hesternus*
Cassin's sparrow	*Aimophila cassinii*
Checkered whiptail lizard	*Cnemidophorus tesselatus*
Chicken	*Gallus gallus*
Chihuahua raven	*Corvus corax*
Chihuahuan grasshopper mouse	*Onychomys arenicola*
Chihuahuan whiptail lizard	*Cnemidophorus exanguis*
Chipping sparrow	*Spizella passerina*
Chiricahua leopard frog	*Rana chiricahuensis*
Coachwhip snakes	*Masticophis flagellum lineatulus, M. flagellum piceus, M. flagellum testaceus*
Colorado chub	*Gila robusta*
Common raven	*Corvus cryptoleucus*
Cotton rats	*Sigmodon* spp.
Couch's spadefoot toad	*Scaphiopus couchi*
Coyote	*Canis latrans*
Deer	*Odocoileus* spp.
Deer mouse	*Peromyscus maniculatus, Peromyscus* spp.
Desert cottontail	*Sylvilagus auduboni*
Desert grassland box turtle	*Terrapene ornata luteola*
Desert grassland kingsnake	*Lampropeltis getulus splendida*
Desert grassland massasauga	*Sistrurus catenatus edwardsii*
Desert grassland whiptail lizard	*Cnemidophorus uniparens*
Desert pocket gopher	*Geomys arenarius*
Desert pocket mouse	*Chaetodipus penicillatus*

Desert shrew	*Notiosorex crawfordi*
Desert spiny lizard	*Sceloporus magister*
*Dire wolf	**Canis dirus*
Duck	*Anser boschas*
Elk	*Cervus elephus*
*Flat-headed peccary	**Platygonus compressus*
Fulvous harvest mouse	*Reithrodontomys fulvescens*
*Giant short-faced bear	**Arctodus simus*
Gila monster	*Heloderma suspectum*
Gila spotted whiptail lizard	*Cnemidophorus flagellicaudus*
Glossy snake	*Arizona elegans*
Golden eagle	*Aquila chrysaetos*
Gopher snake	*Pituophis melanoleucus affinis*
Grasshopper mice	*Onychomys* spp.
Grasshopper sparrow	*Ammodramus savannarum*
Greater earless lizard	*Cophosaurus texanus*
Great Plains ratsnake	*Elaphe guttata emoryi*
Great Plains skink	*Eumeces obsoletus*
Great Plains toad	*Bufo cognatus*
Green ratsnake	*Elaphe triaspis*
Green toad	*Bufo debilis*
Ground squirrels	*Ammospermophilus* spp., *Spermophilus* spp.
Gunnison prairie dog	*Cynomys gunnisoni*
Harvest mice	*Reithrodontomys* spp.
Hispid cotton rat	*Sigmodon hispidus*
Hispid pocket mouse	*Perognathus hispidus*
Horned lark	*Eremophila alpestris*
Horned lizards	*Phrynosoma* spp.
*Horses	**Equus conversidens, *E. niobrarensis, *E. occidentalis, Equus* spp.
House mouse	*Mus musculus*
Jackrabbits	*Lepus alleni, L. californicus*
Javelina (collared peccary)	*Tayassu tajacu*
Kangaroo rats	*Dipodomys* spp.
Lark bunting	*Calamospiza melanocorys*

Lark sparrow	*Chondestes grammacus*
Leopard lizard	*Gambelia wislizeni*
Lesser earless lizard	*Holbrookia maculata*
*Llama	**Hemiauchenia* spp.
Little striped whiptail lizard	*Cnemidophorus inornatus*
Lowland leopard frog	*Rana yavapaiensis*
*Mammoth	**Mammuthus columbi*
*Mastodon	**Mammut americanum*
Meadow larks	*Sturnella magna*, *S. neglecta*
Merriam kangaroo rat	*Dipodomys merriami*
Mexican hook-nosed snake	*Gyalopion quadrangulare*
Mexican pocket gopher	*Pappogeomys castanops*
Mexican wolf	*Canis lupus baileyi*
Mohave rattlesnake	*Crotalus scutulatus*
*Mountain deer	**Navahoceros fricki*
Mourning dove	*Zenaidura macroura*
Mule deer	*Odocoileus hemionus*
New Mexican whiptail lizard	*Cnemidophorus neomexicanus*
New World rats and mice (family)	Cricetidae
Northern grasshopper mouse	*Onychomys leucogaster*
Old World rats and mice (family)	Muridae
Ord kangaroo rat	*Dipodomys ordi*
Packrat	*Neotoma* spp.
Pinyon mouse	*Peromyscus truei*
Plains harvest mouse	*Reithrodontomys montanus*
Plains leopard frog	*Rana blairi*
Plains pocket gopher	*Geomys bursarius*
Plains pocket mouse	*Perognathus flavescens*
Plains spadefoot toad	*Scaphiopus bombifrons*
Plateau whiptail lizard	*Cnemidophorus velox*
Pocket mice	*Chaetodipus* spp., *Perognathus* spp.
Porcupine	*Erethizon dorsatum*
Prairie rattlesnake	*Crotalus viridis viridis*
Pronghorns	*Antilocapra americana*, **Capromeryx* spp.
Rabbits	*Lepus* spp., *Sylvilagus* spp.

Regal horned lizard	*Phrynosoma solare*
*Rhinoceros	**Teleoceros major*
Rock squirrel	*Spermophilus variegatus*
Rock pocket mouse	*Perognathus intermedius*
Rock wren	*Salpinctes obsoletus*
Round-tailed ground squirrel	*Spermophilus tereticaudus*
Round-tailed horned lizard	*Phrynosoma modestum*
Rufous-crowned sparrow	*Aimophila ruficeps*
Rufous-winged sparrow	*Aimophila carpalis*
Sandill crane	*Grus canadensis*
Scaled quail	*Callipepla squamata*
*Shasta ground sloth	**Nothrotheriops shastense*
Short-horned lizard	*Phrynosoma douglassi*
*Shrub ox	**Euceratherium collinum*
Side-blotched lizard	*Uta stansburiana*
Silky pocket mouse	*Perognathus flavus*
Snow goose	*Chen hyperborea*
Sonoran whiptail lizard	*Cnemidophorus sonorae*
Southern grasshopper mouse	*Onychomys torridus*
Southern plains packrat	*Neotoma micropus*
Southern prairie lizard	*Sceloporus undulatus consobrinus*
Southwestern Woodhouse's toad	*Bufo woodhousei australis*
Spadefoot toads	*Scaphiopus* spp.
Spotted ground squirrel	*Spermophilus spilosoma*
Squirrels (family)	Sciuridae
*Stock's pronghorn	**Stockoceros onusroagris*
Tawny-bellied cotton rat	*Sigmodon fulviventer*
*Tapir	**Tapirus* sp.
Texas antelope squirrel	*Ammospermophilus interpres*
Texas horned lizard	*Phrynosoma cornutum*
Thirteen-lined ground squirrel	*Spermophilus tridecemlineatus*
Tiger salamander	*Ambystoma tigrinum*
Trans-Pecos ratsnake	*Elaphe subocularis*
Tree lizard	*Urosaurus ornatus*
Turkey vulture	*Cathartes aura*

Valley pocket gopher	*Thomomys bottae*
Vesper sparrow	*Poocetes gramineus*
Vole	*Microtus* spp.
Western diamondback rattlesnake	*Crotalus atrox*
Western harvest mouse	*Reithrodontomys megalotis*
Western hognose snake	*Heteropogon nasicus kennerlyi*
Western hook-nosed snake	*Gyalopion canum*
Western spadefoot toad	*Scaphiopus hammondi*
Western whiptail lizard	*Cnemidophorus tigris*
Whiptail lizards	*Cnemidophorus* spp.
White-footed mouse	*Peromyscus leucopus*
White-tailed antelope squirrel	*Ammospermophilus leucurus*
White-throated packrat	*Neotoma albigula*
Yellow-nosed cotton rat	*Sigmodon ochrognathus*
Yuma antelope squirrel	*Ammospermophilus harrisi*
Zebra-tailed lizard	*Callisaurus draconoides*

Contributors

Conrad Bahre, Department of Geography, University of California, Davis

Sharon H. Biedenbender, School of Renewable Natural Resources, University of Arizona, Tucson

Tony L. Burgess, Desert Laboratory, Department of Geosciences, University of Arizona, Tucson

Gregory S. Forbes, Department of Biology, New Mexico State University, Las Cruces

Graham I. Kerley, Department of Zoology, University of Port Elizabeth, Port Elizabeth, South Africa

Joseph R. McAuliffe, Desert Botanical Garden, Phoenix, Arizona

Mitchel P. McClaran, School of Renewable Natural Resources, University of Arizona, Tucson

Guy R. McPherson, School of Renewable Natural Resources, University of Arizona, Tucson

Robert R. Parmenter, Department of Biology, University of New Mexico, Albuquerque

Bruce A. Roundy, Department of Botany and Range Science, Brigham Young University, Provo, Utah

Thomas R. Van Devender, Arizona-Sonora Desert Museum, Tucson, Arizona

Walter G. Whitford, U.S. Environmental Protection Agency, Jornada Experimental Range, New Mexico State University, Las Cruces

Index

Abernathy, G. H., 283
Acacia (*Acacia* spp.), 248–49
  catclaw (*A. greggii*), 14, 88, 108, 124, 140
  Chihuahuan whitethorn (*A. neovernicosa*), 80, 81(fig.), 86, 90
  huisache (*A. schaffneri*), 54
  thumor (*A. senegal*), 50
  whiteball (*A. angustissima*), 294
  whitethorn (*A. constricta*), 108
Acarina. *See* Mites, soil
Adams, Lytle, 277–78
Aedes spp. *See* Mosquitoes
Africa, 45, 49–50, 69
Agave (*Agave* spp.), 40, 108, 140, 233
  lechuguilla (*A. lechuguilla*), 80, 81(fig.), 84(fig.)
Agricultural Research Service (USDA), 5
Agriculture, 233, 243, 250, 253
Agropyron spp. *See* Wheatgrass
Agrostis spp. *See* Bentgrass
Aimophila spp. *See under* Sparrows
Alamos (Sonora), 69
Alfilaria. *See* Filaree
Allthorn (*Koeberlinia spinosa*), 84(fig.), 86, 87, 88
Alluvial fans, 107, 110
  erosion of, 114–18
  on Santa Rita Experimental Range, 101–4
  soil horizons on, 104–6
Altar Valley (Ariz.), 248, 284
Altithermal, 88–89
Amaranthus spp. *See* Pigweed
Ambrosia spp. *See* Bursage, triangle leaf; Ragweed
Ambystoma tigrinum. *See* Salamander, tiger
Americans. *See* Anglo-Americans

Amerindians, 131
  in southeastern Arizona, 232–35, 255
Ammodramus spp. *See under* Sparrows
Amnospermophilus spp. *See* Ground squirrels
Amoebae, 153, 160(fig.)
Amphibians, 91–92, 197, 207(table), 210
Amphispiza bilineata. *See* Sparrows, black-throated
Anderson, D., 53
Andropogon gerardii. *See* Bluestem, big
Anglo-Americans, 238–39, 240–41
Animas Valley (N.M.), 212, 213
Anser boshas. *See* Ducks
Antelope. *See* Pronghorn
Antelope Wells (N.M.), 213
Antevs, Ernst, 88–89
Antilocapra americana. *See* Pronghorn
Ant lions (Myrmeliontidae, Neuroptera), 155
Ants, 153, 189
  black desert harvester (*Pogonomyrmex rugosus*), 164, 184, 185, 186
  crazy (*Conomyrma insana*), 162–63, 186
  desert fire (*Solenopsis xyloni*), 163, 164–65
  desert harvester (*P. desertorum*), 164, 186
  desert long-legged (*Aphaenogaster cockerelii*), 186
  ecological roles of, 184–86, 187
  hairless honeypot (*Myrmecocystus depilis*), 162, 186
  harvester (*P.* spp.), 163
  on Jornada del Muerto Experimental Range, 161–65
  red piss (*Iridomyrmex pruinosum*), 186
  seed-harvesting (*Pheidole* spp.), 163
Apache Pass (Ariz.), 240
Apaches, 58, 236, 237, 239, 240, 241
Aphaenogaster cockerellii. *See* Ants, desert long-legged
Aphids, 156
Appleton-Whittell Research Ranch Sanctuary, 10–11, 225
Aquila chrysaetos. *See* Eagles, golden
Arachnids, 155 *See also* Scorpions; Spiders; Sun spiders; Whip scorpions
Araeoschizus decipiens. *See* Beetles, darkling
Aravaipa Creek, 240
Aravaipa Valley, 248, 252
Archaic period, 232. *See also* Cochise Culture
Arctodus simus. *See* Bear, giant short-faced
Arenivaga. *See* Cockroaches, desert
Argentina, 49
Argillic horizons
  development of, 104–6
  water and, 106–7
Aristida spp. *See* Threeawn
Arivaca (Ariz.), 236
Arizona, 1, 68, 89, 114, 208, 255–56, 267
  alluvial fans in, 101–4
  bison in, 214–15
  Hispanics in, 235–40
  nonnative plants in, 47, 252–53
  Pleistocene in, 76–77, 78, 79, 87–88, 94
  prairie dogs in, 211, 212
  precipitation in, 3, 4, 33
  revegetation experiments in, 277–83
Arizona Agricultural Experiment Station, 244, 268–69
Arizona Agricultural Experiment Station Range Reserve Tract, 269
Arizona elegans. *See* Snakes, glossy
Arizona Game and Fish Department, 212

Arizona State Department of Transportation, 253
Arizona State Land Office, 10
Arizona Strip, 278
Arnold, J. F., 270–71
Artemisia. See Sage
Arthropods. *See also by type*
grass canopy, 167–69
ground-dwelling, 165–67
Arundo spp. *See* Cane, giant
Ash, velvet (*Fraxinus pennsylvanica* var. *velutina*), 12
Ashburn Brothers, 245
Astrebla spp. *See* Mitchell grasses
Athene cunicularia. See Owls, burrowing
Atriplex canescens. See Saltbush, four-wing
Australia, 50, 69
Avoiders, 156–57

Babocomari Creek, 215, 236, 239, 241, 245
Babocomari Grant, 237, 245
Babocomari Village (Ariz.), 215
Bac (Ariz.), 235
Backswimmers, 179
Bacteria, 158–59
Badgers (*Taxidea taxus*), 202(fig.), 218
Bailey, Vernon, 211, 212–13
Barley grass (*Hordeum* spp.), 233
Barrel cactus, fishhook, (*Ferocactus wislizenii*), 35(fig.), 40
Barrow, J. R., 284
Bartlett, John Russell, 239
Basin and Range Province, 32–33, 43, 101
Beans (*Phaseolus vulgaris*), 233
Bear
black (*Ursus americanus*), 198
giant short-faced (*Arctodus simus*), 78
Beardgrass (*Bothriochloa* spp.), 17, 269
cane, *B. barbinodis*, 49, 84
Beargrass (*Nolina microcarpa*), 11, 40
Bees, 155
Beetles, 155, 156
click, 221
darkling (*Araeoschizus decipiens, Eleodes* spp.; Tenebrionidae), 165, 216, 221
detritivore (*Eleodes* spp.), 165, 167
dung (*Canthon* spp., *Copris* spp., *Onthophagus* spp., *Phanaeus* spp.), 167
ground or carabid, 167, 216
pinacate (stink, *Eleodes* spp.), 167
predaceous diving, 179
scarab, 167
Bentgrass (*Agrostis* spp.), 72
Bermuda grass (*Cynodon dactylon*), 50, 252, 275
Bermuda high-pressure cell, 3–4
Berriochloa, 75
Beutner, E. L., 55
Big Bend National Park (Texas), 83, 84(fig.), 86, 88
Bimble box (*Eucalyptus populnea*), 50
Biodiversity, 24, 292, 294–95
Biomass, 1, 6–7
of insects, 155–56, 167–68
of prairie dogs, 213–14
and trees, 9–10
Birds, 215, 223. *See also by type*
fire and, 141–42
and grasshoppers, 11, 216
species richness of, 197, 198, 201–3, 204–5(table)
Bison (*Bison*), 78
antiquus, 214, 232
bison, 79, 214–15
latifrons, 79
Blattodea. *See* Cockroaches
BLM. *See* Bureau of Land Management
Bluestem, 17, 269
big (*Andropogon gerardii*), 84

King Ranch (*Bothriochloa ischaemum*), 275
Kleberg (*Dichanthium annulatum*), 50
little (*Schizachyrium scoparium*), 84, 254, 272
yellow, 283
Bombacaceae, 74
Bombyliidae. *See* Flies, bee
Bosque del Apache National Wildlife Refuge, 198, 201
Bothriochloa spp., 17. *See also* Beardgrass; Bluestem
Bouteloua gracilis Province, 54, 55
Bouteloua spp. *See* Chino grass; Grama
Boyce Thompson Southwestern Arboretum, 275
Bridges, J. O., 271, 275
Bristlegrass (*Setaria* spp.), 72, 254, 276
Plains, 281, 285
British Columbia, 173–74
Brittlebush (*Encelia farinosa*), 88
Brome, red (*Bromus rubens*), 47
Broomweed, annual (*Xanthocephalum dracunculoides*), 135
Brown, Clara Spalding, 251
Brown, D. E., 55
Brown, Joel S., 213
Buchloë dactyloides. *See* Buffalograss
Buckwheat, Wright (*Eriogonum wrightii*), 38
Buenos Aires Wildlife Refuge, 294
Buffalograss (*Buchloë dactyloides*), 56, 136
Buffelgrass (*Pennisetum ciliare*), 121, 136, 143, 253, 275
Bufo spp. *See* Toads
Bugs, true, 155
cactus (*Chelinidea vittiger*), 139
lace (*Corythucha morrilli, C. venusta, Gargaphia opacula*), 168
mirid, 156
seed (Lygaeidae), 168
Bullfrog (*Rana catesbeiana*), 210
Bullsnakes (*Pituophis melanoleucus*), 208
Bunchgrass, 8, 32, 136, 137, 163. *See also by type*
Bunting, lark (*Calamospiza melanocorys*), 201
Bureau of Land Management (BLM), 12, 113, 242, 247, 250, 278, 279
Bureau of Plant Industry (USDA), 5
Burrograss (*Scleropogon brevifolius*), 5, 6, 49, 158, 159(fig.), 162
Burros, 211
Burroweed (*Isocoma tenuisecta*), 8, 9(fig.), 57, 268, 282
distribution of, 110, 248
fire and, 137, 140
growth of, 14, 110
root system of, 35(fig.), 38–39
Burrow systems, vertebrate, 218–21
Bursage, triangle leaf (*Ambrosia deltoidea*), 53
Butterflies, 155, 156

Cabeza de Vaca, Alvar Nuñez, 235
Cacti, 8, 137, 139–40, 267–68. *See also by type*
Calabasas (Ariz.), 236–37
Calamospiza melanocorys. *See* Bunting, lark
Calcic horizons, development of, 104–6
Calcium, 134, 221
Caliche, 4, 44
Calliandra eriophylla. *See* Fairy duster
Callipepla squamata. *See* Quail, scaled
Callisaurus draconoides. *See* Lizards, zebra-tailed
Calvin-Benson cycle, 15
Camels (*Camelops hesternus*), 78, 222, 232

Cameron Brothers, 245
Camp Crittenden, 241
Camphorweed (*Heterotheca subaxillaris*), 13(fig.)
Camp Wallen, 241
Canada, 68, 73–74
Cane, giant (*Arundo* spp.), 72
Canis. *See* Coyotes; Wolves
Canthon spp. *See* Beetles, dung
Capromeryx sp. *See* Pronghorn
Carabidae. *See* Beetles, ground
Carbonates, 44–45
CO_2, 14–15
Carnegiea gigantea. *See* Saguaro
Carnegie Desert Laboratory, 52
Carpenter, J. R., 53
Carrion, 218, 221–22
Carrizo (*Phragmites*), 72
Cassaday, J. T., 271
Catclaw (*Acacia greggii*), 14, 88, 108, 124, 140
Cathartes aura. *See* Vultures, turkey
Cattle, 188, 211, 241, 242(fig.). *See also* Livestock
  impacts of, 236, 246–51
  Mexican, 239–40
  overgrazing, 266–69
  wild, 238–39
Cave Creek (Ariz.), alluvial fans, 114–16
Cenchrus ciliaris. *See* Buffelgrass
Centipedes, 167
  giant desert (*Scolopendra heros*), 155
Ceratoides lanata. *See* Winterfat
Cercidium spp. *See* Paloverde
Cervus elaphus. *See* Elk
Chaetodipus penicillatus. *See* Pocket mice, desert
Chamaephytes, 32
Chamiza. *See* Saltbush, fourwing
Chaparillo, 54
Chaparral, 4
Chelinidea vittiger. *See* Bug, cactus
Chen hyperborea. *See* Geese, snow
Chenopodiaceae, 75
Chickens (*Gallus gallus*), 236
Chihuahua, 75, 208, 213, 276
Chihuahuan Desert, 54, 68, 75, 86–87, 89, 94, 139, 216
  invertebrates of, 157–79
  during Pleistocene, 80–85, 88
Chilipoda. *See* Centipedes
Chino grass (*Bouteloua ramosa*), 84(fig.)
Chiricahua Mountains, 241
Chironomidae. *See* Flies, bloodworm
Chirothrips spp. *See* Thrips
Chloris virgata. *See* Fingergrass, feather or fine
Cholla (*Opuntia* spp.), 40, 108, 140, 273
  cane (*O. spinosior*), 8, 9(fig.)
  chainfruit (*O. fulgida*), 8, 9(fig.), 41
  Christmas (*O. leptocaulis*), 118
Chondestes grammacus. *See* Sparrows, lark
Chrysothamnus spp. *See* Rabbit brush, 83, 85
Chub, Colorado (*Gila* cf. *robusta*), 91, 92
Cicadas, 156
Cicadellidae. *See* Leafhoppers
Cienega Creek (Ariz.), 12, 252
Cienega Ranch (Ariz.), 246
Cienegas, 230
Cities, 243–44
Cladocerans, 189
Clays, 44, 105–6
Clements, Frederick, 52–53, 56, 85
Climate, 1, 3–4, 5, 10, 33, 46, 93, 197
  and fire, 144–45
  and grasslands, 49–50
  Holocene, 88–89
  Pleistocene, 75–77, 78–79, 85, 94
  Tertiary, 72–73, 74–75
  vegetation and, 12–14, 41–43, 47–48, 52–53

Clovis big game hunters, 78, 232
Cnemidophorus spp. *See* Whiptail lizards
Coachwhip (*Masticophis flagellum*), 208
Cochise, Lake, 76–77
Cochise County, 246, 248, 254, 256, 267
Cochise Culture, 232–33
Cockroaches, 155
  desert (*Arenivaga*, Polyphagidae), 167, 221
Coleoptera. *See* Beetles
Colinus virginianus ridgewayi. See Quail, masked bobwhite
Collembolans, 156, 159, 160(fig.), 189
Colonial period, Spanish, 198, 235–37
Colorado, 72, 74, 173
Colorado River, lower, 80, 87
Communities, 55
  fire and, 130, 138–39(fig.)
  individualistic theory of, 85–86
Composites, 75
Condor, California (*Gymnogyps californica*), 92(fig.)
Conomyrma insana. See Ants, crazy
Cooke, Philip St. George, 238–39
Cophosaurus texanus. See Lizard, greater earless
Coppice mounds, 119–20
Copris spp. *See* Beetles, dung
Coronado, Francisco Vásquez de, 235
Coronado National Forest, 241–42
Corvus spp. *See* Ravens
Corythucha spp. *See* Bugs, lace
Cotta grass (*Cottea pappophoroides*), 49
Cotton (*Gossypium* spp.), 233
Cotton rats (*Sigmodon* spp.), 211
  Arizona (*S. arizonae*), 225
Cottontails (*Sylvilagus auduboni*), 211
Cottontop, Arizona (*Digitaria californica*), 8, 9(fig.), 14, 49, 136, 272, 276, 285, 291
Cottonwood (*Populus fremonti*), 12
Coyotes (*Canis latrans*), 216, 218, 219(fig.)
Cranes, sandhill (*Grus canadensis*), 198
Creosotebush (*Larrea divaricata*), 5, 14, 52, 81(fig.), 83, 84(fig.), 86, 90, 157
  distribution of, 7(fig.), 80, 87–88
  increases in, 6, 248
Cretaceous period, 69, 72, 93
Crickets, 155, 156
  camel (Gryllacrididae), 167, 168(fig.)
  common (Gryllidae), 167
Crider, F. J., 275
Crops, Old World, 235, 237
Crotalus spp. *See* Rattlesnakes
Croton Springs (Ariz.), 248
Cryptostigmatids, 155
Cuchujaqui, Río, 85
Cucurbita spp. *See* Squash
Cudweed (*Gnaphalium* sp.), 13(fig.)
Cycads, 72
Cynodon dactylon. See Bermuda grass
Cynomys spp. *See* Prairie dogs

Dalea (*Dalea* spp.)
  broom (*D. scoparia*), 80
  feather (*D. formosa*), 38
Damselflies, 179, 189
Dasylirion wheeleri. See Sotol
Date Creek Mountains (Ariz.), 114, 278
  piedmont, 116–18, 119(fig.), 120
Daubenmire, R., 54–55
Decomposition, 180
  by termites, 181–84, 189
Deer, 233
  mountain (*Navahoceros fricki*), 78, 223
  mule (*Odocoileus hemionus*), 198, 211
Defoliation, 20–23, 24
Desert grasslands, 1–2, 31, 50–51
  concept of, 51–56
  features of, 32–33

Deserts, 75
  during Pleistocene, 79–85
Desertscrub, 4, 32, 53, 80, 81(fig.), 84(fig.), 152
  Chihuahuan, 86, 277
Dichanthium annulatum. *See* Bluestem, Kleberg
Dick-Peddie, W. A., 56
Digitaria californica. *See* Cottontop, Arizona
Di Peso, Charles, 233, 234, 236
Diplopoda. *See* Millipedes
Dipodomys spp. *See* Kangaroo rats
Diptera. *See* Flies
Disturbance, 48, 49, 86, 240
  by vertebrates, 218–22
Dixon, R. M., 283
Dodonaea viscosa var. *angustifolia*. *See* Hop bush
Douglas (Ariz.), 244
Dove, mourning (*Zenaidura macroura*), 141, 142
Dragonflies, 179, 189
Dragoon Mountains, 241, 255
Dropseed (*Sporobolus* spp.), 157
  mesa (*S. flexuosus*), 5, 23, 80, 167
  sand (*S. cryptandrus*), 43, 81, 271
Droughts, 6, 89, 287
  impacts of, 266–67
  and plant growth, 33, 38, 40, 41, 42–43, 48
  and shrub increase, 92–93
Ducks (*Anser boshas*), 236
Dunelands, 6
Durango (Mexico), 33, 54, 75
Dytiscidae. *See* Beetles, predaceous diving

Eagles, golden (*Aquila chrysaetos*), 218
Ecosystems, 292–93
Efferia spp. *See* Robber flies
Elaphe spp. *See* Ratsnakes
Elapidae. *See* Snakes, Arizona coral
Elateridae. *See* Beetles, click
Eleodes. *See* Beetles, detritivore
Elgin (Ariz.), 243
Elk (*Cervus elaphus*), 198, 211
Elymus spp. *See* Rye; Squirreltail
Empire-Ciénega Resource Conservation Area, 12, 13(fig.)
Encelia farinosa. *See* Brittlebush
Eocene, 73, 93
Ephedra spp. *See* Joint fir; Mormon tea
Equus spp. *See* Horses
Eragrostis. *See* Lovegrass
Eremophila alpestris. *See* Larks, horned
Erethizon dorsatum. *See* Porcupine
Eriogonum wrightii. *See* Buckwheat, Wright
Erodium cicutarium. *See* Filaree, Texas
Erosion, 46, 255
  of alluvial piedmonts, 113–14
  from grazing, 246, 250, 251
  lithology and, 114–18
  revegetation and, 121–25, 275
  and vegetative cover, 118–20
Escapule (Ariz.), 232
Estevanico, 235
Eucalyptus populnea. *See* Bimble box
Euceratherium collinum. *See* Ox, shrub
Eulimnadia texana, Conchostraca. *See* Shrimp, clam
Euryops multifidus. *See* Resinbush
Evergreens, 13–14
Exploiters
  extensive, 39–40, 46
  intensive, 34, 36, 38–39, 45, 48
Extinction, Pleistocene, 46, 78–79, 222–23

Fairy duster (*Calliandra eriophylla*), 38, 108, 110, 124
Feathergrass, New Mexico (*Stipa neomexicana*), 14, 42

Felis atrox. See Lion, American
Ferocactus wislizenii. See Barrel cactus, fishhook
Fescue (*Festuca*), 72
Filaree, 274–75
  Texas (*Erodium cicutarium*), 47, 185, 252
Fingergrass, feather or fine (*Chloris virgata*), 84, 254
Fir, Douglas (*Pseudotsuga menziesii*), 87
Fires
  animals and, 141–43
  behavior of, 132–33, 143–44
  frequency of, 131–32, 144–45, 230
  grasslands and, 130–31
  human-caused, 223, 234–35, 240
  impacts of, 45–46
  management of, 251–52
  physical environment and, 133–34
  plants and, 134–41
  on Santa Rita Experimental Range, 110, 112
  suppression of, 242, 248, 249–50
Fish, 197
Flameflower (*Talinum aurantiacum*), 40
Fleas, water (*Moina wierzejskii*, Cladocera), 178, 179
Flies, 155, 156. *See also* Robber flies
  bee, 188
  bloodworm, 179
Floral formula, 17, 19
Florets, 16, 17, 18(fig.)
Florida Canyon Wash (Ariz.), 102, 104
Florrisant (Colo.), 74
Flourensia cernua. See Tarbush
Forage, 142, 242, 273–74, 276
Forbs, 2, 8, 216
Forest Service (USDA), 8, 10, 241–42, 247, 251
Fort Bowie, 240
Fort Breckenridge, 240
Fort Buchanan, 240
Fort Grant, 254
Fort Huachuca, 212
Fossils
  grass, 69, 72, 93
  Tertiary, 73–75
Fouquieria splendens. See Ocotillo
Fraxinus pennsylvanica var. *velutina. See* Ash, velvet
Frogs (*Rana* spp.)
  Chiricahua leopard (*Rana chiricahuensis*), 210
  lowland leopard (*R. yavapaiensis*), 210
  plains leopard (*R. bairi*), 210
Frye Mesa (Ariz.), plant introduction on, 121–25
Fungi, 158–59, 161, 180

Gadsden Purchase, 239
Galleta (*Hilaria* spp.)
  H. rigida, 38
Gallus gallus. *See* Chickens
Gálvez, Bernardo de, 236
Gambelia wislizeni. See Lizards, leopard
Gargaphia opacula. See Bugs, lace
Geese, snow (*Chen hyperborea*), 198, 201
Gentry, H. S., 54
Geography, 1–2
Geology, 74
Ghaf (*Prosopis spicigera*), 50
Gila cf. *robusta. See* Chub, Colorado
Gila River, 243
Gila-San Simon area, 250–51
Gila Valley, 250–51, 278
Gileños, 234
Glendening, G. E., 271
Gnaphalium sp. *See* Cudweed
Gnathamitermes perplexus. See Termites, subterranean

Goats, 46, 211
Goosefoot family. *See* Chenopodiaceae
Gophers. *See* Pocket gophers
Gossypium spp. *See* Cotton
Government Draw (Ariz.), 255
Graham County (Ariz.), 246
Grama (*Bouteloua* spp.), 17, 72, 250, 254, 255, 269
  black (*B. eriopoda*), 5, 6(fig.), 16, 18(fig.), 23, 35(fig.), 38, 42, 46, 52, 53, 57, 81, 121, 136, 157, 160(table), 167, 169, 283
  blue (*B. gracilis*), 10, 11(fig.), 13(fig.), 32, 36, 38, 53, 54, 56, 81, 271, 272
  crowfoot, Rothrock (*B. rothrockii*), 8, 23, 71(fig.), 271, 276
  hairy (*B. hirsuta*), 10, 11(fig.), 13(fig.)
  sideoats (*B. curtipendula*), 13(fig.), 14, 23, 81, 84, 124, 271, 276, 283, 288
  six-weeks needle (*B. aristidoides*), 9(fig.), 137
  slender (*B. repens*), 8, 84
  sprucetop (*B. chondrosioides*), 13(fig.), 84
Graminidites, 69
Graminites, 72
Graminocarpon, 69
Granivores, granivory, 215–16
Grants, land, 237, 238(fig.)
Grasses, 2, 3, 6, 32, 48, 68, 134, 216
  canopy arthropods, 167–69
  climate and, 42–43
  and erosion, 118, 120
  evolution of, 69–72
  fire and, 45–46, 136–37
  growth and regrowth of, 20–23, 24
  humans and, 233, 234
  morphology of, 15–19
  nonnative, 47, 225, 265–66
  perennial, 9–10, 36, 39
  revegetation and, 121, 269–72, 278–83, 285–86, 293–94
  water use of, 36–38, 108
  See also by type
Grasshoppers, 143, 155, 156, 169, 172(table), 189
  bandwinged (*Trimerotropis pallidipennis*), 143, 170–71
  Gladstone spur-throat (*Melanoplus gladstoni*), 143
  predation on, 11, 188, 216
Grassland Biome of North America, 53
Grasslands, 32, 45. *See also* Desert grasslands
  climate and, 49–50
  defining, 50–56
  fire and, 130–31
  formation of, 72–75
  human impacts on, 230–31
  Pleistocene, 75–79
  soil moisture and, 48–49
Grazing, 12, 23, 168
  on Appleton-Whittell Research Ranch Sanctuary, 10–11
  evolution of, 73–74
  and fire, 132, 136, 143
  impacts of, 46, 224(fig.), 244–47, 294
  management of, 5, 242, 247–51
  research on, 8–9
Grazing Service, 242. *See also* Bureau of Land Management
Great Basin, 55
Great Plains, 55, 68, 73, 75, 76, 84, 93, 94
Green Valley (Ariz.), 244
Griffiths, D. A., 244, 269, 275
Groundsel, thread-leaf (*Senecio longilobus*), 248
Ground sloth, Shasta (*Nothrotheriops shastense*), 223
Ground squirrels, 198, 218

Amnospermophilus spp., 211
spotted (*Spermophilus spilosoma*), 81, 141, 198, 211, 215
Grus canadensis. *See* Cranes, sandhill
Gryllacrididae. *See* Crickets, camel
Gryllidae. *See* Crickets, common
Guévavi, 236
Gutierrezia sarothrae. *See* Snakeweed
Gyalopion spp. *See* Snakes, hook-nosed
Gymnogyps californica. *See* Condor, California

Half-shrubs, 32
Haplothrips haplophilus. *See* Thrips
Hawks, 216
Hay harvesting, 248, 254–55
Heffner, Harry L., 275
Helianthus annuus. *See* Sunflower
Hemiauchenia. *See* Llama
Hemiptera. *See* Bugs, true
Herbel, C. H., 283
Herbivores, herbivory
impacts of, 46–49
vertebrate, 211–14
Heterodon nasicus kennerlyi. *See* Snakes, western hognose
Heteropogon spp. *See* Robber flies
Heteropogon contortus. *See* Tanglehead
Heterotermes aureus. *See* Termites, subterranean
Heterotheca subaxillaris. *See* Camphorweed
Hidalgo County (N.M.), 208
Hilaria spp. *See* Galleta; Mesquite grass; Tobosa
Hispanics, impacts of, 235–40, 255–56
Hohokam, 214–15, 233–34
Holbrookia maculata. *See* Lizards, lesser earless
Holocene, 232
plant communities of, 85–89, 94
Hooker, H. C., 245
Hop bush (*Dodonaea viscosa* var. *angustifolia*), 50
Hordeum spp. *See* Barley grass
Horned lizards (*Phrynosoma* spp.)
regal (*P. solare*), 208
round-tailed (*P. modestum*), 92, 208
short-horned (*P. douglassi*), 208
Texas (*P. cornutum*), 208
Horses (*Equus* spp.), 73–74, 78, 222, 236
cf. *conversidens*, 232
cf. *niobrarensis*, 232
cf. *occidentalis*, 232
Howell's Ridge Cave (N.M.), vertebrate assemblage from, 89–92, 94, 223–24
Huachuca Mountains, 241, 248, 255
Hueco Bolson, 80, 81, 83
Hueco Mountains (Texas), 80–81, 83, 86–87
Huisache (*Acacia schaffneri*), 54
Humans, impacts of, 5, 222–25, 230. *See also various cultures*; *time periods*
Humphrey, R. R., 54
Hymenoptera. *See* Ants; Bees; Wasps

IBP. *See* International Biological Programme
India, 50, 69
Indians. *See* Amerindians
Insects, 139. *See also by type*
and birds, 202–3
predators on, 188–89
in soils, 159–61
International Biological Programme (IBP), 6–7
Invertebrates, 143. *See also by type*
adaptations of, 156–57
biomass of, 155–56
diversity of, 152–55

ecological roles of, 179–90, 250
on Jornada del Muerto Basin Study Area, 157–79
Iridomyrmex pruinosum. See Ants, red piss
Irrigation, 243, 250, 253
Irwin, J. N., 246
Isocoma tenuisecta. See *Burroweed*
Isoptera. *See* Termites

Jackrabbits (*Lepus alleni, L. californicus*), 211, 216, 250
Jackson Hole (Wyo.), 74
Janos (Chihuahua), 213
Jatropha macrorhiza. See Purgeroot
Javelina (*Tayassu tajacu*), 28, 211
Jensen, Ben F., 277
Johnson, H. B., 85–86
Johnson grass (*Sorghum halepense*), 275
Joint fir (*Ephedra trifurca*), 86, 90
Jordan, Gilbert L., on revegetation, 277, 278–81, 287
Jornada del Muerto Plain (N.M.), 271
Jornada Experimental Range, 23, 121
characteristics of, 5–7
invertebrates on, 157–79
plant growth on, 14, 42
revegetation on, 275–76
Joshua tree (*Yucca brevifolia*), 87
Juglans regia. See Walnut, English
Juniper (*Juniperus* spp.), 39, 80, 85, 247, 248
California (*J. californica*), 87
oneseed (*monosperma*), 87, 140
Rocky Mountain (*scopulorum*), 87
Utah (*osteosperma*), 87

Kangaroo rats (*Dipodomys* spp., 46, 141, 214, 216, 217(fig.)
bannertail (*D. spectabilis*), 81, 198, 215, 218–21
Merriam (*D. merriami*), 215, 250
Ord (*D. ordi*), 198, 215, 219(fig.)
Kansas, 72, 76
Kansas Settlement (Ariz.), 243
Kingsnakes, desert grassland (*Lampropeltis gelutus splendida*), 208
Kino, Eusebio Francisco, 235
Kleingrass (*Panicum coloratum*), 275
Koeberlinia spinosa. See Allthorn

Lampropeltis gelutus splendida. See Kingsnakes, desert grassland
Land and Hays Company, 245
Land imprinting, 283–84
Landscape, 32–33
erosion and, 113–14
evolution of, 100–106
fragmentation of, 5, 231, 252
Larks, 215
horned (*Eremophila alpestris*), 142, 201
Larrea divaricata. See Creosotebush
Las Colinas (Ariz.), 215
Lasiacis ruscifolia. See Negrito
Leafhoppers, 168
Lechuguilla (*Agave lechuguilla*), 80, 81(fig.), 84(fig.)
Lehner Ranch (Ariz.), 78, 232
Lepidoptera. *See* Butterflies; Moths
Leporidae. *See* Rabbits
Leptochloa dubia. See Sprangletop, green
Lepus spp. *See* Jackrabbits
Lion, American (*Felis atrox*), 78
Lithology, erosion and, 114–18
Little, E. A., 269
Little Hatchet Mountains (N.M.), vertebrates, 89–91
Livestock, 5, 31, 47, 57, 210
on Appleton-Whittell Research Ranch Sanctuary, 10–11

feed for, 254–55
impacts of, 46, 132, 136, 143, 168, 211, 224(fig.), 244–51, 294
introduction of, 235–36, 237
management of, 267–68
Mexican, 239–40
termites and, 181, 183, 189
Lizards, 203, 216. *See also* Horned lizards; Whiptail lizards
desert spiny (*Sceloporus magister*), 208
greater earless (*Cophosaurus texanus*), 205
leopard (*Gambelia wislizeni*), 208
lesser earless (*Holbrookia maculata*), 205
side-blotched (*Uta stansburiana*), 205, 208, 221
southern prairie (*S. undulatus consobrinus*), 205
tree (*Urosaurus ornatus*), 208
zebra-tailed (*Callisaurus draconoides*), 205, 208
Llama (*Hemiauchenia*), 78
Llano Estacado, 76
Loma Los Ratones (Chihuahua), 213–14
Long Term Ecological Research (LTER), 7, 197
Lordsburg Playa, 91(fig.)
Lovegrass (*Eragrostis* spp.), 265
Boer (*E. curvula* var. *conferta*), 143, 225, 275–76, 281, 283, 287
Cochise (Atherstone) (*E. lehmanniana x E. trichophora*), 275, 276, 281, 288, 289
Lehmann (*E. lehmanniana*), 8, 10, 14, 47, 50, 110, 121, 137, 143–44, 160(table), 161, 216, 225, 253, 271, 275, 276, 280–81, 283, 285, 286, 287, 288, 289–90, 291, 293
plains (*E. intermedia*), 10, 11(fig.), 13(fig.), 14, 84, 135, 287
weeping (*E. curvula*), 275
Wilman (*E. superba*), 50, 275, 281, 285
LTER. *See* Long Term Ecological Research
Lycurus setosus. See Wolftail
Lygaeidae. *See* Bugs, seed
MacDougal, Daniel, 52
Madera Canyon Wash, alluvial fan of, 101–4, 106, 111–12(fig.)
Magnesium, 134
Maize (*Zea mays*), 233
Mammals. *See also by type*
fire and, 141–42
hypsodont, 73–74
Pleistocene, 46, 77–78
species richness of, 197, 198, 199–201(tables)
Mammoths, Columbian (*Mammuthus columbi*), 78, 215, 223, 232
Mammut americanum. See Mastodon
Mammuthus columbi. See Mammoths, Columbian
Maravillas Canyon (Texas), 83, 84
Mariola (*Parthenium incanum*), 80, 81(fig.), 86, 87, 90
Marshes, 230
Martin, Paul Schultz, 89, 232
Massasauga (*Sistrurus catenatus edwardsi*), 208, 221, 223
Masticophis flagellum. See Coachwhip
Mastodon (*Mammut americanum*), 78
Mayeux, H. S., 85–86
McGinnies, W. G., 270–71
Meadowlarks (*Sturnella magna, S. neglecta*), 201
Meadow Valley Flat (Ariz.), 212
Mearns, Edgar A., 212
Melanoplus gladstoni. See Grasshoppers, Gladstone spur-throat
Melicgrass (*Melica* spp.), 72
Meristems, 20, 22, 23

Mescal (*Agave* spp.), 40
Mesquite (*Prosopis*), 14, 52, 57, 140, 223, 233, 268, 285
  biomass production and, 9–10
  honey (*P. glandulosa*), 5, 6, 80, 83, 84(fig.), 86–87, 88, 90, 140, 157, 214
  invasion of, 247, 248–50
  ghaf (*P. spicigera*), 50
  velvet (*P. velutina*), 8, 9(fig.), 12, 13(fig.), 35(fig.), 39, 45, 56, 87, 108, 110, 112–13, 117(fig.), 124, 132, 136, 140, 224
  vine (*Panicum* cf. *obtusum*), 81, 157, 254
Mesquite grass, curly (*Hilaria belangeri*), 32, 53, 55, 124, 269, 272
Mesquite, vine (*Panicum* cf. *obtusum*), 81
Mexican Plateau, 53–54, 68, 74, 75
Mexicans, in southern Arizona, 237–40
Mexico, 1, 2, 3, 33, 47, 51, 53–54. *See also various states*
Mice, 198, 211. *See also* Pocket mice
  deer (*Peromyscus* spp.), 215
  grasshopper (*Onychomys* spp.), 215, 216
  harvest (*Reithrodontomys megalotis, R. montanus*), 215
  white-footed (*P. leucopus*), 141
Microarthropods, 159–61
Microbes, 180
Microtus spp. *See* Voles
Micruroides euryxanthus. See Snakes, Arizona coral
Middens, packrat, 78–85, 86–88, 94
Millipedes, 156
  desert (*Orthoporus ornatus*, Spirostepidae), 155
Mimosa spp. *See* Wait-a-minute bush
Miocene, 72, 73, 74–75, 93, 134
Mitchell grasses (*Astrebla* spp.), 50
Mites, soil, 154–55, 159, 161, 180, 181, 189
Mogollon Culture, 234
Mohave Desert, 85
Moina wierzejskii, Cladocera. *See* Fleas, water
Monoporites, 69
Monsoons, 88–89, 94
Montana, 188
Mormon tea (*Ephedra* spp.), 35(fig.), 124
Mosquitoes (*Aedes* spp.), 178, 179, 189
Moths, 155, 156
  yucca (*Tegeticula yuccasella*), 186–88, 189
Muhly, bush (*Muhlenbergia porteri*), 5, 9(fig.), 14, 32, 38, 272, 283, 285
Mule Mountains (Ariz.), 255
Murray Springs (Ariz.), 78, 215, 232
Muscotah Marsh (Kan.), 76
Museum of Southwestern Biology (Univ. of New Mexico), 197–98
Myrmecocystus depilis. See Ants, hairless honeypot
Myrmeliontidae. *See* Ant lions

Naco (Ariz.), 232
Nanophanerophytes, 38
National Audubon Society, 11
National forest reserves, 241–42
National Science Foundation, 7
Native Americans. *See* Amerindians
Navahoceros fricki. See Deer, mountain
Nebraska, 75
Needlegrass (*Stipa*), 72
  New Mexican (*S. neomexicana*), 14, 42
Negrito (*Lasiacis ruscifolia*), 69
Nematodes, 153, 154, 155, 156, 159, 160(fig.), 189
Neotoma spp. *See* Packrats
New Mexico, 1, 56, 72, 105(fig.), 214, 223, 267

climate in, 3, 4, 5
grasslands in, 68, 87
kangaroo rats in, 218–19
Pleistocene in, 76, 77–78, 79, 94
prairie dogs in, 211–13
revegetation experiments, 274, 275–76, 283
vertebrate assemblages from, 89–91, 197–210
New Mexico Agricultural Experiment Station, 269
Nicotiana spp. *See* Tobacco
Nitrogen, 133–34, 180
Niza, Marcos de, 235
Nogales (Ariz.), 244
Nolina microcarpa. See Beargrass
Nopal cenizo (*Opuntia durangensis*), 54
Notiosorex crawfordi. See Shrew, desert
Notonectidae. *See* Backswimmers
Nutrients, 180, 183
ants and, 184–85
fire and, 133–34
trees and, 9–10, 285
vertebrates and, 221–22

Oaks (*Quercus*), 247, 248
Arizona white (*Q. arizonica*), 10
Emory (*Q. emoryi*), 10, 11(fig.)
shrub (*Q. pungens*), 80
shrub live (*Q. turbinella*), 87
Ocotillo (*Fouquieria splendens*), 40, 80, 81(fig.), 83, 84(fig.), 86, 87, 88, 90, 110, 140
Odocoileus hemionus. See Deer, mule
Odonata. *See* Damselflies; Dragonflies
Ogallala formation, 72
Oligocene, 69, 72, 74, 93
Omninablautus. See Robber flies
Onthophagus spp. *See* Beetles, dung
Onychomys spp. *See* Mice, grasshopper
O'otam, 233–34
Oplismenus hirtellus. See Zacate barbón
Opuntia. See Cholla; Prickly pear
Orthoporus ornatus, Spirostepidae. *See* Millipedes, desert
Orthoptera. *See* Crickets; Grasshoppers
Ospirocerus abdominalis. See Robber flies
Overgrazing, impacts of, 53, 244–48, 266–68
Overhunting, Paleo-Indian, 222–23, 232
Ovis canadensis. See Sheep, bighorn
Owls, 216
burrowing (*Athene cunicularia*), 202
Ox, shrub (*Euceratherium collinum*), 78, 223
Oxybelis aeneus. See Snakes, brown vine

Packrats (*Neotoma* spp.), 78, 141, 211, 250
middens of, 79–85, 86–88, 94
Paleocene, 69, 72–73, 93
Paleo-Indians, 78, 222–23, 232
Palms, 72, 74
Paloverde (*Cercidium* spp.)
blue (*C. floridum*), 8, 9(fig.), 88, 140
foothill (*C. microphyllum*), 8, 53
Panicoideae, 69
Panicgrass (*Brachiaria* spp., *Panicum* spp.), 72, 287
blue (*P. antidotale*), 275, 281, 283, 288
browntop (*B. fasciculata*), 234
panizo caricillo (*P. trichoides*), 69
vine mesquite (*P.* cf. *obtusum*), 81
Panicum spp. *See* Kleingrass; Panicgrass
Papago. *See* Tohono O'odham
Pappogeomys castanops. See Pocket gopher, Mexican
Pappusgrass (*Pappophorum* spp.)
P. vaginatum, 9(fig.)
spike, *Enneapogon desvauxii*, 49, 50

Parthenium incanum. *See* Mariola
Parthenogenesis, 156, 209
Pastizales, 75
Peaches (*Prunus persica*), 236
Peccary, flat-headed (*Platygonus compressus*), 78, 222. *See also* Javelina
Pennisetum ciliare. *See* Buffelgrass
Pentzia (*Pentzia incana*), 50
Perognathus spp. *See* Pocket mice
Peromyscus spp. *See under* Mice
Petryszyn, Yar, 213
Phanaeus spp. *See* Beetles, dung
Pharus cornutus, 70(fig.)
Phaseolus vulgaris. *See* Beans
Pheidole spp. *See* Ants, seed-harvesting
Phleum asperum. *See* Timothy, annual
Phosphorus, 133, 221
Photosynthetic pathways, 14–15, 24, 42, 69, 72
Phragmites spp. *See* Carrizo
Phrynosoma spp. *See* Horned lizards
Phytomers, 16, 17(fig.), 20–22
Piedmonts, alluvial, 113–14, 116, 118, 119(fig.), 120
Pigweed (*Amaranthus* spp.), 75
Pima, 233
Pima County, 246
Pimería Alta, 235
Pinaleño Mountains, 241
Pines (*Pinus*)
  border pinyon (*P. discolor*), 247
  P. crossii, 74
  pinyon (*P. edulis, P. remota*), 80
  ponderosa (*P. ponderosa*), 87, 132
  single-leaf pinyon (*P. monophylla*), 87
Piptochaetium. *See* Ricegrass, pinyon
Pituophis spp. *See* Bullsnakes
Plants, 5
  environmental responses of, 287–90
  and fire, 132–33, 134–41
  growth of, 41–43, 47–49
  native, 293, 294–95
  nonnative, 47, 121–25, 252–53, 274–76, 285–86
  production by, 12–14, 24, 33–34, 136
Platygonus compressus. *See* Peccary, flat-headed
Playas, 90(fig.), 91–92, 94, 178–79, 189
Playas Valley (N.M.), 90, 212, 213
Pleistocene, 94, 46, 232
  alluvial fan development, 101–4, 121–22
  grasslands of, 75–76
  vertebrate fauna of, 77–79
  woodlands during, 79–85
Pliocene, 121
Poaceae. *See* Grasses; *grasses by type*
Poacites, 72
Pocket gophers, 218
  Mexican (*Pappogeomys castanops*), 211
  valley (*Thomomys bottae*), 211, 219(fig.)
Pocket mice
  desert (*Chaetodipus penicillatus*), 215
  hispid (*Perognathus hispidus*), 141, 215
  silky (*Perognathus flavus*), 215, 219(fig.)
Pogonomyrmex spp. *See under* Ants
Pollination, 16, 186–88
Polyphagidae. *See* Cockroaches
Ponds, 178–79, 189
Pooecetes gramineus. *See* Sparrows, vesper
Populus fremonti. *See* Cottonwood
Porcupine (*Erethizon dorsatum*), 198
Portal (Ariz.), 223
Poston, Charles D., 237
Prairie dogs (*Cynomys* spp.), 218
  Arizona (*C. ludovicianus arizonensis*), 212, 213

blacktail (*C. ludovicianus*), 81, 198, 202(fig.), 211, 212, 231
Gunnison's (*C. gunnisoni*), 211, 213(fig.)
Precipitation, 5, 10, 47
patterns of, 3–4, 33
plant growth and, 41–42
plant production and, 14, 33–34
and soil types, 43–45
Predation, 11, 188, 216
Prickly pear (*Opuntia* spp.), 32, 40, 108, 119(fig.), 140, 273
blind (*O. rufida*), 84(fig.)
nopal cenizo (*O. durangensis*), 54
pancake (*O. chlorotica*), 117(fig.)
tuna blanca (*O. megacantha*), 54
variable (*O. phaecantha*), 8, 9(fig.), 81(fig.), 86, 88
Proctacanthus milbertii. See Robber flies
Promachus nigralbus spp. *See* Robber flies
Pronghorn
Antilocapra americana, 78, 142, 185, 198, 203(fig.), 230–31
Capromeryx sp., 78, 223
Stockoceros onusroagris, 78, 223
Prosopis. See Mesquite
Prostigmatids, 155
Protozoans, 153, 156, 159, 189
Prunus persica. See Peaches
Pseudoscorpions (Pseudoscorpiones), 155
Pseudotsuga menziesii. See Fir, Douglas
Psilocurus spp. *See* Robber flies
Purgeroot (*Jatropha macrorhiza*), 40

Quail
masked bobwhite (*Colinus virginianus ridgewayi*), 141, 142, 294
scaled (*Callipepla squamata*), 201
Quercus. See Oaks
Quiburi (Ariz.), 235, 236, 237

Rabbit brush (*Chrysothamnus* spp.), 83, 85
Rabbits, 211, 233, 236
Ragweed (*Ambrosia* spp.), 75
Rainfall. *See* Precipitation
Rain forests, 72–73
Rana spp. *See* Frogs
Ranching, 12, 52
American, 241, 247–51
Mexican, 239–40
Range sites, 113, 285
Rats. *See* Cotton rats; Kangaroo rats; Packrats
Ratsnakes (*Elaphe* spp.)
Great Plains (*E. guttata emoryi*), 208
green (*E. triaspis*), 208
trans-Pecos (*E. subocularis*), 208
Rattlesnakes (*Crotalus* spp.)
Mohave (*C. scutulatus*), 208, 223
prairie (*C. viridis viridis*), 202, 208, 221, 223
western diamondback (*C. atrox*), 208, 221
Ravens (*Corvus*)
Chihuahua (*C. corax*), 218
common (*C. cryptoleucus*), 218
Reclamation, 292
Regasilus. See Robber flies
Rehabilitation, 292
Reithrodontomys spp. *See* Mice, harvest
Reproduction, of grasses, 16–18
Reptiles, 91, 221. *See also by type*
species richness of, 197, 203, 205, 207–9
Rescuegrass (*Bromus unioloides*), 274
Resinbush (*Euryops multifidus*), introduction of, 123–25
Resprouting, 140–41
Restoration, 292–93, 294–95
Revegetation
with cacti and shrubs, 273–74
goals of, 291–95

with native grasses, 269–72
with nonnative plants, 47, 121–25, 252–53
with pelleted seed, 277–78
plant trials in, 274–76
research on, 265–66, 284–91
seedbed preparation and, 278–84
and soil erosion, 121–25
Rhinoceros (*Teleoceras major*), 75
Rhizomes, 16, 20, 137
Rhus spp. *See* Sumac
Ricegrass, pinyon (*Piptochaetium* spp.), 72
Rich Lake (Tex.), 76
Rillito Creek (Ariz.), 255
Rio Grande, 198
Rio Grande Valley (N.M.), 235
Rio Grande Village (Tex.), 83, 84(fig.)
Riparian areas, 198, 201
Robber flies, 174–75(table)
Asilidae, 176(fig.)
Efferia spp., 172–73, 175, 177, 178, 188, 189
Heteropogon spp., 175, 177, 188
Omninablautus spp., 177
Ospiriocerus abdominalus, 175
Proctocanthus milbertii, 188
Promachus nigralbus spp., 178
Psilocurus spp., 175
Regasilus spp., 177
Scleropogon spp., 178, 188
Wilcoxia spp., 177
Rock types. *See* Lithology
Rocky Mountains, 74, 93
Rodents, 141, 198, 215. *See also by type*
soils and, 218–21
Root systems, 13, 35(fig.), 137, 289
burroweed, 38–39
extensive exploiters, 39–40
of intensive exploiters, 34, 36
of water storers, 40–41
and water use, 108, 110
Rose, wild (*Rosa stellata*), 87
Rosewood, Arizona (*Vauquelinia californica*), 87
Rotifers, 179, 189
Rural areas, 243, 256
Rye (*Elymus* spp.), 274
Sacaton (*Sporobolus* spp.), 17
alkali (*S. airoides*), 5, 91(fig.), 158, 162, 283
big (*S. wrightii*), 10, 11(fig.), 56, 136, 234, 254
Sacramento Mountains (N.M.), 87
Sage (*Artemisia* spp.)
big (*A. tridentata*), 83, 85, 87, 275
sand (*A. filifolia*), 80, 83
Saguaro (*Carnegiea gigantea*), 8, 53, 233
Salamander, tiger (*Ambystoma tigrinum*), 91, 92(fig.), 210
Salpinctes obsoletus. *See* Wrens, rock
Salsola australis. *See* Tumbleweed
Saltbush (*Atriplex* spp.)
fourwing (*A. canescens*), 81, 83, 86, 87, 88, 90, 273–74, 276, 283
A. halimoides, 274
San Andres Mountains (N.M.), 87
San Agustin Plains (N.M.), 76
San Augustín de Tucson, 236
San Bernardino Grant, 237
San Bernardino Valley, 238, 239, 256
Sand, soil moisture and, 43–45
San Pedro River Valley (Ariz.), 212, 215, 232, 235, 236, 238, 239, 243, 245, 248, 249, 253, 256
San Rafael Ranch (Ariz.), 245
San Rafael Valley (Ariz.), 212
San Simon Valley (Ariz.), 248, 250, 278, 279(fig.), 280–82
Santa Catalina Mountains (Ariz.), 241
Santa Catarina (Ariz.), 235, 237
Santa Cruz County (Ariz.), 246
Santa Cruz de Terrentate (Ariz.), 237
Santa Cruz Valley, 243, 245, 253, 256

Anglo-Americans in, 240–41
Mexicans in, 237–40
Spanish in, 235–37
Santa Rita Experimental Range, 23, 41, 211, 224, 248–49(fig.), 269, 271
characteristics of, 7–10
soil types on, 101–6
vegetation types on, 110–12
Santa Rita Mountains, 241, 245, 248, 255
San Xavier del Bac, 236, 237
Savanna, 49, 50, 51, 53, 73–74, 93
Apacherian mixed shrub, 32, 58
mixed shrub, 57–58
Pleistocene, 77–78
Scaphiopus spp. *See* Spadefoot toads
Sceloporus. *See under* Lizards
Schizachyrium scoparium. *See* Bluestem, little
Schmutz, E. M., 55
Sciuridae. *See* Ground squirrels
Scleropogon spp. *See* Robber flies
Scleropogon brevifolius spp. *See* Burrograss
Scolopendra heros. *See* Centipedes, giant desert
Scorpions, true, 155, 156, 167. *See also* Whip scorpions
whip. *See* Vinegaroons
Scrub, 50, 51
Seedbeds, preparation of, 278–84, 291
Seeding
environmental responses and, 287–90
native grasses, 269–72
with pelleted seed, 277–78
revegetation trials, 274–76
Seedlings, 42
establishment models for, 290–91
and water use, 39–40, 41, 270–71
Senecio longilobus. *See* Groundsel, thread-leaf
Setaria. *See* Bristlegrass
Sevilleta Long-Term Ecological Research Program (LTER), vertebrates of, 197–210
Sheep, 46, 211, 236, 241, 246
bighorn (*Ovis canadensis*), 233
Sherbrooke, Wade, x
Shreve, Forrest, 53–54
Shrew, desert (*Notiosorex crawfordi*), 216
Shrimp
clam (*Eulimnadia texana*, Conchostraca), 178, 179, 189
fairy (*Streptocephalus texanus, Thamnocephalus platyurus*, Anostraca), 178, 179, 189
tadpole (*Triops longicaudatus*, Notostraca), 178, 179, 189
Shrublands, 50, 51, 56
Shrubs, 1, 2, 7(fig.), 8, 13, 39, 55, 108, 224
dwarf, 38, 45
erosion and, 118–20
fire and, 132, 140
as forage, 273–74
increases in, 6, 89–91, 92–93, 94–95, 247–48
Sidr (*Ziziphus nummularia*), 50
Sierra Madre Occidental, 53–54, 74, 93, 208
Sierra Madre Oriental, 74, 93
Sierra Vista (Ariz.), 243, 244
Sigmodon spp. *See* Cotton rats
Silver Bell (Ariz.), 255
Silver City (N.M.), 212
Skink, Great Plains (*Eumeces obsoletus*), 221
Snakes, 203, 216. *See also* Coachwhips; Ratsnakes; Rattlesnakes
Arizona coral (*Micruroides euryxanthus*), 208

brown vine (*Oxybelis aeneus*), 208
glossy (*Arizona elegans*), 223
gopher (*Pituophis melanoleucus*), 208
hook-nosed (*Gyalopion* spp.), 208
western hognose (*Heterodon nasicus kennerlyi*), 208, 221
Snaketown (Ariz.), 214
Snakeweed (*Guitierrezia sarothrae*), 5, 6(fig.), 8, 32, 35(fig.), 38, 85, 108, 124, 140, 206(figs.), 248, 268, 282
disturbance and, 86, 140, 220(fig.)
Sobaipuri, 234, 235–36
Socorro County (N.M.), vertebrates from, 197–210
Soil Conservation Service, 53, 113, 253, 271–72, 285
Soil moisture, 42
plant types and, 47–49, 108, 110
root systems and, 33–40
soil types and, 43–45
Soils, 3, 4, 189, 203(fig.)
ants and, 162–63, 184–86, 187
biota of, 158–61, 180
fire and, 133–34
and landscape evolution, 101–6
and moisture, 43–45
revegetation and, 280–81
termites and, 181–84
and vegetation types, 6(fig.), 7(fig.), 8, 33, 86, 100, 108–13
and vertebrates, 218–22
water storage of, 106–7
Solenopsis xyloni. *See* Ants, desert fire
Solifugae. *See* Sun spiders
Sonoita (Ariz.), 243, 246
Sonoita Creek (Ariz.), 240, 241
Sonora, 68, 69, 85, 208, 276
Sonoran Desert, 8, 53, 55, 75, 87–88, 94, 121
during Pleistocene, 80, 85
Sopóri, 236, 237
Sorghum halepense. *See* Johnson grass
Sotol (*Dasylirion wheeleri*), 10, 11(fig.), 32, 40, 52, 57, 84(fig.), 86–87, 88, 140
South Africa, 44–45
South America, 49
Southern Hemisphere, 69
Southern Pacific Railroad, 241, 245
Southwestern Forest and Range Experiment Station, 269, 271
Spadefoot toads (*Scaphiopus*)
Couch's (*S. couchi*), 92, 210
plains (*S. bombifrons*), 210
western (*S. hammondi*), 210
Spaniards, impacts of, 235–37
Sparrows, 216
Baird's (*Ammodramus bairdii*), 201
black-throated (*Amphispiza bilineata*), 201, 206(figs.)
Botteri's (*Aimophila botterii*), 141, 225
Cassin's (*Aimophila cassinii*), 141, 201
chipping (*Spizella passerina*), 141
grasshopper (*Ammodramus savannarum*), 141
lark (*Chondestes grammacus*), 142
rufous-crowned (*Aimophila ruficeps*), 225
vesper (*Pooecetes gramineus*), 141
Spermophilus spilosoma. *See* Ground squirrels, spotted
Spiders, 155, 156, 167
Spikelets, 16–17, 18(fig.)
Spizella passerina. *See* Sparrows, chipping
Sporobolus spp. *See* Dropseed; Sacaton
Sprangletop, green (*Leptochloa dubia*), 49, 81, 276
Squash (*Cucurbita* spp.), 233
Squirreltail (*Elymus elymoides*), 14, 17
Steppe, 51, 54–55

Stipa spp. *See* Feathergrass; Needlegrass
Stipideum, 72
Stockoceros onusroagris. *See* Pronghorn
Stock Raising Act, 247
Stolons, 16, 18(fig.), 20, 137
Streptocephalus texanus. *See* Shrimp, fairy
Sturnella spp. *See* Meadowlarks
Subshrubs, 32, 38–39, 42, 45
Succulents, 2, 13, 32, 40, 45, 137, 139–40. *See also by type*
Sulphur Springs Valley (Ariz.), 212, 245, 252, 253, 254
Sumac (*Rhus* spp.)
  African (*R. lancea*), 50
  little-leaf (*R. microphylla*), 86
Sunflower (*Helianthus annuus*), 135
Sun spiders, 155, 156, 221
Sylvilagus auduboni. *See* Cottontails

Talinum aurantiacum. *See* Flameflower
Tanglehead (*Heteropogon contortus*), 22, 23, 50, 84–85, 87, 271, 276
Tapir (*Tapirus* spp.), 78, 223, 232
Tarantulas, 155, 156
Tarbush (*Flourensia cernua*), 5, 6, 7(fig.), 14, 39, 52, 86, 90(fig.), 158
Taxidea taxus. *See* Badgers
Tayassu tajacu. *See* Javelina
Taylor Grazing Act, 242, 247, 249
Tegeticula yuccasella. *See* Moths, yucca
Teleoceras major. *See* Rhinoceros
Tenebrionidae. *See* Beetles, darkling
Tennessee, 69, 93
Termites, 153, 181–82, 189
  subterranean (*Gnathamitermes perplexus, Heterotermes aureus*), 183–84
Terrapene ornata luteola. *See* Turtle, desert grassland box
Tertiary period, 69, 72, 73, 74, 93
Texas, 1, 47, 76, 86, 89, 208, 267
  grasslands in, 54, 68
  during Pleistocene, 80–81, 83–84
  precipitation in, 3, 4
Thamnocephalus platyurus, Anostraca. *See* Shrimp, fairy
Thatcher (Ariz.), 122(fig.)
Thelyphonidae, Uropygi. *See* Whip scorpions
Thomomys bottae. *See* Pocket gopher, valley
Thornber, J. J., 267, 273, 275
Thornscrub, Sinaloan, 68
Threeawn (*Aristida* spp.), 14, 17, 136, 157, 269
  red (*A. longiseta*), 23, 254
  Santa Rita (*A. glabrata*), 137
  six-weeks (*A. adscensionis*), 50, 137, 216
Thrips, 156
  Chirothrips spp., 168, 169
  Haplothrips haplophilus, 168
Thumor (*Acacia senegal*), 50
Thysanoptera. *See* Thrips
Tillers, tillering, 16, 17(fig.), 20–22
Timothy, annual (*Phleum asperum*), 274
Tinajas Altas Mountains (Ariz.), 87
Tingidae. *See* Bugs, lace
Toads, 210, 216. *See also* Spadefoot toads
  Great Plains (*Bufo cognatus*), 210
  green (*B. debilis*), 210
  southwestern Woodhouse's (*B. woodhousei australis*), 210
Tobacco (*Nicotiana* spp.), 233
Tobosa (*Hilaria mutica*), 5, 7(fig.), 16, 36, 46, 55, 72, 114, 158, 159(fig.), 162, 254, 269
  on alluvial fans, 115(fig.), 116–18, 119(fig.)
  drought and, 6, 43
  and grassland definition, 52, 53, 56

Tohono O'odham, 233, 234
Tombstone, 241
Toumey, J. W., 244, 273
Trees, 1, 2, 8, 39, 47, 247–48. *See also by type*
Tres Alamos (Ariz.), 235, 237
Tridens, slim (*Tridens muticus*), 81
Trimerotropis pallidipennis. See Grasshopper, bandwinged
Triops longiaudatus, Notostraca. *See* Shrimp, tadpole
Tritle, F. A., 245
Trophic interactions, 210–11
Tubac (Ariz.), 236, 237, 241
Tucson, 71(fig.), 236, 237, 241
Tumacácori, 235, 236, 237, 240
Tumacácori Mountains, 241
Tumbleweed (*Salsola australis*), 220(fig.)
Tuna blanca (*Opuntia megacantha*), 54
Turtle, desert grassland box (*Terrapene ornata luteola*), 202(fig.), 203, 211

U-Bar Cave (N.M.), 77–78
United States Department of Agriculture, 5, 8. *See also* Forest Service; Soil Conservation Service
United States Department of the Interior, 12. *See also* Bureau of Land Management
United States Predator and Rodent Control, 211
University of Arizona, 8, 268, 277–78
Urban areas, 231, 243–44
Urosaurus ornatus. See Lizards, tree
Ursus americanus. See Bear, black
Uta stansburiana. See Lizards, side-blotched

Vail, W. L., 245
Vegetation
  classification of, 52–53
  and climate, 12–14
  diversity of, 4–12, 100, 292–93
  water storage and, 108–13
Vertebrates, 225–26. *See also by type*
  from Socorro County, N.M., 197–210
  soils and, 218–21
  species richness of, 196–98
  trophic interactions of, 210–18
Vinegaroons. *See* Whip scorpions
Voles (*Microtus* spp.), 91
Voorhies, Charles T., 212
Vultures, turkey (*Cathartes aura*), 218

Wait-a-minute bush (*Mimosa* spp.)
  M. biuncifera, 108, 140
  M. dysocarpa, 140
Walnut, English (*Juglans regia*), 236
Walnut Gulch Experimental Watershed, 119–20
Wasps, 155, 156
Waterman Mountains (Ariz.), 87–88
Water potentials, 36, 37(fig.)
Water storage
  soils and, 106–7
  vegetation patterns and, 108–13
Water storers, 40–41, 46
Wetlands, 198, 201
Wheatgrass (*Agropyron* spp.)
  bluebunch (*A. spicatum*), 274
  crested (*A. cristatum*), 188
  western (*A. smithii*), 42
Whetstone Mountains (Ariz.), 248
Whip scorpions, 155, 156
Whiptail lizards (*Cnemidophorus* spp.), 205, 221
  checkered (*C. tesselatus*), 209
  Chihuahuan (*C. exanguis*), 209
  desert grasslands (*C. uniparens*), 209
  Gila spotted (*C. flagellicaudus*), 209
  little striped (*C. inornatus*), 209
  New Mexican (*C. neomexicanus*), 209

plateau (*C. velox*), 209
Sonoran (*C. sonorae*), 209
western (*C. tigris*), 208
Whitethorn
acacia (*Acacia constricta*), 108
Chihuahuan (*A. neovernicosa*), 80, 81(fig.), 86, 90
Whitfield, C. J., 53, 55
Wilcox formation, 69
Wilcoxia. *See* Robber flies
Willcox Playa (Ariz.), 77
Wilson, C. P., 269, 270
Winterfat (*Ceratoides lanata*), 83, 273
Wolftail (*Lycurus setosus*), 23, 49
Wolves (*Canis*)
dire (*C. dirus*), 78
Mexican (*C. lupus baileyi*), 231
Woodlands, 4, 10, 50, 78–79, 86, 89, 94
Wood rats. *See* Packrats
Woody plants, 13, 32. *See also by type*
as cattle feed, 267–68
fire and, 45, 134, 138–39(fig.), 140, 141
increases in, 247–48, 255
water use of, 39–40, 108, 110
Wrens, 216
rock (*Salpinctes obsoletus*), 225
Wright, L. Neal, 279, 286–87
Wyoming, 74

Xanthocephalum dracunculoides. *See* Broomweed, annual

Yucca (*Yucca* spp.), 35(fig.), 40, 108, 140
Joshua tree (*Y. brevifolia*), 87
soaptree (*Y. elata*), 5, 6(fig.), 80, 83, 86, 90(fig.), 155, 186–88
Torrey (*Y. torreyi*), 80, 81(fig.)

Zacate barbón (*Oplismenus hirtellus*), 69
Zacatecas, 54, 75
Zenaidura macroura. *See* Dove, mourning
Ziziphus nummularia. *See* Sidr